Fire in Archaeology

Papers from a session held at the European Association of Archaeologists Sixth Annual Meeting in Lisbon 2000

Edited by

Dragos Gheorghiu

BAR International Series 1089

2002

Published in 2016 by
BAR Publishing, Oxford

BAR International Series 1089

Fire in Archaeology

ISBN 978 1 84171 467 7

BAR Publishing is the trading name of British Archaeological Reports (Oxford) Ltd. British Archaeological Reports was first incorporated in 1974 to publish the BAR Series, International and British. In 1992 Hadrian Books Ltd became part of the BAR group. This volume was originally published by Archaeopress in conjunction with British Archaeological Reports (Oxford) Ltd / Hadrian Books Ltd, the Series principal publisher, in 2002. This present volume is published by BAR Publishing, 2016.

Printed in England

BAR
PUBLISHING

BAR titles are available from:

BAR Publishing
122 Banbury Rd, Oxford, OX2 7BP, UK
EMAIL info@barpublishing.com
PHONE +44 (0)1865 310431
FAX +44 (0)1865 316916
www.barpublishing.com

Acknowledgements

The editor wants to thank all those who contributed to the present volume, for their long patience.

As always thanks to Maria Catuna for her invaluable help in editing the texts and to Florin Alexa and Corina Sarbu who contributed to the processing of illustrations.

D. G.

CONTENTS

LIST OF CONTRIBUTORS

Phyllis Anderson Ambrosiani – Swedish National Heritage Board, Stockholm, Sweden <phyllis.anderson-ambrosiani@raa.se>

Michal Artzy -Institute for Maritime Studies, University of Haifa, Haifa, Israel, 31905. <martzy@research.haifa.ac.il>

Massimo Cardosa – Research Institute for Prehistory and Archaeology, Milan, Italy <mcardosa@tin.it>

Richard Carlton - Department of Archaeology, University of Newcastle upon Tyne, Newcastle upon Tyne, UK <R.J.Carlton@newcastle.ac.uk>

Nuccia Negroni-Cataccio – University of Milan, Department of Antiquities, Section of Archaeology, via Fratell Casiraghi 491, Sesto San Giovanni, 20090 Milano, Italy <negroni@cilea.it>

Laura Domanico – "The Old West Manse", 25 Butte Terrace, Millport Isle of Cumbrae, Scotland, Millport KA 28 OAE, Scotland, UK
<CStrut@aol.com>

Esther Gatto – Laboratoire d'Anthropologie, Universite de Bordeaux 1, Av des Facultes, 33405, Talence, France <e.gatto@anthropologie.u-bordeaux.fr>

Dragos Gheorghiu – University of Arts Bucharest, 19 Budisteanu, Bucharest, Romania <dgheorghiu@digi.ro>

Laura Guidetti - Research Institute for Prehistory and Archaeology, Milan, Via Marconi 13, 28010 Gargallo, Italy, <laura_guidetti@yahoo.com>

Monica Miari – Archaeological Superintendence for the Region Emilia Romagna, Italy <miari@icil64.cilea.it>

Lars Larsson – Institute of Archaeology and the Historical Museum, Department of Archaeology, Sandgatan 1, S223 50 Lund, Sweden <lars.larsson@ark.lu.se>

Vasilica Lungu – Institute for South-East European Studies, Casa Academiei, 13, Calea 13 Septembrie, Bucharest, Romania.

Marcel Otte – Universite de Liege, Service de Prehistoire, Place du XX Aout, 7, bat A1, Liege, Belgium <prehist@ulg.ac.be>

Nicholas Petrov – St.Petersburg State University Faculty of History, Department of Archaeology, Mendeleyevskaya linia, 5, St.Petersburg 199034, Russia <nickpetrov@infopro.spb.su>

Christophe Pottier - Museum National d'Historie Naturelle, Laboratoire de Prehistorie – UMR 6569 du CNRS, Abri Pataud, F-24620, Les Eyzies-de-Tayac, France <cpottier@mnhn.fr>

Edina Rudner ¯ Institute of Geography, Hungarian Academy of Sciences, Budapest, XI., Budaörsi u. 43–45., 1112, Hungary
/
CIMO scholar, Department of Geography, University of Helsinki, P.O. Box 64, 00014 Helsinki University, Finland <Edina.Rudner@helsinki.fi>

Pal Sumegi ¯ Department of Geology and Palaeontology, University of Szeged, Szeged, Egyetem u. 2., 62, Hungary

Mirjana Stefanovic- ARF Anthropology Department, 232 Kroeber Hall, University of California at Berkeley, Berkeley, CA 94720, USA <mirjana@sscl.berkeley.edu>

Bruce Veitch – Archae-Aus Pty Ltd, PO Box 117, South Fremantle WA 6162, Australia <bruce@archae-aus.com.au>

Christer Westerdahl - Vanermuseet, SE 53117 Lidkoping, Sweden <christer.westerdahl@malmo.se>

Towards Pyroarchaeology

The present book is a collection of papers inspired by the sessions on "The Archaeology of Fire" held at the 6th and 7th European Association of Archaeologists Meetings in Lisbon and Esslingen in September 2000 and 2001.

The choice of this topic was made based on the fact that in all archaeological literature the number of the studies on fire is minimal. In archaeological research fire seems to have been the forgotten phenomenon, all attention being focussed on material culture. This neglect may be due to an anthropological shift that occurred along with the process of industrialization, when fire took on new associations, being perceived in almost all the cases as an "accident", or risk factor. One can infer that modernity produced an "oblivion of fire", by transforming it into a negative phenomenon, related to punishment and disaster.

Fire was central in the anthropology of traditional societies, being a phenomenon whose transforming and relational nature structured pre-Modern cultures. Memories from epochs when fire constituted a daily reality are present in all mythologies, tales, as well as in the current language. In Romanian there still remains a saying expressing the superlative of beauty through a comparison with the beauty of fire.

Fire as an artefact-phenomenon

The nature of fire is essentially relational, the existence of the phenomenon being due to its relationship to material world. Its perception is therefore both as artefact and phenomenon. For that reason, an approach to fire could cover, from a micro to a macro perspective, the study of the relationship of the artefact-phenomenon with materials, human body, built space, environment. Being an ephemeral phenomenon, that is studied only on its (indirect) material effects, it deserves complex modes of analysis, such as the direct ones that involve the replication of the process of combustion, to the indirect, as anthracology, the study of charcoals.

In the study of traditional societies fires involves the study of relationships with almost all technologies and rituals. For this reason and because its ephemeral nature, it is suggested that, every archaeologist should be aware in their research of this invisible presence in material culture. There is no archaeologist that has not confronted with the indexes of fire during the excavations, but until today, material artefacts are central to archaeological research than the process of making and controlling fire. Archaeological studies on fire have concentrated on the origins of humankind, rather than on the prehistoric and protohistoric epochs.

An archaeological study of fire becomes necessary to understand human development. Behind almost every object or organization of living and funerary spaces hides a pyrotechnology that is waiting to be discovered, for that reason the purpose of the present book is to endorse a pyroarchaeology that will promote a complex study of fire.

The popularity of the two sessions on the archaeology of fire, organized at EAA in 2000 and 2001, confirms that the subject is attractive to archaeologists, that it was referred to in their studies, but as a minor subject, not closely examined or explained.

The present book is therefore an invitation to reflection on the approaches to the study of fire, as an essential phenomenon in human evolution. The book includes studies of anthracology, ethnoarchaeology, field archaeology, symbolism, technology and experimental archaeology, whose ideas converge to some universals, such as the relationship of fire with environment, materials, human body, its quality of transformability and its anthropological centrality.

To underline the relational character of fire and the above-mentioned areas of study, I have renamed the titles of each contribution in the introduction.

Fire and civilization

Marcel Otte considers fire as being a metaphysical motor and the evolutionary factor towards modern humanness, whose role was to modify the mechanical characteristics of materials, to produce physical light, to gather people and to generate behavioural cults all around the world.

He notes that the production light and its origin sometimes from the sky, imposed communication between terrestrial and celestial worlds. Additionally, the property of fire to communicate with the outer world, expressed in the case of the transformation of the human body by means of cremation, imposed a relationship with memory.

Fire and environment

Edina Rudner and Pal Sumegi approach fire by analysing one of its material effects: the charcoal. A particular science, anthracology, subject to the dangers of misinterpreting results, tries to identify the differences between a set fire and a spontaneous forest fire, or between the fireplace remains and the remains of a trunk. In such cases data collecting is crucial for interpretation, the temperature of charcoal burning being an index of the type of fire. If, as is the case of the charcoal collected from caves the anthropic source is certain, for open-air sites, the differences between the traces of natural fires and human fires are more difficult to identify.

Fire and centrality

After being "domesticated" and daily utilized in hearts, fire became a structuring factor of human activities, its centrality in daily life being fully established. The example from Upper Palaeolithic from Abri Pataud (France), a site discovered at the end of 19th century, and excavated in several campaigns in the 20th century, demonstrates, beside the pyro-centrism mentioned, a diversification in the utilization of fire. Christophe Pottier discusses the Middle Gravettian level by analysing the spatial organization of combustion structures delimited from the rest of the settlement by a semicircle of stones. Through the analysis of the concentration of objects one could re-enact the intentional movements of objects in relationship with fire, the principal activities gravitating around the combustion zones.

Fire and human body (Neolithic)

In Late Neolithic France there is evidence of a simultaneous treatment of the dead, cremation and inhumation, as well as that of a relationship between cremated bones and different structures as pits, dolmens, passage-graves, tholoi, tumuli, caves and rock shelters. According to Esther Gatto, to understand the special relationship with fire, mixed approaches, both anthropological and archaeological, are necessary. The action of fire on the human body or on the skeleton are difficult to identify in the archaeological records, for the understanding of the funerary ritual experiments and different methods of study of cremated remains are necessary. Primary burials must be differentiated from secondary ones. Particular rituals such as the selective collection of bones must be identified.

Fire and transformation

The role of fire in transforming materials is discussed by Lars Larsson in the context of Scandinavian Neolithic. Here, its ritual role was to "sacrifice" prestige objects such as axes or houses. Sometimes objects were "killed" before cremation, this legitimising act impressing the audience, and conserving the value of objects through their metamorphosis through fire. Larson equates this custom of destruction with one of sacral "transformation", the intentional selected objects being burned in a sacral destruction. This equality seems to have had anthropomorphic analogies, since the burning of axes was perceived as a sort of cremation of the human body, because the colour of the fired flint is similar to that of cremated human bones. For Larsson fire had a simultaneous destructive and creative nature.

Fire was positioned in special places, being protected by architectural structures with a possible visual dominating role. In time, the role of fire changed. In the Iron Age it was used mainly as a medium of passage, being connected to the rites of passage, and ceremonial purification near the grave.

Fire and cultivation

The example of the utilization of fire as a regulator for the plants' cycles of growth and food managed by fire, discussed by Bruce Veitch, could be used as an inspirational model for plant management in prehistory. Veitch discusses two models of utilization of fire for plant management in Australia and New Zealand, his paper examining the case of *Braken fern* plant and ecosystem management.

Through ethnographic and ecological evidence, he concludes that plants can be enhanced by fire, re-establishing fronds after being burned. The firing of the inhibiting litter cover became a sort of removal of competition for the *Bracken fern* plant. By experimenting on a large scale, Veitch demonstrated a stimulation of plant growth and apparent vigour resulting from fire. A comparative analysis between Australian hunter-gatherers and Neo-Zealand horticulturalists reveals that, in spite of great different cultural ecologies, plant management using fire was performed in a similar way.

Fire and house sacrifice

A strategy in house building with the interest of destruction by fire at the end of their life-cycle existed in the Balkan Neolithic and Chalcolithic. Experiments demonstrate the difficulty of firing wattle and daub structures at high temperatures, this being possible only by adding a great amount of fuel. According to Mirjana Stevanovic, the intentional firing of architectural structures was a way of preserving the house and its objects and its place within the settlement, the role of fire being of fixing in memory the objects, the place of the house and the event itself. She believed she had identified the intentional firing of wattle and daub buildings by analysing fragments of architecture with vitrifying points and the fire path. Making use of archaeological examples from Vinca culture, as well as ethnological examples from different cultures, Stevanovic illustrates the house's life and death, discussing the diverse modes of house abandonment and the symbolical meanings of the process of construction and deconstruction by fire.

Fire and ceramics

One of the most constant utilizations of fire through time has been the process of making ceramics. Richard Carlton describes the contemporary process of production of calcite-tempered pottery on the Dalmatian coast, as a model to understand the Iron Age or Romano-Illyrean technology of ceramic production in the Western Balkans.

Using extensive ethnological data from the Island of Iz, in the Adriatic Sea, Carlton describes some technological traditions of paste composition, construction methods, and firing of ceramic objects. The causal relationship between the nature of the paste (determined by the clay quality and the quantity of temper used), the temperature of firing and the final appearance of the ceramic object, is insisted upon. Complementary to the ethnological study, experiments validated the abovementioned relationship, as well as the post firing process of colouring the vases.

Fire and air

There is a relationship that has been practically neglected in current studies on prehistoric fires, due to its "virtual" character, that is the relationship with air.

Dragos Gheorghiu's study focuses on the immaterial or 'virtual" relationship with air of pyroinstruments, such as pits, kilns, portable kilns, braziers, and intentionally fired houses.

The pyroinstruments, or draught objects, function with the help of airflow circulating through them. One cannot study the virtual (in archaeological record) element "fire" without studying the virtual element "air", experimentation being the main instrument of study. The paper contains many examples of study-cases of pyroinstruments.

Fire and human body (Middle and Late Bronze Age)

The widespread custom of cremation at the beginning of Late Bronze Age in North Western Italy is approached by Laura Domanico through examples from necropoli from Piedmont, Canegrate and Premeno. She reconstructs parts of the funerary rituals after the examination of the internal organization of the burial sites, the spatial relation between tombs, grave structure, funeral pyres, and the analysis of the burned personal belongings of the deceased. The structure of the cremation ritual, as well as the social status of the deceased, are examined through the tombs' contents and fired objects.

Domanico devotes a special attention is given to intentional fragmentation and to the traces of fire on the objects cremated with the dead.

Fire and ritual food (Late Bronze Age)

Nuccia Negroni-Catacchio, Massimo Cardosa, Laura Guidetti and Monica Miari's study on the ritual uses of fire in everyday life (ovens, domestic hearth, etc.), and in cult and religion (animal sacrifice, holy meals), focuses on Proto-Etruscan sites from Etruria. A particular case of fire utilization, the Protovillanovan grotto from the settlement of Sorgenti della Nova (Viterbo), revealed new data on holy fires used for cremating animal offerings. The grotto's function, was related to fertility rituals in which fire was used to roast animals' meat.

Historical texts describing the Thesmophoric cults in honour of Demeter are used to interpret these fertility rituals, and different devices in relation to fire, such as fireplaces and kilns, are also analysed.

Fire and stone

From the beginning of the last century the significance of the firing of stones in Bronze Age Sweden as a cultural trait has been disputed.

Phyllis Anderson-Ambrosiani's new perspective on the subject, as the distribution of materials or status (i.e. the visibility in the landscape and the relationship with grave monuments or rock carvings), offers new explanations for this old problem. Cracked stones are perceived as a physical manifestation of man's use of fire, as " a human activity and energy in interaction with raw materials and resources"; this offers a new insight into prehistoric cultures, especially concerning the significance and technology of fire.

Fire and human body (Iron Age)

Cremation of the human body was central to Iron Age Greek funerary rituals, and it seems that the phenomenon was developed during the growing complexity of the Greek polis and society, and due to the process of colonization of "barbarian" territories that started in the 8th century B.C.

Vasilica Lungu's study of the necropolis of Orgame on the Black Sea coast reveals interesting links between power and uses of fire in the organization of funerary architecture.

Lungu's approach to the study of fire combines the analysis of archaeological remains, ancient texts, and ceramic iconography. These reveal the social implications in the transformation of the body of the "hero-founder of the city" cremated on a funerary pyre, as well as the importance of the memory of the event in creating colonists' identity.

Fire and human body (Pre-Christian)

The location of funerary piles in pre-Christian North -Western Russia was outside the grave.

Nicholas Petrov's excavation of a Pskov long barrows culture pit and his study of this in relation to ritual fire raised a number of questions: what ritual actions connected a funeral pile to a grave? how were the cremated bones selected and later transferred? Relating the information about pagan funerary rituals from Medieval texts to these archaeological finds, Petrov considers the possible variants of the cremation ritual.

Fire and symbol (Iron Age and Middle Ages)

Christer Westerdahl and Michal Artzy's hermeneutic approach to identify the representations of fire on Iron Age Gotland picture-stones is based on information offered by the Nordic Medieval oral tradition as well as by archaeological record. In the perspective of the symbolism of the myths of origin,

Westerdhal and Artzy, demonstrate that there could be a relationship between cremation, gender, and ships on one side and between fire and limestone on the other.

In this perspective, the patterns on Gotland picture-stones are interpreted as representing fire and not the sun, this inferring to picture-stones a particular identity.

Concluding remarks

Some preliminary conclusions of the present book on the nature of fire, as it is revealed by the archaeological record, could be the following:

Fire was central to human existence;
It acted as a socializing instrument;
It had a utilitarian characteristic;
It had a public characteristic;
It was in a relationship with dominant locales;
It was a symbolic instrument;
It was in a relationship with graves and funerary rituals;
It was in a relationship with the human body or with the skeleton;
It possessed the quality of transforming materials and their colour;
It possessed a quality of cleansing, particularly in funerary rituals;
It possessed a characteristic of separation;
It possessed a sacral characteristic of rebirth and purification;
It possessed a sacral dual character of construction and deconstruction (for objects, houses, human bodies, environment), alone, or in relationship with liquids;
It possessed a spectacular characteristic, being in relation to dominant monuments;
It had a nocturnal role, impressing the audience;
It possessed an ephemeral role, its temporary presence being put in relation to a special event or monument;
It possessed the character of fixing important events in memory;
It possessed a social role of identity construction;
It was an instrument for writing history.

It is hoped that the present book will stimulate complex approaches that will study the dual nature of fire, as artefact and as phenomenon, emphasizing the complex nature of this subject.

Dragos Gheorghiu March 2002

Fascinating fire. Experiment with up-draught kiln in Vadastra, summer 2001.
Photo Florin Alexa.

FIRE AS AN EVOLUTIONARY FACTOR

Marcel Otte

To my son, family pyromaniac, last priest in the arts of fire

Introduction

The comments presented here aim at clarifying the crucial role played by the mastery of fire, along with other elements, in the development of humanity. In my view, it is not simply one of many technological victories, but a veritable agent of spiritual, mythical and social development forming the basis of our current values. These different aspects will be discussed here in chronological order of their appearance and of their importance: there is thus a logical succession whose organization reveals the meaning. Certain material data mark this process, but the essence of the phenomenon remains of metaphysical nature, such as the mysterious and profound attraction that fire exercises even today in our "secular" spirits.

Origins

In order to leave the original environment, common to the higher primates (gorillas, chimpanzees, orangutans), there must first have operated a profound modification in the dietary regime. A primarily vegetarian diet (leaves, fruits) shifted to a carnivorous diet at an early date (2-3 million years ago). This passage was necessary in order to survive in areas lacking sufficient vegetal resources: savannahs and steppes (Delluc 1995). The contribution of protein and calories, from then coming from meat, transformed the relationship between man and his environment, even though our digestive system was better suited to vegetal foods. In order to facilitate the preparation and digestion of meat, cooking was required and in this way allowed liberation from the original environment (Brain and Sillen 1988). In this sense, hunting and consumption of meat and the use of fire formed the necessary conditions for expansion into northern regions, where evidence of human occupation exists from between 1 and 2 million years ago. Yet, traces of hearths or of combustion are found from 1.5 million years ago in Kenya and South Africa, apparently corresponding to this accomplished and successful "departure" to open areas (Oakley 1954, Gowlett *et al.* 1981, Barbetti 1986, James 1996). We can thus consider that the origin, the use of fire offered man at least the "biological possibility" for expansion out of the forests.

Tools

From a technological viewpoint, man used fire to modify the mechanical characteristics of certain materials. For this, the literature is rich and abundant, illustrating the order and the nature of these different successive applications (e.g., Collin *et al.* 1991, Collina-Girard 1998). Fire-hardened wood, used as a spear point, was recovered at Lerhingen (Oakley 1955). Similar actions have been described for rare wood found at different early sites including Krapina, Abric Romani and Ambrona. This method of hardening vegetal materials has also been often described in ethnographic studies. Glues (resin, betuline, various tars) were first observed during the Mousterian and were widespread during the Upper Paleolithic (for "composite tools"). Through time, the more obvious applications of fire were successively firing of statuettes, then pottery and finally, metallurgy. The mastery of man over the elements can be seen as progressively broadened and increasingly complex. It contributed to the metaphysical feeling of a power of creator of all things. This feeling of power placed man in a position of "demiurge", between the gods and the world, and supported his conviction of escaping the laws of destiny, applicable to all other creatures.

Even in our days, in both primitive and modern societies, during the Middle Ages and modern times, smiths and other metalworkers were feared and respected. They possessed the "secrets" necessary to transform raw material, secrets that were jealously guarded and transmitted only within the family or clan. The occult power of medieval alchemists, masters of the laws of matter, is another example. They were the precursors of modern chemistry, when science was substituted for religion. Classical mythology presents other evidence by the Greek (Hephaistos) and Roman (Vulcan) gods and not least the famous Titan Prometheus, who stole the fire from the gods in order to offer it to man, for better or worse ...

The symbol

Fire principally provided *light* which constituted an incontestable and visible product also mastered by man, who could then see in the shadows, contrary to the natural laws. This physical characteristics of fire finds an obvious analogy in the light (and heat) of the sun. This force from the sky was soon harnessed by the megaliths which picked up the movements of the sun and created significant shadows. The divine action thus "descends" to earth as rays which constitute the cosmic link between heaven and earth. All subsequent architecture used the play of celestial light, modified by season, latitude and orientation. Our most beautiful churches, with enflamed windows, continually remind us of this.

In contrast, the ascent of man, addressing the heavens and the gods, occurred by the fire of pyres and sacrificial altars. Heat, light, smoke and ash entered into celestial communication by fires. The purification of the dead, in particular, was frequently carried out by the incineration of the body which thus entered in contact with the divinities and guaranteed its resurrection, symbolically contained in the cremation. We return to this idea in the "flames of memory", illustrating a fact, a memory, a presence or an idea, powerfully respectable and that we would like to keep "alive", beyond death and physical disappearance.

Between these two worlds, celestial and terrestrial, between gods and man, are found paths of exchange and reciprocity. As the celestial light descends on earth to define the seasons, calibrate the calendar of human actions, ceremonies and festivals, man manifests in his turn this relationship by fire of joy. The changing of the seasons is marked by the fires of St. John, the burning of Old Man Winter and the fireworks of St. Sylvester. All are reunited to "ascend to heaven" what remains of human faith in a derisory "sacrifice" which today is nothing more than that of our wallets. Uncountable cults of fire, throughout the world, perpetuate this original fascination and this willingness to maintain the lines of exchange, by sacrifice, with the divinity (Frazer 1931).

Figure 1. The conquest of fire, engraving by Vitruve, 1521 (after Perlès 1977).

The demiurge

The force represented by the mastery of fire also has an effect on the animal world. This domination by controlled fear suggests to man the notion of influence, similar to that of the gods, extended this time to their living counterparts. A symbolic force is thus acquired as a result of consciousness, innovation and knowledge of man in comparison with physically superior beings. The influence, first mechanical (by materials transformed), then physical (by light) finally became behavioural, substituting for the animal will.

By analogy, the use of fire was felt to be a protection against the spirits, repelled to the shadows and cold by the light and heat produced. It is in this way symbolic of the victory of consciousness ("light") on the feared unknown (invisible spirits). The spiritual force of intelligence is thus incarnated by the light, controlled and directed by man. These exchanges with the spirits, by the intermediary of fire, are abundant in the rituals of ancient Japan and China.

The group

Finally, fire constitutes a gathering agent of the social group, to exchange ideas, transmit knowledge and maintain inter-individual cohesion. In the common meaning, the domestic "hearth" has more symbolic meaning than technological. This hearth is the place where one is reassured, protected and among friends. This "radiant" location around the fire constitutes from now on the basis of human society, where it is formed, consolidated and transformed.

The passion for fire is so profound and so powerful that it fascinates our children and still provokes spontaneous assembly "burning with passion".

Bibliography

BARBETTI, M.
1986 Traces of Fire in the Archaeological Record, Before One Million Years Ago? *Journal of Human Evolution* 15:771-781.

BRAIN, C.K. and SILLEN, A.
1988 Evidence from the Swartkrans cave for the earliest use of fire. *Nature* 336(1):464-466.

COLLIN, F., MATTAR, D., PIRNAY, L. and SPECKENS,J.
1991 L'obtention du feu par percussion: approche expérimentale et tracéologique. *Bulletin des Chercheurs de la Wallonie*, tome XXXI:19-49.

COLLINA-GIRARD, J.
1998 *Le feu avant les allumettes*. Paris, Editions de la Maison des sciences de l'homme.

DELLUC, G.
1995 *La nutrition préhistorique*. Périgueux, Pilote.

FRAZER, J.
1931 *Mythes sur l'origine du fe*u. Paris, Editions Payot.

GOWLETT, J., HARRIS, J., WALTON, D. and WOOD, B.
1981 Early archaeological sites, hominid remains and traces of fire from Chesowanja, Kenya. *Nature* 294(12)):125-129.

JAMES, S.
1996 Early Hominid Use of Fire: Recent Approaches and Methods for Evaluation of the Evidence. In *The Lower and Middle Palaeolithic*, Series Colloquia, Volume 5, XIII International Congress of Prehistoric and Protohistoric Sciences, Forlí, Italy, 8/14 September 1996, edited by O. Bar-Yosef, L. Cavalli-Sforza, R. March et M. Piperno. Forlí, A.B.A.C.O. Edizioni.

OAKLEY, K.P.
1954 Evidence of Fire in South African Cave Deposits. *Nature* 174:261.

OAKLEY, K.P.
1955 Fire as Palaeolithic Tool and Weapon. *Proceedings of the Prehistoric Society* 4:36-48.

PERLÈS, C.
1977 *Préhistoire du Feu*. Paris, Masson.

CHARCOAL AS A REMAIN OF NATURAL AND HUMAN-SET FIRES OF PALAEOLITHIC TIMES – CASE STUDY FROM HUNGARY

Edina Z. Rudner and Pál Sümegi

Introduction

Anthracology is used to reconstruct vegetation elements from charcoal, charred wood or from wooden remains. It may mean that the remains were subjected to high pressure and they are charred due to this process. However, in the case of younger, let's say Pleistocene samples, we come across charcoal as the result of burning. During so short period of time charring would lead only to lignite phase. In this paper we would like to highlight amongst others the possible causes of fires during the Upper Pleistocene, or from an archaeological point of view, during the Upper Palaeolithic.

Charcoal represents local arboreal vegetation elements. The macrocharcoal fragments are less likely to have been wind-born (Clark 1988), carried over long distances and so, are the remains of a plant *in situ* (Sümegi and Rudner 2001). It can survive even very dry conditions during several tens of thousands of years (Lowe and Walker 1997). So it provides irreplaceable information on vegetation at sites where other types of vegetation remains (e.g. pollen) hardly can be found. Thorough examination of charcoal from different point of views is called anthracology. It has traditions in Quaternary research in Hungary (see Stieber 1968; Horváth 1958, 1966, 1979, 1982, 1987, 1990; Sárkány 1938; Sárkány and Stieber 1950, 1952; Simoncsics 1955; Szalai 1995; Valkó and Stieber 1970), but in other countries, like France and Belgium, as well (Vernet ed. 1992; Damblon 1996, 1997a, 1997b). Regarding Hungarian anthracological research, it has been started in connection with archaeological excavations, namely in the 19th century. The first report came to light from Békés Bronze Age excavation in 1890-1891 by Deininger (1890, 1891). Since then till the beginning of the 1970s anthracology (called anthracotomy at that time) had been continuously following archaeologist's work. As a summary on anthracological research so far as 1970s Stieber presented a complete work in 1968 where lots of the data are based on archaeological charcoal (see also Stieber 1957, 1967, 1969). However, the investigations nearly totally stopped later on, and were re-started around the mid-1990s, but at this time not on archaeological samples. That arouses the question that not all charcoal samples of younger times are connected to man's activity. So not all prehistoric charcoals are the result of human-set fires. Many archaeologists also use macrocharcoal data in their work. It means they interpret results of macroscopic charcoal examinations sometimes provided by wood anatomists. Here we also would like to highlight the possible misleading interpretations of charcoal finds. As anthracologists and geologists, we look at this fossil material differently as archaeologists. Here we would like to draw archaeologists' attention to consider more aspects while interpreting data from fossil charcoal, and drawing conclusion about fires of prehistoric times.

Although the anthracological school of Montpellier (Vernet ed. 1992) and the few other anthracologists (Damblon 1996, 1997a, 1997b; Gale 1997) have rather detailed if not the most detailed knowledge of interpreting charcoal anthracologically from archaeological sites, many archaeologists, who only use the data obtained from charcoal perhaps by wood anatomists seems to have some problems in interpretation, and not all conclusions that could be are drawn at most of the times (Rolando and Thinon 1988).

The main aspects what we would like to examine here closely are the decision at site excavation (both geological and archaeological) whether the found charcoal was the result of human set fires, or just spontaneous forest fires. Also, it can be important if we find a concentrated charcoal spot to decide whether it was a fireplace or it is just the remains of a tree trunk. As we are not archaeologists we only can imagine how important it could be to decide that human impact on an excavated phenomena had been traced or not.

Material and Methods

Altogether 33 charcoal bearing sites are mentioned in this paper (Table 1). The sites can be divided into three main groups, according to their former relation to man: 1. cave-sites, 2. open-air sites: a. archaeological excavations, b. geological sampling sites (T. Dobosi 1998-1999).

During the geological excavations we followed fine-stratigraphical methods (Sümegi 1996).

Regarding sampling for charcoal, very different kinds of methods were used, as only some of the samples were collected by the writers. Many samples collected by archaeologists were taken from bulk charcoal samples. That means there is a large spot of charcoal but only some part of it is taken to bring to the laboratory. Although, appropriate charcoal sampling methods are known from the early nineties (Vernet ed. 1992), archaeologists in Hungary starts just nowadays to collect charcoal samples according to these directives (after Rudner 2000, in press). Some samples from Holocene collected in the '70s – '90s are also good, as the archaeologist collected every piece they found on the excavation (Rudner and Jerem 2002, in press). However, in this paper only samples from Pleistocene are mentioned.

Charcoal samples (Schoch 1986) from loess profiles and lake sediments are compared to charcoal samples from archaeological sites. The basis of their comparability is that radiocarbon measurements were done on the charcoal itself, after that they were examined xylologically. Loess layers were sampled for charcoal in dense intervals (10 to 20, maximum 25 cm), with fine-stratigraphy according to Krolopp (1983) and Szanyi and Braun (1995). Samples were obtained from the same quantities of sediments, from 1 dm3, and were washed for shells and charcoal. From lake-sediments, together with pollen, charcoal was obtained from 4-8 cm thick subsections of the cores. After that we could identify specimens. For wood anatomical investigations: 1 dm3 (~2.65 kg) of sediment was dried at 100°C, then sieved for charcoal with a metal screen with mesh size of 0.5 mm, in water. Then charcoal remains were left to dry at room temperature, measured, then put into labelled glasses. The size of our charcoal remains varies between 0.05–2.0 cm, but we took under examination only the fraction sized of 0.2–2.0 cm. To identify them we used stereomicroscope (up to magnification ×100), microscope with reflected light (×600–800) (Stieber 1957, 1967), SEM and N-SEM (to take photos at magnification ×400–800), and we have examined three fracture faces of them, i.e. transversal, longitudinally radially and tangentially fractured surfaces. The books of Greguss (1944–47, 1955, 1959, 1972) and Schweingruber (1978, 1990) were applied to help identification, besides, Babos (1970, 1981), Hollendonner (1913), Greguss und Varga (1950), Schmidt (1941), Wagenführ (1966) and Wilhelm (1925) also.

We measured 25 samples of the layers with a high precision radiocarbon dating system (Hertelendi 1990; Hertelendi *et al.*, 1987, 1989; Csongor *et al.*, 1982; Sümegi and Hertelendi 1998) using macrocharcoal fragments and Mollusc shells (Hertelendi *et al.*, 1992). Measurements were done in the Nuclear Research Institute of Hungarian Academy of Sciences in Debrecen, Hungary (lab code begins with deb-). The rest of radiocarbon data were taken from the references.

What I would like to emphasise here is, that it is very important to make detailed macroscopic description of the charcoal-bearing layer regarding the embedding circumstances of the sample material. It is important to observe as carefully as possible the size of the charcoal spot, whether it belongs to one tree branch or how many pieces seem to belong to the same "log", etc. It is also worthy to note where exactly the sample was taken from, and also to take photos of the sampled spot. These all play an important role during the interpretation of data. As charcoal can survive after collection even for hundreds of years, it is worthy to sample correctly during excavation and analysis of material and data evaluation can take place at the most suitable time later on.

Results and Discussion

Anthracology

The macrocharcoal we have investigated represents the local arboreal vegetation of the sites, that means, vegetation maximum about a hundred meter far from the site. They are thin-section sized (Clark 1988), or "*in situ*" (Sümegi and Rudner 2001) macroscopic charcoal fragments, or charcoal from fireplaces of early humans (Vernet 1992). The anthracomass contained species of *Picea* sp. (fir), *Larix* sp. (larch), *Pinus sylvestris, Pinus cembra* (pines), *Carpinus betulus* (hornbeam), *Salix* sp. (willow), *Betula* sp. (birch), *Juniperus* sp. (juniper).

We made distinction between *Larix* and *Picea* fragments, for details see Rudner and Sümegi (2001).

In the Norhtern part of Hungary in the Middle Mountain ranges *Picea* occurred together with *Larix. Pinus cembra* also occurred in the coldest regions. In the Great Hungarian Plain mainly *Pinus sylvestris* could be identified, *Betula* also appeared. At the border of these two regions nearly all type of founded species could be observable. *Salix* followed the river system. This can be applicable for the recurring forest spots (Rudner and Sümegi 2001).

All the charcoal material proved to be the result of fire, this is supported by microscopic examinations, as well as by derivatographical analyses (Szöőr 2000). The latter show that the firing temperature was around 5-600 °C, this suggest forest fires (MacDonald, *et al.* 1991; Wein *et al.* 1987).

We examined charcoal from 8 archaeological sites, among which 4 cave-sites and 4 open-air sites were analysed. From cave sites it is clear, that charcoal derives from fireplaces. Very rarely it can also be washed down from the surface, but at most of the sites also human traces can be found connected to the charcoal. Regarding open air-sites, all of the samples came to light through archaeological excavations. The samples can either belong to former fireplaces, or to remains of the fired vegetation around the archaeological settlement. Concerning the firing of these samples it is not very easy to decide whether it was the result of fire on purpose, or of a spontaneous forest fire. Or, even, these remains can be charcoal particles, which derived from the fireplaces.

22 geological sampling sites have also been analysed anthracologically. The division between the open-air sites and geological sampling sites can be difficult sometimes from the charcoal-point of view, if only charcoal was found. In most of the cases where a big charcoal spot is found it is

Table 1. The sites mentioned in the paper, with radiocarbon data and references.

No.	Site	*Type*	Level (m under-ground)	Sediment type of charcoal-bearing layer	*Age*	References
1.	Basaharc-téglagyári vető II.	Loess deposit	4,0–4,25	Paleosol and subsequent layer(loess) border	24,030±317	Pécsi, 1977; Willis *et al.*, 2000; Vértes, 1965
2.	Csillaghegy	Loess deposit		Loess	15,935±142	Gábori-Csánk, 1984; T. Dobosi, 1996, 1998-1999, 1999, 2000b; Vértes, 1965; Sümegi *et al.*, 1998
3.	Dunaszekcső	Loess deposit	6,5–7,0	Loess, embrional soil	21,740±320	Krolopp, 1977; Willis *et al.*, 2000; Pécsi and Schweitzer, 1995
4.	Jankovich-barlang	Archaeological site		?	Upper Palaeolithic	Gábori-Csánk, 1993; Vértes, 1965
5.	Kardoskút	Lake sediment coring	6,0–6,2	sand	23.370±300	Magyari, 1998, Sümegi *et al.*, 1999; Willis *et al.*, 2000
6.	Katymár (2 szint)	Loess deposit	9,0–9,25 6,0–6,25	Paleosol subsequent layer(loess) border Loess, embrional soil	29.828±554 23,749±494	Molnár és Geiger, 1981; Willis *et al.*, 2000
7.	Kecskés-galya	Archaeological site		?	Upper Palaeolithic	Vértes, 1965
8.	Kiskő-oldal	Archaeological site		?	Magdalénien	Vértes, 1965
9.	Lakitelek I.	Loess deposit	3,0–3,2	Paleosol	22,110±300	Sümegi, 1995; Willis *et al.*, 2000
10.	Lakitelek II.	Loess deposit	3,0–3,5	Paleosol subsequent layer(loess) border	26.736±629	Willis *et al.*, 2000
11.	Madaras	Loess deposit	10,0–10,25	Loess, embrional soil	21937±252	Dobosi, 1967; Molnár and Krolopp, 1978; Krolopp, 1995; Willis *et al.*, 2000; ; T. Dobosi, 1998-1999, 1999, 2000a, 2000b
12.	Megyaszó-Szelestető	Archaeological site	1,10–1,30	fireplace	27,070 ±680	Vértes, 1965; T. Dobosi and Simán, 1996; T. Dobosi, 1996; 1998-1999, 2000a, 2000b
13.	Mende	Loess deposit	10,0–10,2	Paleosol subsequent layer(loess) border	29,800±600	Geyh *et al.*, 1969; Willis *et al.*, 2000
14.	Nagy-Mohos (2 szint)	Lake sediment coring	4,22–4,26 3,94–3,99;	Peaty layers	21,756±267 18,159±247	Magyari *et al*, 1999, 2000; Willis *et al.*, 2000
15.	Parassa-III. Orgonás	Archaeological site	1,3–1,4	fireplace	27,350 ±610	Vértes, 1965; T. Dobosi, 1998-1999, 2000a, 2000b; T. Dobosi and Simán, 2000
16.	Püspökhatvan-Diós	Loess deposit	1,1–1,3	Paleosol subsequent layer(loess) border	27.700±300	Csongrádi-Balogh and T. Dobosi, 1995; T. Dobosi, 1998-1999, 2000a, 2000b; Willis *et al.*, 2000
17.	Ságvár	Loess deposit		Paleosolés subsequent layer(loess) border	18,900±1001 7,760±150	Vogel and Waterbolk, 1964; Gábori and Csánk, 1960; Vértes, 1965; T. Dobosi, 1998-1999, 2000a, 2000b
18.	Susak	Loess deposit	 21,40–21,65	fireplace Paleosol-subsequent layer(loess) border	Würm	Schweitzer (ed) 2000
19.	Süttő	Archaeological site		?	Upper Palaeolithic	Vértes, 1965; Krolopp, 1995
20.	Szeged-Öthalom	Loess deposit	4,25–4,5	Paleosol subsequent layer(loess) border	25.200±300	Vértes, 1965; T. Dobosi, 1999, 2000b; Krolopp *et al*, 1996
21.	Tiszaalpár-magaspart	Loess deposit		Paleosol	Upper Würm	Sümegi *et al.*, 1992, Frechen and Pécsi, 1999
22.	Tokaj 1:Bodrogkeresztúr-téglagyár	Loess deposit	5,0–5,25	Paleosol subsequent layer(loess) border	26.851±398	Sümegi, 1996; Sümegi and Hertelendi 1998; Willis *et al.*, 2000
23.	4: Kereszt-hegy I.	Loess deposit	4,75–5,0	"	27.323±644	Sümegi, 1996; Sümegi and Hertelendi 1998; Willis *et al.*, 2000
24.	3: Kereszt-hegy II.	Loess deposit	5,25–5,5	"	26.962±657	Sümegi, 1996; Sümegi and Hertelendi 1998; Willis *et al.*, 2000
25.	5: Csorgó-kút I.	Loess deposit	4,75–5,0	"	26.618±532	Sümegi, 1996; Sümegi and Hertelendi 1998; Willis *et al.*, 2000
26.	6: Csorgó-kút II.	Loess deposit	5,0–5,25 5,75–6,0	" Loess, embrional soil	28.225±360	Sümegi, 1996; Sümegi and Hertelendi 1998; Willis *et al.*, 2000

27.	7: Patkó-bánya (2 szint)	Loess deposit	4,0–4,25 5,0–5,25	Loess, embrional soil Paleosol subsequent layer(loess) border	27,491±362 23.519±494	Geyh *et al*, 1969; Sümegi, 1996; Sümegi and Hertelendi 1998; Willis *et al*., 2000
28.	9: Terem-tető	Loess deposit	4,75–5,0	Paleosol	30.174±1101	Sümegi, 1996; Sümegi and Hertelendi 1998; Willis *et al*., 2000
29.	10: Tarcal	Loess deposit	6,75–7,0	Paleosol subsequent layer(loess) border	27.251±288	Sümegi, 1996; Sümegi and Hertelendi 1998; Willis *et al*., 2000
30.	11: Bodrogkeresztúr Henye-tető	Archaeological site	0,4–0,6	Paleosol subsequent layer(loess) border	26.318±365	Sümegi, 1996; Krolopp, 1977; Sümegi and Hertelendi 1998; Willis *et al*., 2000; Sümegi *et al*., 2000
31.	Vác	Archaeological site		Loess	Upper Palaeolithic	Geyh *et al*, 1969

assumed that it was a fireplace connected to early humans. However, it also can be remain of a tree trunk, even if the form of a trunk cannot be identified at all. We reconstructed forestations and forest fires in at least 4 phases during the Upper Palaeolithic (Rudner and Sümegi 2001) which seems to be the result of natural processes. Derivatographical analyses (Szöőr 2000) showed that the firing temperature was around 5-600 °C and this suggests forest fires (MacDonald, *et al*. 1991; Wein *et al*. 1987). Also extraterrestrial causes are thought to be responsible for these firing events. Lightening is also a probable cause.

References

ALLEN, J. R. L.
1998 Windblown trees as a palaeoclimate indicator: regional consistency of a mid-Holocene wind field. *Palaeogeography, Palaeoclimatology, Palaeoecology*: 144.

BABOS, K.
1970 Különböző termőhelyi eredetű, papírfa méretű erdeifenyők évgyűrűszélességeinek, rosthosszának és térfogatsúlyának vizsgálata, *Faipari Kutatások 1970*: 269–281.
1981 A szibériai vörösfenyő (*Larix sibirica* Ledeb.) néhány anatómiai jellemzője, Bot. Közlem. 68, 29–35. Bartholin, Th., 1979, The Picea-Larix problem. *IAWA Bulletin* 1979(1): 7–10.

BEHRE, K.-E.
1989 Biostratigraphy of the last glacial period in Europe. *Quaternary Science Reviews* 8: 25–44.

CARCAILLET, C. and M. THINON
1996 Pedoanthracological contribution to the study of the evolution of the upper treeline in the Maurienne Valley (North French Alps): methodology and preliminary data. *Rev. Palaeobot. Palyn.* 91: 399–416.

CSONGOR, É., I. SZABÓ and E. HERTELENDI
1982 Preparation of counting gas of proportional counters for radiocarbon dating. *Radiochemical and Radioanalytical Letters* 55: 303.

CSONGRÁDI-BALOGH É. and V. T. DOBOSI
1995 Palaeolithic settlement traces near Püspökhatvan. *Folia Archaeologica* 44: 37–59.

DAMBLON, F.
1996 Les depots tourbeaux et l'histoire de la vegetation sur le plateau des Hautes-Fagnes (Belgique). *Annales de la Société géologique de Belgique* 117: 259–276.

DAMBLON, F.
1997a Anthracology and past vegetation reconstruction, Pavlov I. – Northwest. The Upper Palaeolithic burial and its settlement context. In Svoboda, J. (ed.) *The Dolni Vestonice Studies* 4, Brno: 437–439.
1997b Palaeobotanical study of representative Upper palaeolithic sites in the Central European Plain: a contribution to the SC-004 project. *Préhistoire Européenne* 11: 245–253.

DAMBLON, F. and P. HAESAERTS
1997 Radiocarbon chronology of representative Upper Palaeolithic sites in the Central European Plain: a contribution to the SC-004 project. *Préhistoire Européenne* 11: 255–276.

DAMBLON, F., P. HAESAERTS and J. VAN DER PLICHT
1996 New datings and considerations on the chronology of Upper Palaeolithic sites in the Great Eurasiatic Plain. *Préhistoire Européenne* 9: 177–231.

DEININGER, I.
1890 Pflanzenreste der prähistorischen Fundstatte von Lengyel. In Wosinszki (ed.): *Das prähistorische Schanzwerk von Lengyel*. III. Bp.
1891 Adatok kultúrnövényeink történetéhez. A lengyeli őskori telep növénymaradványai. *A keszthelyi Magyar Királyi Gazdasági Tanintézet* 1891-i évkönyve. Nagykanizsa.

GALE, R.
1997 Charcoal analysis, Medieval leads: Archaeological excavation and conservation of a lead working site, Howden Clough, Bradfield, South Yorkshire. In Bevan, B. (ed.) *Peak District National Park Archaeology Service*: 16–19.

GÁBORI, M. and V. GÁBORI
1957 Études archéologiques et stratigraphiques dans les stations de loess paléolithiques de Hongrie. *Acta Arch.* 8: 3–117.

GÁBORI-CSÁNK, V.
1960 A ságvári telep abszolút kormeghatározása. *Archeológiai Értesítő* 87: 125–129.
1978 Une oscillation climatique à la fin du Würm en Hongrie. *Acta Archaeologica Hungarica* 30: 3–11.
1984 A felső paleolitikum nyoma Budapesten. *Budapest régiségei* XV, Budapest: Budapesti Történeti Múzeum: 7–14.
1993 Le Jankovichien. *Études et Recherches Archéologiques de l'Université de Liège et Musée Historique de Budapest.*

GENCSI, L. and R. VANCSURA
1997 *Dendrológia*, Budapest: Mezőgazda Kiadó, 728.
GEYH, M. A., F. SCHWEITZER, L. VÉRTES and I. C. VOGEL
1969 A magyarországi würmi eljegesedés új kronológiai adatai. *Földrajzi Közlemények* 18: 5–18.
GREGUSS, P.
1944–1947 A fontosabb fenyőgenuszok meghatározása szövettani alapon. *Acta Universitatis Szegediensis* 3: 1–65.
1945 A középeurópai lomblevelű fák és cserjék meghatározása szövettani alapon. *Országos Magyar Természettudományi Múzeum*, Budapest.
1955 *Xylotomische Bestimmung der Heute Lebenden Gymnospermen*. Budapest: Akadémiai Kiadó.
1959 Holzanatomie der Europäischen Laubhölzer und Sträucher. Budapest: Akadémiai Kiadó.
1972 *Xylotomy of the living Conifers*. Budapest: Akadémiai Kiadó.
GREGUSS, P. and I. VARGA
1950 Xylotomische Bestimmungsschlüssel der Pinus Arten. *Acta Biologica*.
HAESAERTS, P., I. BORZIAK, J. VAN DER PLICHT and F. DABLON
1998 Climatic events and Upper Palaeolithic chronology in the Dniester Basin: new 14C results from Cosautsi. *Radiocarbon* 40: 649–657.
HAESAERTS, P., F. DABLON, M. BACHNER and G. TRNKA
1996 Revised stratigraphy and chronology of the Willendorf II sequence, Lower Austria. *Archaeologica Austriaca* 80: 25–42.
HERTELENDI, E.
1990 Sources of random error in the Debrecen radiocarbon laboratory. *Radiocarbon* 32: 283.
HERTELENDI, E., J. GÁL, A. PAÁL, S. FEKETE, M. GYŐRFFY, I. GÁL, Zs. KERTÉSZ and S. NAGY
1987 Stable isotope mass spectrometer, Fourth working meeting isotopes in nature. Wand, U. and G. Strauch (eds.) *Akademie der Wissenschaften der DDR Zentralinstitut für Isotopen und Strahlenforschung*, Leipzig 1: 323.
HERTELENDI, E., E. CSONGOR, L. ZÁBORSZKY, J. MOLNÁR, J. GÁL, M. GYŐRFFY and S. NAGY
1989 Counting system for high precision C-14 dating. *Radiocarbon* 31: 399–408.
HOLLENDONNER, F.
1913 A fenyőfélék fájának összehasonlító szövettana, *Országos Erdészeti Egyesület*, Budapest.
HORVÁTH, E.
1958a Pleisztocén gerinces és faszénmaradványok, valamint langyosvíz feltörési helyek Gencsapátiban. *Vasi Szemle* 12: 60–67.
1966 Pleisztocén famaradványok Szombathely környékéről, Savaria, *A Vas m. Múz. Ért.* 3: 9–28 (1965).
1973 Jahrringanalyse und dendroklimatologische Untersuchung von Bäumen, III. *Magyar Növényanatómiai Szimpózium előadásainak kivonata*. Visegrád, 42–43.
1975 Későglaciális és óholocén növénymaradványok Szombathelyről. *Bot. Közlem.* 62: 312.
1979 Dendroklimatologische untersuchungen an Holzstämmen aus Ungarn, Savaria. *A Vas m. Múz. Ért.* 7–8: 11–54 (1973–74).
1982 A famaradványok helye, szerepe és jelentősége a régészeti kutatásokban. Régészeti Továbbképző Füzetek. A régészet és természettudomány kapcsolata, *Budapest Magyar Nemzeti Múzeum* 1: 39–51.
1987 Wooden remains from the Roman villa at Kővágószőlős. *Janus Pannonius Múzeum Évkönyve* 30–31: 237–241, Pécs (1985–1986).
1990 Charcoal remains of the upper loess section, Vértesszőlős. In Kretzoi, M. and T. Dobosi, V. (eds.) *Site, Man and Culture*. Budapest: Akadémiai Kiadó: 137–143.
KROLOPP, E.
1977 Absolute chronological data of the Quaternary sediments in Hungary. *Földrajzi Közlemények* 26: 228–232.
1983 Biostratigraphic division of Hungarian Pleistocene formations according to their Mollusc fauna. *Acta Geologica Hungarica* 26: 62–89.
KROLOPP, E. and P. SÜMEGI
1995 Palaeoecological reconstruction of the Late Pleistocene, based on Loess Malacofauna in Hungary. *GeoJournal* 36: 213–222.
KROLOPP, E., P. SÜMEGI, L. KUTI, E. HERTELENDI and L. KORDOS
1996 Szeged-Öthalom környéki löszképződmények keletkezésének paleoökológiai rekonstrukciója. *Földtani Közlöny* 125: 309–361.
KUKULAK, J.
2000 Sedimentary record of early wood burning in alluvium of mountain streams in the Bieszczady range, Polish Carpathians, Palaeogeography, Palaeoclimatology. *Palaeoecology* 164: 167–175.
LAGEARD, J. G. A., P. A. THOMAS and F. M. CHAMBERS
2000 Using fire scars and growth release in subfossil Scots pine to reconstruct prehistoric fires. *Palaeogeography, Palaeoclimatology, Palaeoecology* 164: 87–99
LANGOHR, R.
1993 Types of tree windthrow, their impact on the environment and their importance for the understanding of archaeological excavation data. *Helinium* 33: 36–49.
LOWE, J. J. and M. J. C. WALKER
1997 *Reconstructing quaternary environments*. Essex: Addison Wesley Longman Ltd.
MacDONALD, G. M., C. P. S. LARSEN, J. M. SZEICZ and K. A. MOSER
1991 The reconstruction of boreal forest fire history from lake sediments: a comparison of charcoal, pollen, sedimentological, and geochemical indices. *Quaternary Science Reviews* 10: 53–71.
MAGYARI, E., G. JAKAB, E. RUDNER and P. SÜMEGI
1999 Palynological and plant macrofossil data on Late Pleistocene short-term climatic oscillations in NE-

Hungary. Proceedings of the 5 EPPC. *Acta Palaeobotanica, Supplement* 2: 491–502.

MAGYARI, E., G. JAKAB, P. SÜMEGI, E. RUDNER and M. MOLNÁR

2000 Paleoökológiai vizsgálatok a keleméri Mohos-tavakon, In Szurdoki E. (ed.) *Tőzegmohás lápok Magyarországon: kutatás, kezelés, védelem*. pp. 101–131. Miskolc: CEEWEB Munkacsoport.

MASON, S. L. R.

2000 Fire and Mesolithic subsistence – managing oaks for acorns in northwest Europe? *Palaeogeography, Palaeoclimatology, Palaeoecology* 164: 139–150.

MEDZIHRADSZKY, Zs. and M. JÁRAI-KOMLÓDI

1996a Az ember természetformáló tevékenysége a holocén folyamán a Kárpát-medencében. In Hably, L. (ed.) *Emlékkötet Andreánszky Gábor (1895-1967) születésének 100. Évfordulójára*. pp. 147–154. Budapest: Magyar Természettudományi Múzeum.

1996b Late Holocene vegetation history and the activity of man in the Tapolca Basin. *Annales Historico-naturales Musei Nationalis Hungarici* 88: 21–29.

MENDOZA ANAYA, D., L. CARAPIA MORALES, V. RODRÍGUEZ LUGO, L. TORRES MONTES, F. FRANCO VELÁZQUEZ and T. As. LEE WHITING

2000 Analysis by MEB, EDS and XRD of Samples of Wall Bricks, Slag and Minerals Belonging to Remains of a Possible Metal Smelting Furnace of the Colonial Period from Jalenton, Chiapas, Mexico. *Absracts of 32nd Archaeometry Symposium*: 201.

MOORE, J.

2000 Forest fire and human interaction in the early Holocene woodlands of Britain. *Palaeogeography, Palaeoclimatology, Palaeoecology* 164: 124–138.

NICHOLS, G. J., J. A. CRIPPS, M. E. COLLINSON and A. C. SCOTT

2000 Experiments in waterlogging and sedimentology of charcoal: results and implications. *Palaeogeography, Palaeoclimatology, Palaeoecology* 164: 43–56.

PÉCSI, M.

1975 Lithostratigraphical subdivision of the loess sequences in Hungary. *Földrajzi Közlemények* 23: 228–239.

1977 A hazai és az európai löszképződmények paleogeográfiai kutatása és összehasonlítása, Geonómia és Bányászat, MTA X. *Osztályának Tudományos Közleményei* 10: 183–221.

1993 Negyedkor és löszkutatás. Budapest: Akadémiai Kiadó, 375 p.

PÉCSI, M. and F. SCHWEITZER

1995 The lithostratigraphical, chronostratigraphical sequence of Hungarian loess profiles and their geomorphological position. In Pécsi, M. and F. Schweitzer (eds.), *LOESS inFORM 3*, pp. 31–62.

RINGER, Á.

1988 Possible correlations between loess and cave deposit stratigraphies for the Upper Pleistocene in Hungary, Paleogeography of Carpathian regions. In Pécsi, M. and L. Starkel (eds.), Geogr. Res. Inst., Budapest. *Theory – Methodology – Practice* 47: 65–86.

RINGER, Á.

2000 Upper Plesitocene chronostratigraphy of North-Eastern Hungary. In Mester, Zs. and Á. Ringer (eds.) *A la recherche de l'Homme Préhistorique*. pp. 95–102.Liège: ERAUL 95.

ROLANDO, C. and M. THINON

1988 Perspectives offertes par l'identification de petits fragments de charbon de bois. *PACT* 22.

RUDNER, Z. E.

2001a Faszénvizsgálatok régészeti lelőhelyen, Százhalombatta Oktatónapok 2, (közlésre elfogadva, nyomtatás alatt).

RUDNER, Z. E. and E. JEREM

2001 Anthracological investigations at Sopron-Krautacker, Proceedings of 31st Archaeometry Symposium, Budapest, Hungary. *British Archaeological Reports-Archaeolingua, East European Series* 1: 53–62.

RUDNER, Z. E. and P. SÜMEGI

2001 Recurring taiga forest-steppe habitats in the Carpathian Basin. *Quaternary International* 76-77: 177–189.

SÁRKÁNY, S.

1938 A Szeleta-barlang faszénmaradványai. *Bot. Közl.* 35.

SÁRKÁNY, S. and J. STIEBER

1950 A Szelim-barlang újabb faszénmaradványainak vizsgálata, *Annales Universitatis Scientarium Budapestiensis de Rolando Eötvös nominatae Sect. Biologica.*

1952 A Remete szurdoki alsó barlang alluviális faszenei, MTA *Biol. Oszt. Közlem.* 1: 239–254.

SCHMIDT, E.

1941 *Mikrophotographischer Atlas der mitteleuropäischen Hölzer*. Neudamm: Verlag von J. Neumann.

SCHOCH, W.

1986 Wood and charcoal analysis. In Berglund, B. E. (ed.), *Handbook of Holocene Palaeoecology and Palaeohydrology*, pp. 619–626. New York: John Wiley & Sons.

SCHWEINGRUBER, F. H.

1978 *Microscopic Wood Anatomy*. Zug: Zürcher AG, CH-6301.

SCHWEINGRUBER, F. H.

1990 *Anatomy of european woods*. Zug: Zürcher AG, CH-6301.

SCOTT, A. C., J. A. CRIPPS, M. E. COLLINSON and G. J. NICHOLS

2000 The taphonomy of charcoal following a recent heathland fire and some implications for the interpretation of fossil charcoal deposits. *Palaeogeography, Palaeoclimatology, Palaeoecology* 164: 1–31.

SIMONCSICS, P.

1955 Investigation of charcoal remains of the

palaeolithic limonit mine of Balatonlovas. *Acta Archaeologica Academiae Scientarium Hungaricae* 5: 293–295.

SMART, T. L. and E. S. HOFFMAN

1988 Environmental Interpretation of Archaeological Charcoal, In Hastorf, C. A. and B. S. Popper (eds.) *Current Paleoethnobotany*, pp. 167–205. Chicago and London: The University of Chicago Press.

STIEBER, J.

1957 A hazai felsőpleisztocénből származó faszénmaradványok anthrakotómiai vizsgálata, Kandidátusi disszertáció, Budapest, kézirat.

1967 A magyarországi felsőpleisztocén vegetáció-története az anthrakotómiai eredmények (1957-ig) tükrében, *Földtani Közlöny* 97(3): 305–317.

1968 *Anthrakotómia, kvarter kronológia és a hazai pleisztocén vegetáció, (Athracotomy, quarterchronology and the Hungarian Pleistocene vegetation)*, Thesis for Acad. Doct. Department of Plant Anatomy, Eötvös Lóránd University, Budapest.

1969 A hazai későglaciális vegetációtörténet anthrakotómiai vizsgálatok alapján. *Földtani Közlöny* 99/2: 188–193.

SÜMEGI, P.

1989 *Upper pleistocene evolution history of the Hajdúság (Hungary) region, on the basis of stratigraphical, palaeontological, sedimentological and geochemical investigations.* Thesis for D.Univ., Department of Mineralogy and Geology, Kossuth Lajos University, Debrecen.

1995 Quartermalacological analysis of late-Pleistocene loess sediments of the Great Hungarian Plain, Quaternary malacostratigraphy in Hungary, Fűköh, L. (ed.), Gyöngyös: *Malacological Newsletter Suppl.* 1: 17–78.

1996 *Az észak-magyarországi löszterületek összehasonlító őskörnyezeti rekonstrukciója és rétegtani értékelése*, Thesis for PhD, Department of Mineralogy and Geology, Kossuth Lajos University, Debrecen.

SÜMEGI, P. and E. HERTELENDI

1998 Reconstruction of microenvironmental changes in the Kopasz hill loess area at Tokaj (Hungary) between 15 and 70 ka BP. *Radiocarbon* 40: 855–863.

SÜMEGI, P. and E. KROLOPP

1995 A magyarországi würm korú löszök képződésének paleoökológiai rekonstrukciója Mollusca-fauna alapján. *Földtani Közlöny* 125: 125–148.

2000 Palaeoecological reconstruction of the Ságvár-Lascaux interstadial. In , Mester, Zs. and Á. Ringer (eds.) *A la recherche de l'Homme Préhistorique*. pp. 103–112. Liège: ERAUL 95.

SÜMEGI, P. and Z. E. RUDNER

2001 In situ charcoal fragments as remains of natural wild fires of Upper Würm in the Carpathian Basin. *Quaternary International* 76-77: 165–176.

SÜMEGI, P., R. KERTÉSZ and E. HERTELENDI

2002 Environmental change and human adaptation in the Carpathian basin at the late glacial/postglacial transition. Proceedings of 31st Archaeometry Symposium 1998, Budapest, Hungary, *British Archaeological Reports, East European Series 1.*

SÜMEGI, P., E. KROLOPP and E. HERTELENDI

1998 A Ságvár-Lascaux interstadiális paleoökológiai rekonstrukciója. *Acta Geographica Debrecina* 34, 165–180.

SÜMEGI, P., Z. E. RUDNER and I. BESZEDA

2000 *Stratigaphical and palaeoecological examinations, Bodrogkeresztúr-Henye (NE Hungary) Upper Palaeolithic site.* T. Dobosi, V. (ed.), Hungarian National Museum, Budapest, 217–220.

SÜMEGI P., E. HERTELENDI, E. MAGYARI and M. MOLNÁR

1998a Evolution of the environment in the Carpathian Basin during the last 30,000 BP years and its effects on the ancient habitats of the different cultures. In Költő, L. and L. Bartosiewicz (eds.), *Archimetrical Research in Hungary*. pp. 183–197. Budapest.

SÜMEGI, P., E. MAGYARI, P. DANIEL, E. HERTELENDI and E. RUDNER

1999 A kardoskúti Fehér-tó negyedidőszaki fejlődéstörténetének rekonstrukciója. *Földtani Közl.* 129: 479–519.

SZALAI, G.

1995 *A fás vegetáció változása a Duna-Tisza Köze déli részén a holocén kezdetétől napjainkig*, Kézirat, diplomadolgozat, ELTE Növényszervezettani Tanszék, Bp., 47 p.

SZANYI, J. and M. BRAUN

1995 Kémiai elemek viselkedésének vizsgálata lápokban félvariogramok segítségével. In Bárdossy, Gy. (ed.), *A IV. Geomatematikai Ankét előadásai*. pp. 77–83. Szeged: MFT kiadványa.

SZÖŐR, Gy.

2000 Felsőpleisztocén erdőtüzek termoanalitikai bizonyítása és feltételezett paleoklimatológiai szerepük. In Fábián, SZ. Á. és J. Tóth (eds.) *Geokronológia és domborzatfejlődés.* pp. 167–186. Pécs: Pécsi Tudományegyetem.

T. DOBOSI, V.

1967 Új felső-paleolit telep az Alföldön. *Archeológiai Értesítő* 94: 184–193.

1994 Contribution to the Upper Palaeolithic topography, *Acta Archaeologica Academiae Scientarium Hungaricae* 46: 3–20.

1996 The Hungarian Upper Palaeolithic. *ERAUL* 76: 77–85.

1998–1999 Upper Palaeolithic open-air sites. Settlement and function. *Archaeologica Austriaca* 82–83: 57–63.

1999 Postplenigalcial repeopling of the Hungarian Plain. *Folia Quaternaria* 70: 297–315.

2000 Upper Palaeolithic research in Hungary – A situation report from 2000. *Praehistoria* 1: 149–159.

2000 Interior parts of the Carpathian Basin between

30,000 and 20,000 BP, Hunters of the Golden Age, In Roebroeks, W., Mussi, M., Svoboda, J. and K. Fennema (eds.), *The mid Upper Palaeolithic of Eurasia 30,000 – 20,000 BP*. pp. 231–239.University of Leiden.

T. DOBOSI, V. and K. SIMÁN

1996 New Upper Plaeolithic site at Megyaszó-Szelestető, *Communicationes Archaeologicae Hungariae*, 5–22.

2000 Upper Palaeolithic in the Region of Hont villageIn Mester, Zs. and Á. Ringer (eds.), *A la recherche de l'Homme Préhistorique*. pp. 321–337. Liège: ERAUL 95.

THÉRY, I., J. GRIL, J.-L. VERNET, L. MEIGNEN and J. MAURY

1996 Coal used for Fuel at Two Prehistoric Sites in Southern France: Les Canalettes (Mousterian) and Les Usclades (Mesolithic). *Journal of Archaeological Science* 23: 509–512.

VALKÓ, E. and J. STIEBER

1970 Anthrakotómiai vizsgálatok a veszprémi würmi löszből. *Földtani Közlöny* 100: 209–210.

VAN ANDEL, T. H. and P. C. TZEDAKIS

1996 Palaeolithic landscapes of Europe and environs, 150,000–25,000 years ago: an overview, *Quat. Sci. Rev.* 15: 481–500.

VERNET, J.-L. (ed.)

1992 Les charbons de bois, les anciens écosystèmes et le rôle de l'homme. *Bull. Soc. Bot. France* 139, 725 p.

VERNET, J.-L. and T. S. THIEBAULT

1987 An approach to northwestern Mediterranean recent prehistoric vegetation and ecologic implications. *Journal of Biogegraphy* 14: 117–127.

VERNET, J.-L., C. PACHIAUDI, F. BAZILE, A. DURAND, L. FABRE, C. HEINZ, M.-E. SOLARI and S. THIEBAULT

1996 Le 13C de charbons de bois préhistoriques et historiques méditerranéens, de 35,000 BP à l'actule. Premiers résultats, *C.R. Acad. Sci. Paris* 323: 319–324.

VÉRTES, L.

1965 *Az őskőkor és az átmeneti kőkor emlékei Magyarországon,* Budapest: Akadémiai Kiadó. 385 p.

1966 The Upper Palaeolithic Site on Mount Henye at Bodrogkeresztúr. *Acta Arch. Hung.* 18: 3–14.

VOGEL, I. C. and H. I. WATERBOLK

1964 Groningen radiocarbon dates V. *Radiocarbon* 6: 349–369.

WAGENFÜHR, R.

1966 *Anatomie des Holzes,* Leipzig: VEB Fachbuchverlag, 377 p.

WEIN, R. W., M. P. BURZYNSKI, B. A. SREENIVASA and K. TOLONEN

1987 Bog profile evidence of fire and vegetation dinamics since 3000 years BP in the Acadian Forest, *Can. J. Bot.* 65: 1180–1186.

WHITEHOUSE, N. J.

2000 Forest fires and insects: palaeoentomological research from a subfossil burnt forest. *Palaeogeography, Palaeoclimatology, Palaeoecology* 164: 247–262.

WILHELM, K.

1925 *Schlüssel zum bestimmen eiheimischer hölzer nach äusseren merkmalen,* Wien: Druck und Verlag von Carl Gerold's sohn.

WILLIS, K. J.

1997 The Impact of Early Agriculture upon the Hungarian Landscape. In Chapman. J. and P. Dolukhanov (eds.), *Landscapes in Flux. Central and Eastern Europe in Antiquity*, pp. 193–207. Oxford: Oxbow Books.

WILLIS, K. J., K. D. BENNETT and H. J. B.BIRKS

1998 The late Quaternary dinamics of pines in Europe. In Richardson, D. M. (ed.), *Ecology and Biogeography of Pinus*, pp. 107–121. Cambridge: Cambridge University Press.

WILLIS, K. J., E. RUDNER and P. SÜMEGI

2000 The Full-Glacial Forests of Central and Southeastern Europe, *Quaternary Research* 53: 203–213.

WILLIS, K. J., Z. E. RUDNER and P. SÜMEGI

2001 Full Glacial Tree Refugia in Europe. Response to Letter to the Editor – C. Carcaillet and J.-L. Vernet, *Quaternary Research* 55: 388–389.

WILLIS, K. J., P. SÜMEGI, M. BRAUN, K. D. BENNETT and A. TÓTH

1998 Prehistoric land degradation in Hungary: who how and why? *Antiquity* 72: 101–113.

UN HABITAT DU GRAVETTIEN MOYEN A BURINS DE NOAILLES DE L'ABRI PATAUD (DORDOGNE, FRANCE)

Christophe Pottier

Abstract

Although fire has been used for a long time, it has been harnessed by man around 500 000 years ago. Since then, fire has become an essential element in the habitat of men through the ages. Numerous excavations on prehistoric sites have shown evidence of this harnessing: around which man organized their living quarters. Hearths have been discovered in Abri Pataud, situated in Eyzies-de-Tayac (Dordogne, France). This site has considerable archaeological stratigraphy dating from the early part of the upper Palaeolithic period. Hearths, of which this is the very first study, were discovered at the base of middle Gravettian with Noailles burins. The quality of the excavation and the availability of the original scientific documents allowed us to carry out a study of the construction and organisation of these combustion structures. The spatial distribution of the lithic archaeological material, corroborated by the archaeozoological studies, enabled us to analyse the relationship between these hearths and the other sectors of this occupation space.

Introduction

Bien que voulant se placer en toute fin de la chaîne de l'Evolution, l'Homme est une espèce animale à part entière. Cependant sa rapide évolution fut jalonnée d'étapes cruciales entraînant une coupure de plus en plus prononcée avec le monde animal, isolement progressif aboutissant à sa situation actuelle. L'une de ces étapes de rupture fut sans conteste la découverte des usages du feu. Mais pour cela, il a fallu en premier lieu que l'Homme maîtrise sa peur vis-à-vis de cet élément de destruction, peur présente instinctivement en toute espèce animale. Cette première emprise est certainement ancienne, plusieurs milliers d'années, probablement avec l'apparition d'*Homo erectus*. La maîtrise de cette peur permit alors à l'Homme les premières utilisations du feu.

Mais ce n'est que bien plus tard, vers 500 000 ans, que le feu est véritablement domestiqué, sa production à volonté étant uniquement un simple problème technique, dont la résolution fut rapide. Le feu fait désormais partie intégrante de la vie quotidienne des hommes.

Avec la domestication du feu, apparaissent les foyers, dont l'organisation est plus ou moins élaborée. Les activités domestiques du quotidien des préhistoriques vont se structurer autour de ces derniers, lesquels deviennent le point central de l'habitation. Outre l'utilisation du feu pour la cuisson des aliments, de nombreux usages vont apparaître tel que le durcissement au feu des armes en bois, technique apparue au Paléolithique moyen. Au Paléolithique supérieur, les usages se diversifient : cuisson de statuettes en argile, chauffe du silex pour la taille. Mais la maîtrise du feu permet aussi le chauffage de l'habitation, l'éclairage par l'intermédiaire de torches ou de lampes en pierre. Les hommes prennent l'habitude de se rassembler autour du foyer, des rites s'installent. Le foyer devient un lieu de d'échanges et de liens pour les membres de la tribu.

Le site de l'abri Pataud est l'un des sites les plus importants pour le début du Paléolithique supérieur. Dans ce gisement de nombreux foyers ont été mis au jour, certains présentant une organisation très structurée. C'est notamment le cas dans le niveau archéologique attribué au Gravettien moyen de l'abri Pataud.

Le site de l'abri Pataud

Le gisement de l'abri Pataud est situé dans le département de la Dordogne (France), au cœur du village des Eyzies-de-Tayac. Il s'agit d'un ancien abri sous roche effondré, ouvert vers l'ouest/sud-ouest dans une barre rocheuse orientée nord-sud, sur la rive gauche de la Vézère, affluent de la Dordogne (fig. 1).

Bien que cité pour la première fois en 1899 sous le nom de « Croze-de-Tayac » par Emile Rivière qui y fit une petite fouille (Rivière, 1899), il faut attendre 1957, après le rachat de la propriété, pour que des campagnes de fouilles soient programmées. Six campagnes auront lieu sous la direction du professeur Hallam L. Movius, de 1958 à 1964 (Movius, 1975 et 1977), mettant au jour une importante stratigraphie archéologique datée de la première moitié du Paléolithique supérieur.

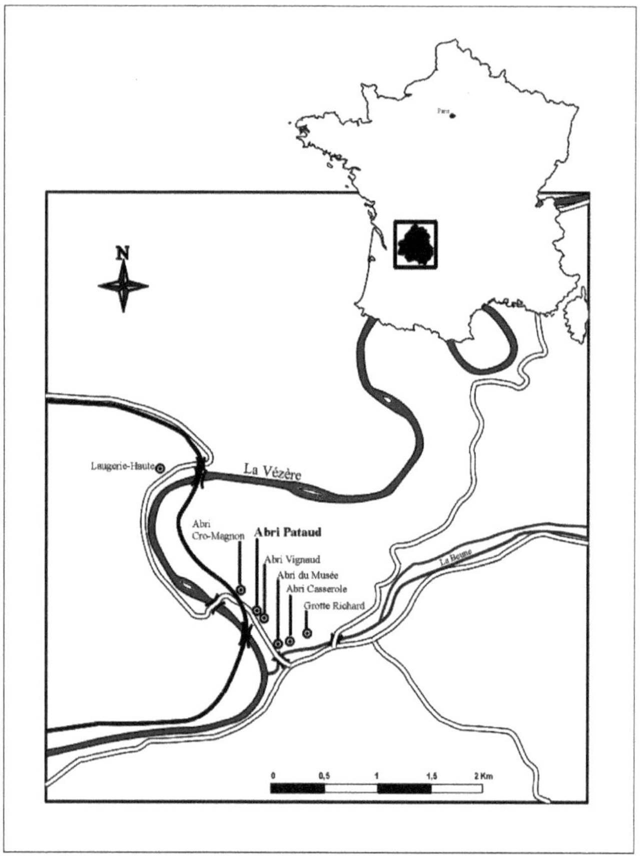

Figure 1: Localisation géographique de l'abri Pataud

Le gisement fut découpé en sept tranchées d'ouest en est (*Square A* à *Square G*) et en huit tranchées du sud au nord (*Trench I* à *Trench VIII*), chaque tranchée ayant 2 m de côté. Suite à la présence de nombreux blocs d'effondrements, la zone principale de fouilles fut réduite aux tranchées II-nord, III, IV, et V-sud, lorsque le niveau gravettien à burins de Noailles fut atteint (fig. 2).

Après quelques années d'abandon, le site renaît en 1985 sous l'impulsion du professeur Henry de Lumley du Muséum national d'Histoire naturelle, à travers la conservation des collections archéologiques, le nettoyage et la présentation du gisement et l'aménagement d'un musée de site afin de présenter au public le résultat des recherches scientifiques menées à l'abri Pataud. Des campagnes de fouilles de sauvetage sont alors organisées conjointement par le Muséum national d'Histoire naturelle et la Direction des Antiquités préhistoriques d'Aquitaine. De 1989 à 1991, trois campagnes sont organisées afin de permettre le nettoyage et le relevé systématique de toutes les coupes stratigraphiques. Tous les niveaux archéologiques décrits lors des campagnes de fouilles franco-américaines sont alors retrouvés.

Stratigraphie du gisement

Quatorze niveaux archéologiques majeurs ont été mis au jour. Ces niveaux sont séparés par des dépôts d'éboulis (constitués de blocs effondrés de la voûte et de la paroi), stériles ou « pauvres » en matériels archéologiques.
Les niveaux 14 à 6 sont attribués à l'Aurignacien, datés de 34 250 ± 675 BP (niveau 14 : Aurignacien ancien) à 24 340 ± 700 BP (niveau 6: Aurignacien évolué). Les niveaux 5 à 2 sont attribués au Gravettien, datés de 28 400 ± 1100 BP (niveau 5 : Gravettien ancien) à 22 000 ± 600 BP (niveau 2 : Gravettien récent). Enfin le niveau 1 est attribué au Solutréen inférieur et est daté de 20 400 ± 450 BP (fig. 3).

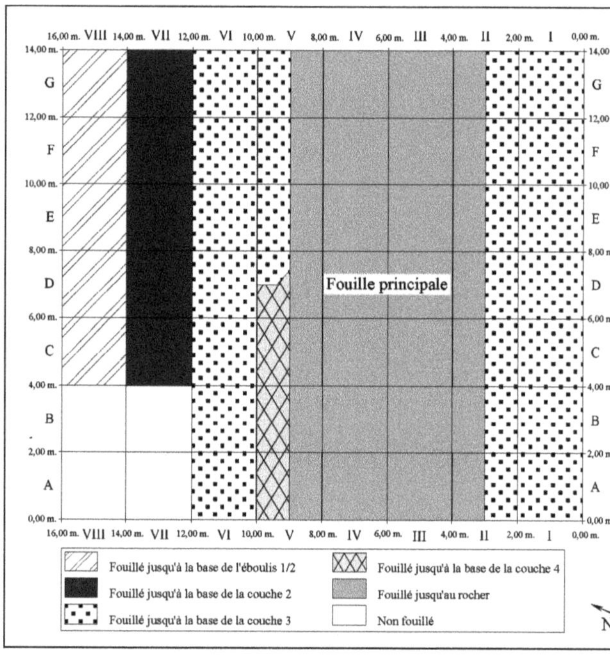

Figure 2: Plan de la zone fouillée (d'après Movius, 1977)

Le Gravettien moyen à burin de Noailles, fouillé au cours des campagnes de 1959 et de 1960, correspond, à l'abri Pataud, au niveau 4. Il est daté de 26 000–27 000 BP. Séparé du niveau 3 par l'éboulis 3/4, attribué au Gravettien moyen final, le niveau 4 fut subdivisé par le professeur Hallam L. Movius Jr. en trois unités principales: *Upper*, *Middle* et *Lower*.

De fait, de nombreux foyers ont été mis à jour, notamment au sommet de l'éboulis 3/4, ainsi que dans la subdivision *Middle* du niveau 4. Mais c'est à la base du niveau 4, dans la subdivision *Lower*, qu'une structure d'habitation a été révélée.

Organisation des structures de combustion

D'énormes blocs d'effondrement en avant du gisement permettent de délimiter un lieu protégé de l'extérieur, entre ces blocs et le fond de l'abri. Les chasseurs-cueilleurs arrivant sur le site vont organiser leur espace d'habitation. Dans cette partie bien protégée de l'abri sous roche, ils vont aménager un ensemble de structures de combustion selon un schéma bien précis (fig. 4).

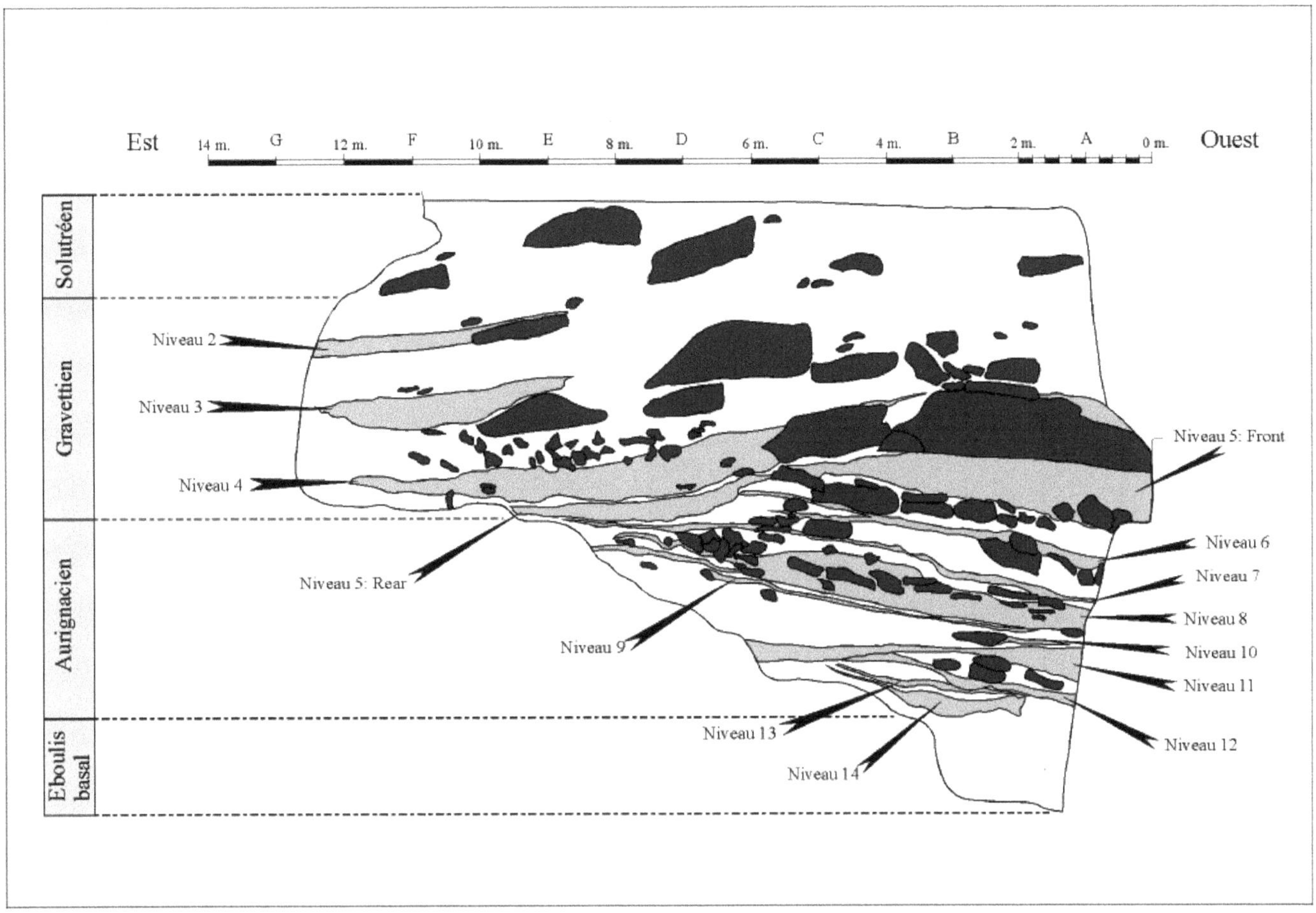

Figure 3: Stratigraphie de l'abri Pataud (d'après Movius, 1977)

Les notes de fouilles disponibles indiquent que ces foyers se présentaient sous la forme de cuvettes peu profondes. Les diamètres variaient entre 0,20 m et 0,50 m. Parmi ces foyers, trois d'entre eux étaient réunis en un ensemble plus complexe, l'ensemble D. Enfin l'aire de ce complexe, composée au total de sept structures de combustion, est délimitée par un demi-cercle partiel constitué par des blocs de calcaire, dont la disposition est sans conteste volontaire. Cet anneau pouvait ainsi délimiter une construction circulaire ou ovalaire, bien qu'aucune trace de celle-ci n'ait été observée sur le sol.

Les notes de fouilles qui nous sont parvenues décrivent les foyers C, D et E (fig. 5).

Le foyer C est une dépression d'environ 0,70 m par 0,50 m. Du matériel carbonisé a été retrouvé au fond de cette dépression, la couche s'épaississant du nord-ouest vers le sud-est. L'ensemble était recouvert d'une fine poussière de couleur grisâtre (cendre). Le mur nord-ouest du foyer C présente une pente, dont l'angle est compris entre 15° et 20° vers le centre de la dépression, tandis que les murs sud-est et sud sont presque verticaux. Le mur sud-ouest est constitué par des blocs. Il semble cependant que ces blocs n'ont pas été placés intentionnellement, mais auraient plutôt glissé dans la cuvette. Enfin le côté nord-est est incliné à environ 50° et présente de nombreux petits fragments de calcaire. Sur le côté nord-ouest, une accumulation de cendre semblent indiquer qu'une personne s'est assise à cet endroit et a ramené la cendre vers elle. Au sein de ce foyer, quelques ossements ont été retrouvés, ainsi que des fragments de calcaire brûlés, quelques silex patinés par le feu, et des os et des dents non brûlés dans la partie ouest. Les quelques pièces brûlées indiquent qu'un feu chaud a été entretenu pendant un certain temps au sein de cette structure.

L'ensemble D correspond à l'association de trois foyers. Le foyer D1, 0,55 m de longueur, 0,40 m de largeur, et d'environ 0,10 m de profondeur, est le foyer le plus organisé de ces trois foyers et le plus important. Les murs sont verticaux, les murs sud-est et sud-ouest étant formés par blocs. Le foyer D2 est le foyer le plus petit de l'ensemble D. Quelques os brûlés y ont été retrouvés, ainsi que quelques os non brûlés. L'aspect des sédiments et des os brûlés indique l'existence d'un feu moyen. Le foyer D3 est le foyer le plus simple. Au fond de celui-ci, de nombreux fragments d'os brûlés ont été retrouvés, de même que des fragments de mandibule et de phalange de renne, mais non brûlés. Ce foyer est limité, dans le coin sud-est, par des blocs.

Le foyer E est un petit foyer, de 0,55 m de longueur et de 0,30 m de largeur. Il présente une forme sub-rectangulaire, en cuvette, mais de faible profondeur, 0,05 m pour sa partie la plus profonde. Dans la séquence finale du Gravettien moyen (Eboulis 3/4), un foyer présentant une structure analogue a été retrouvé. Dans ce foyer, un certain nombre de très petits blocs brûlés, dont nous ignorons s'il s'agit de blocs de calcaire ou de galets, ont été retrouvés notamment dans la partie sud du foyer, c'est-à-dire dans la partie la plus profonde. Ces blocs ont semble-t-il été disposés intentionnellement, et forment une sorte de sol plus ou moins plat. Dans le foyer E, peu de matériel a été retrouvé.

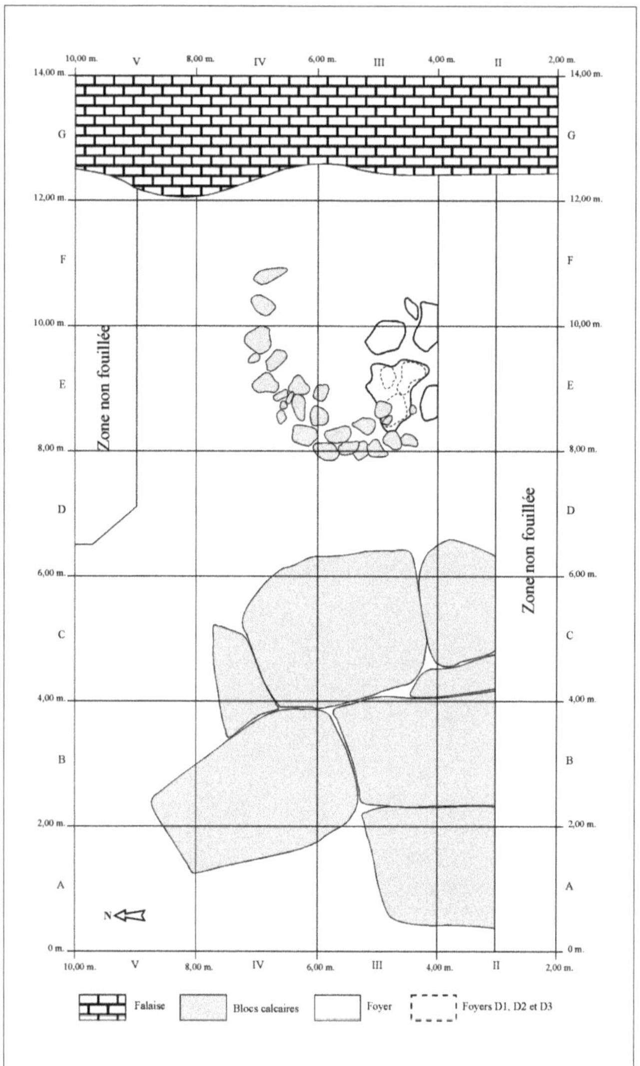

Figure 4: Plan du sol d'habitation de la subdivision Lower *(d'après Movius, 1977)*

Etudes des zones d'activité

Un sol d'occupation présente une organisation spatiale qui dépend des activités se déroulant dans cet espace. Des zones de concentration différentielle des objets archéologiques vont donc tout naturellement s'établir en fonction des activités exercées à tel ou tel endroit de l'habitat. Afin d'appréhender cette organisation spatiale du sol d'occupation, il importe de prendre en compte toutes traces permettant la reconstitution des mouvements de déplacement intentionnel ou non des objets. L'une des approches de cette dynamique réside en l'étude des répartitions spatiales des objets archéologiques.

Figure 5: Organisation des structures de combustion de la subdivision *Lower* (d'après Movius, 1977)

De l'ensemble de la séquence du Gravettien moyen, la subdivision *Lower* est la plus riche en outillage lithique avec quatre mille sept cent huit pièces, dont plus de la moitié correspond aux burins (tous types confondus).

La répartition spatiale de l'outillage lithique montre des concentrations de matériels à l'extérieur de la zone délimitée par le demi-cercle de pierres. Deux concentrations distinctes sont observables. L'une en avant de l'abri, en zone AII-BIII, l'autre au niveau de la tranchée V (fig. 6a). Cependant, une différence très nette apparaît entre la partie sud de la tranchée V et sa partie nord. Cette différence est un artefact de fouille. En effet, la zone principale (tranchées II-nord, III, IV, et V-sud) fut fouillée entre 1959 et 1960, alors que la tranchée V-nord a été fouillée en 1963, suite à une décision d'étendre la fouille à cet endroit. Malgré la présence de cet artefact, nous avons une forte accumulation d'outils en avant de l'abri, le fond semblant peu occupé. Une étude des répartitions spatiales, par types d'outils, montrent une même répartition, quel que soit le type de matériel observé. Ainsi l'utilisation des grattoirs et des burins semble avoir lieu préférentiellement en avant du site, dans la tranchée V et en zone AII-BIII (fig. 6b et 7c). Comparativement, peu de matériel se retrouve au voisinage de l'anneau de blocs. En outre dans ce cas, les grattoirs, ainsi que les burins sont surtout à l'extérieur du cercle de pierre, plutôt qu'à l'intérieur, à proximité des foyers. La répartition spatiale des nucléus est absolument identique à celles observées pour l'outillage (fig. 6d). Il semblerait au premier abord que, soit les activités telles que le débitage ou l'utilisation des outils soient localisées en avant du site, soit les fortes concentrations de matériel indiquent une zone dépotoir. Dans la première hypothèse, les environs immédiats des foyers seraient alors réservés à des activités autres, telle que le repos. Une première approche des produits de débitage non

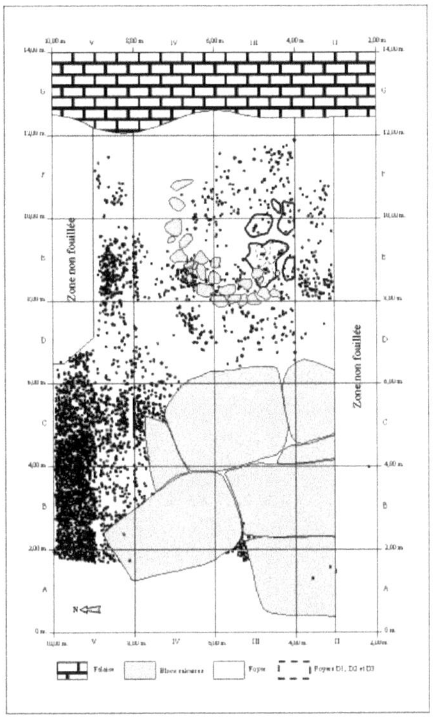

Figure 6 a: répartition spatiale de l'outillage

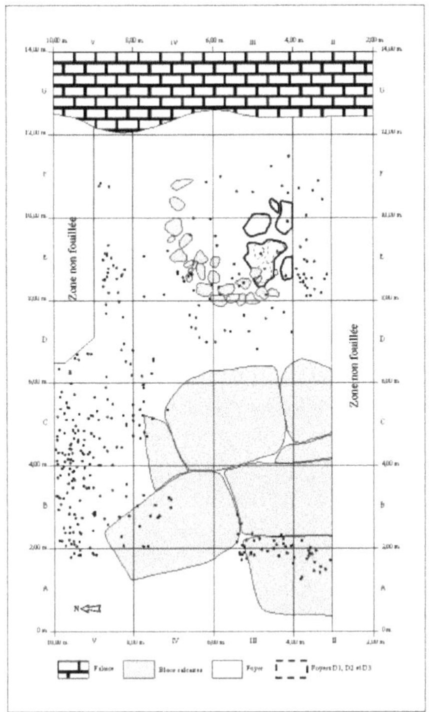

Figure 6 b: répartition spatiale des grattoirs

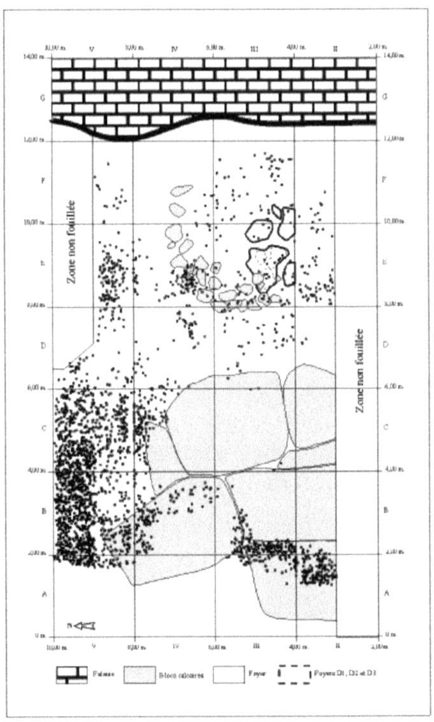

Figure 6 c: répartitions spatiales des burins

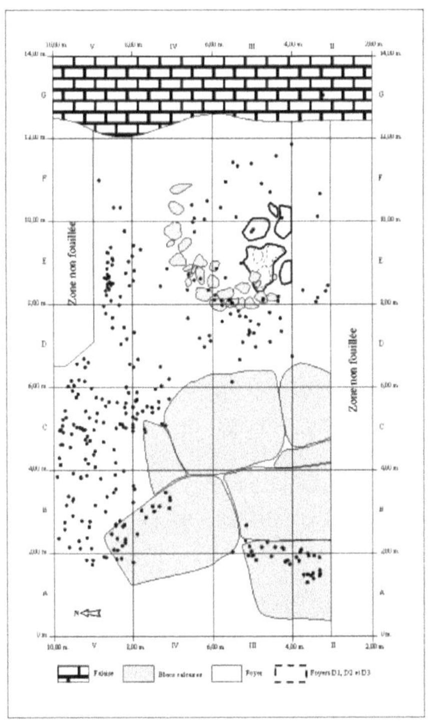

Figure 6 d: répartition spatiale des nucléus

Figure 6: Répartitions spatiales du matériel lithique

retouchés semble vouloir confirmer cette idée de l'existence d'un dépotoir en avant du site. En effet les premières répartitions spatiales des produits de débitage non retouchés montrent une concentration en avant du site, dans la même zone où se concentre la majorité du matériel lithique. Cependant il faut garder à l'esprit que la partie fouillée du site ne correspond qu'à 5 % ou 10 % de la totalité du gisement.

L'étude de la faune et la répartition spatiale des restes osseux totaux (NRT) montrent une forte densité d'ossements en zones BV et CV. La zone CV présente en outre la plus forte densité en ossements déterminés, bien que le taux de détermination soit faible. De même les esquilles brûlées se concentrent essentiellement en zone BV, alors que dans les foyers peu d'esquilles brûlées ont été retrouvées, sans doute dû à des vidanges. Les esquilles non brûlées ont une répartition identique. Enfin les ossements striés ont une dispersion plus large mais la zone CV présente là aussi le taux le plus élevé, ce qui pourrait peut-être indiquer un lieu de traitement des animaux. Cependant l'ensemble des données sur la répartition des restes osseux mènent aux mêmes conclusions que les répartitions spatiales du matériel lithique : soit les activités principales se déroulaient dans les zones BV-CV, soit nous avons affaire ici à une zone dépotoir, résultat de vidanges des foyers et de leurs abords.

Les colorants ont été fortement utilisés pendant ces occupations gravettiennes, et de nombreux fragments de colorants, hématite et manganèse, ont été retrouvés. Ces fragments, au nombre de dix-sept, présentent une double répartition : huit d'entre eux sont localisés en zone BV, et neuf en zone FIII. Il est à noter que de nombreuses plaquettes peintes ont été retrouvées dans ce niveau archéologique, près de la paroi du fond de l'abri, indiquant que celui-ci était peint. Aussi, si l'avant de l'abri correspond à une zone dépotoir, cette présence en zone FIII de nombreux restes de colorants pourrait indiquer une aire de préparation de ces colorants, ceux présents en zones BV et CV résultant du nettoyage des foyers et de leurs abords.

Discussion

L'habitation concentre les activités multiples en un même espace, autour de structures de combustion. La vie du groupe peut être approchée dans sa dynamique quotidienne par l'analyse des répartitions spatiales du matériel archéologique. La qualité des fouilles menées par le professeur Hallam L. Movius Jr., ainsi que la disponibilité des documents relatifs à ces fouilles ont permis la réalisation de ces répartitions spatiales, d'une part pour l'outillage lithique, d'autre part pour la faune. Les données archéozoologiques indiquent un campement occupé tout au long de l'année. L'importante accumulation de matériel lithique confirme cette conclusion. Les répartitions spatiales montrent que la subdivision *Lower* présente une forte concentration aussi bien en restes osseux qu'en industrie lithique en avant de l'abri, dans la tranchée V. En effet nous y retrouvons l'essentiel du matériel lithique (les produits de débitage bruts et retouchés, les nucléus), mais aussi du matériel faunique (ossements déterminés, esquilles brûlées et non brûlées, ossements striés).

L'organisation de l'avant de l'abri indique deux possibilités : soit cette zone de l'abri était réservée aux diverses activités domestiques (traitement des animaux et débitage lithique), soit il s'agit d'une zone d'accumulation résultant du nettoyage de l'espace alentour des foyers et de leur vidange. Les zones d'activités seraient alors localisées à proximité immédiate du complexe de foyers, l'espace libre compris entre ces structures et le fond de l'abri étant davantage des endroits réservés au repos, à la préparation des colorants. Ce complexe correspondrait ainsi au foyer domestique principal constituant le centre de l'habitation, entre la paroi et les blocs d'effondrements. La structure principale a été édifiée dans une zone sans blocs d'effondrements, au fond de l'abri, donc sous la protection du surplomb rocheux. En outre, ce complexe principal montre une organisation élaborée. A côté de cette structure principale, des foyers satellites pouvaient être aménagés dans les zones réservées à des activités plus spécialisées. Cependant il est important d'insister sur le fait qu'une petite partie du site a été fouillée. Il est alors possible que le sol d'occupation dégagé ne corresponde qu'à une extrémité d'une organisation d'habitations plus importante.

Bibliographie

DAVID, N. C.
1985 Excavation of the abri Pataud, Les Eyzies (Dordogne): the Noaillian (Level 4) assemblages and the Noaillian culture in western Europe, *American School of Prehistoric Research*, Peabody Museum, Harvard University, Cambridge, Massachusetts, bull. n°37, 355 p.

DAVID, N. C.
1995 Le Noaillien (« Périgordien Vc ») de l'abri Pataud niveau 4, éboulis ¾: moyen + inférieur, niveau 4a. In Le Paléolithique supérieur de l'abri Pataud, Dordogne. Les fouilles de H. L. Movius Jr., édité par H. M. Bricker. Paris, Maison des Sciences de l'Homme, *Documents d'Archéologie Française*, n°50, pp. 105-131.

MOVIUS, H. L. Jr.
1966 The hearths of the upper Perigordian and Aurignacian horizons at the Abri Pataud, Les Eyzies, Dordogne and their possible significance, *American Anthropologist*, vol. 68, n°2, part. 2, p. 296-325.

MOVIUS, H. L. Jr.
1975 Excavation of the Abri Pataud, Les Eyzies, (Dordogne). Contributors, *American School of Prehistoric Research*, Peabody Museum, Harvard University, Cambridge, Massachusetts, bull. n°30, 305 p.

MOVIUS, H. L. Jr.
1977 Excavation of the Abri Pataud, Les Eyzies, (Dordogne). Stratigraphy, *American School of Prehistoric Research*, Peabody Museum, Harvard

University, Cambridge, Massachusetts, bull. n°31, 167 p.

POTTIER, C.
1999 Le Périgordien Vc ou Noaillien de l'abri Pataud (Les Eyzies-de-Tayac, Dordogne). Etude technologique et typologique de l'outillage du niveau 4: Upper, Paris, Muséum National d'Histoire Naturelle, Institut de Paléontologie Humaine, Mémoire de DEA, 101 p.

RIVIERE, E.
1899 La Croze-de-Tayac, Dordogne, *Comptes Rendus de l'Association Française pour l'Avancement des Sciences*, 28e session, Boulogne-sur-Mer, p. 294.

CREMATION FIRES IN LATE NEOLITHIC FRANCE

Esther Gatto

1. Cremation: an old funerary practice

Cremation is not a recent funeral practice. The Upper Paleolithic can be illustrated by the young woman discovered in the tomb I at Mungo (26500-24500 BP, Australia; Leclerc et Tarrête 1994; Valla 1994). In France, a number of sites, ranging from the Mesolithic (La Vergne; Courtaud and Duday 1995; Duday and Courtaud 1998) to the Middle (Neuvy-en-Dunois; Masset 1968, 1995 or in the south of France, Vaquer 1988), were found showing cremated bones. Although traces of cremation have been discovered in many archaeological sites this funeral practice is not predominant.

Cremation is also very rare in the Late Neolithic. Many types of cremated remains are known: primary and secondary burials, fire in situ or not...

2. Different types of cremated remains in the Late Neolithic

In the Late Neolithic, common burials were frequent in France. The dead were deposited either successively - collective burials- or simultaneously –multiple burials (Duday 1987; Duday *et al.* 1990).

Cremations can be primary deposits (the bones are discovered where the bodies were initially deposited; Duday 1978).
Two types of primary burials are differentiated:

- repeated fires: in collective burials, bodies are burned one after the other. This is the case at "La grotte du Gardon" (Ambérieu-en-Bugey, Ain; Gatto and Buquet 2000; Gatto, in press)
- single combustion: this can be illustrated by the simultaneous burial at Reichstett-Mundolsheim-Souffelweyersheim (Bas-Rhin; Blaizot 2001) in which the part of the fire is clear. Eleven corpses were previously deposited on a wood-shed before cremation. It is not obvious if the last deposit was a consequence of an accidental fire or a cremation.

The last fire could indicate the end of the tomb occupation. This might be the case at "Lacroix-Saint-Ouen" (Oise, Le Goff *et al.* 1995; Billand *et al.* 1995).

Burnt remains can also be secondary deposits. Corpses or bones are cremated on a combustion area and buried in another place where they were discovered. Most of these cases are found in the south-east of France, in the dolmens of Var (Bérard 1954; Courtin 1974; Roudil and Bérard 1981; Sauzade 1983, 1987, 1988; Sauzade *et al.*, 1975, 1988). The collecting of calcined or carbonised bones could have been more or less exhaustive. Two combustion areas have been excavated at "La Bouissière" (Cabasse, Var; Bérard 1954; Roudil and Bérard,1981) and at "La Clauze" (Carnas, Gard; Louis and Peyrolle 1934).

At "Le Rond du Lévrier" (Haute-Loire; Morel and Perrot, 1976), cremated remains were found between layers of unburned bones. At "La Coste" (Saint-Haon, Haute-Loire; Gisclon 1983, 1986), calcined bones were moved to the rock face. At "Peyrolebade" (Allard 1984), they were moved against the walls to leave space for subsequent inhumations.

Cremation is a diversified funerary practice. Moreover, both cremation and inhumation occurred during the whole Late Neolithic.

3. Distribution of cremated burial sites in France

Eighty-four cremation sites are shown on figure 1 and listed in table 1. Two clear concentration areas can be seen: one includes The "Bassin parisien" and "La Marne", the other covers the south of France (Lot, Tarn-et-Garonne, Aveyron, Gard et Var).

This spatial distribution is rather difficult to interpret. Was cremation intensive in these two areas or is research in these regions particularly developed?

4. Different shelters for cremated remains

The second map (Fig. 2) illustrates the diversity of the places where burnt bones were sheltered: simple pits, well-built pits (wooden structure or stone built), dolmens, passage-tombs, tholos, tumulus, cave and rock shelters. These structures are commonly used at that period and reflect regional characteristics. Sepulchral passages are more frequent in the "Bassin parisien". Caves and dolmens are mostly found in the

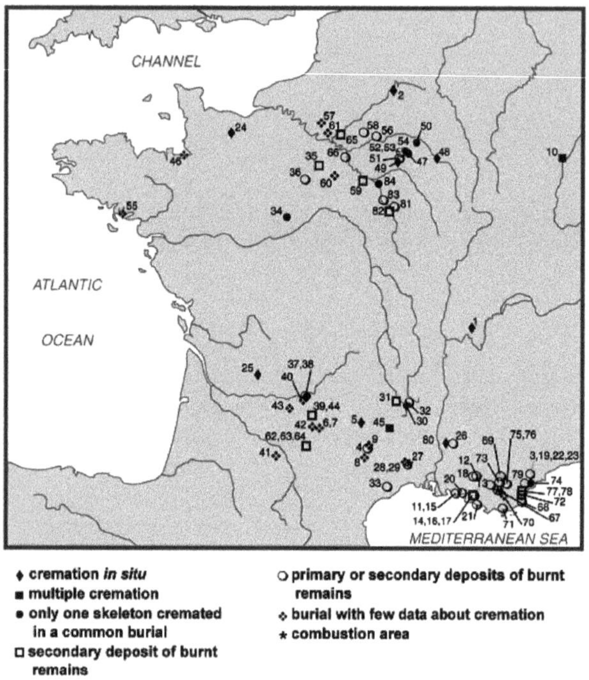

Fig. 1

south of France. We can also notice a high number of pit graves in and around the "Bassin parisien".

Cremated remains are deposited in burial structures quite similar to those used for inhumations.

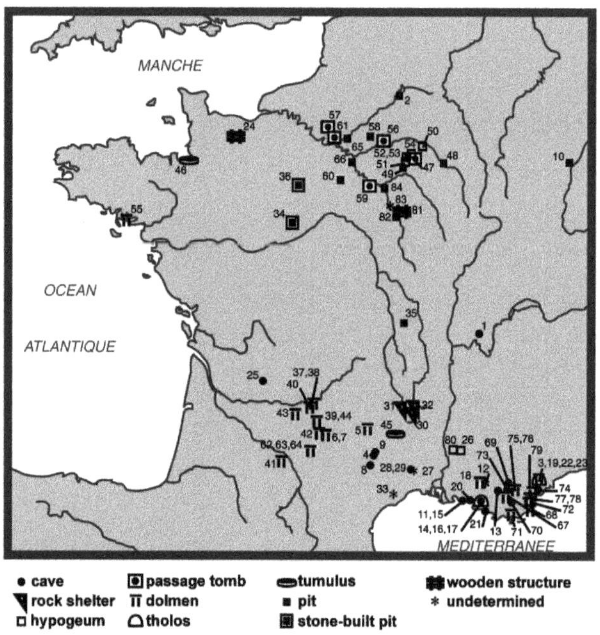

Fig. 2

5. Cremation: Why?

Sanitary reasons are often favoured to explain this particular funerary treatment. Is it a remembrance of the past? In the human mind, fire has often purifying properties. For instance, leprouses were burnt at Lausanne (Morard 1981, cited by Susini 1988: 200). In the Christian world, witches, rebels or heretics were condemned to the stake. In this case, the aim was to purify the soul and to get rid of them (Thomas 1981: 171).

Was this also true in the Neolithic period?
Cremation was not the most prominent funerary treatment. Less than one out of twenty burials were cremations. Does this mean that Neolithic people were particularly resistant to bad sanitary conditions?

Why incinerate the dead during the Neolithic? Was cremation limited to a part of the population? Was cremation related to specific funerary rituals? Only archaeological data can give us some clues. However, the main problem is the elaboration of appropriate archaeological and anthropological methodologies.

6. Who is cremated?

One of the central focuses of anthropological studies is to determine a possible selection of cremated people. Was this treatment reserved to men or women? Nowadays, this problem is difficult to solve. From a morphometrical or visual approach, a reliable sex determination is only possible on coxal bones (Bruzek 1991, 2002). But, coxal bones tend to break and twist excessively during combustion.

The assessment of the age at death is also difficult because of the bad preservation of burnt bones. No reliable method of age at death estimation exists. With the sacro-pelvic surface, which is the best bone indicator, adults can be classified in three large categories with a reliability of 80%: young adults (20-30 years old, middle-aged adults (30-60 years old) and adults deceased after 60 years old (Schmitt 2001). As already mentioned, coxal bones are badly preserved. However, age-at-death assessment of immatures is quite accurate through the calcification of C.F.A. Moorrees, E.A. Fanning and E.E. Hunt, (1963a, 1963b). If mandibles are not preserved, maxillaries can be used (Schour and Massler, 1940). Maturation of bones is also reliable (Birkner 1991). Measurements of long bones correlated to stature and age cannot be used because of contraction of bone during cremation: « about 6-13% with great individual variability » during the combustion and « conditions of cremation process, mainly temperature, are strongly correlated with degree of contraction (higher temperature – greater contraction) » (Strzalko *et al.* 1974). This phenomenon is different depending on the archaeological sites. At "La grotte du Gardon", no restriction is attested.

The information that could be obtained is the demographic profile of the population enabling the detection of possible

Table 1 – List of cremated burial sites in Late Neolithic France.

1 - Gardon, Ambérieu-en-Bugey (Ain), cave
2 - Ribemont (Aisne), pit
3 - Le Coulet du Stramousse, Cabris (Alpes-Maritimes), dolmen
4 - Les Cascades 1, Creissels (Aveyron), cave
5 - Peyrolebade, Espalion (Aveyron), dolmen
6 - Rosier 1, Martiel (Aveyron), dolmen
7 - Rosier 3, Martiel (Aveyron), dolmen
8 - Les Treilles, Saint-Jean et Saint-Paul (Aveyron), cave
9 - Les Truels, Millau (Aveyron), cave
10 - Reichstett-Mundelsheim-Souffelweyersheim (Bas-Rhin), pit
11 - Les Abeilles, Châteauneuf-les-Martigues (Bouches-du-Rhône), cave
12 - La Bastide Blanche, Peyrolles (Bouches-du-Rhône), cave
13 - Baume d'Onze Heures, Trets (Bouches-du-Rhône), cave
14 - Enco-de-Bote, Allauch (Bouches-du-Rhône), tholos
15 - La Marane, Châteauneuf-les-Martigues (Bouches-du-Rhône), cave
16 - Le Pilon du Roy, Allauch (Bouches-du-Rhône), cave D
17 - Pitchoun-Ome, Allauch (Bouches-du-Rhône), cave
18 - La Plaine, Meyrargues (Bouches-du-Rhône), dolmen
19 - Pomeiret, Saint-Vallier (Bouches-du-Rhône), dolmen
20 - Le Rove (Bouches-du-Rhône), cave
21 - Terrevaine (La Ciotat), Bouches-du-Rhône, cave
22 - Tholos 1 de Sainte-Anne, Saint-Vallier (Bouches-du-Rhône), tholos
23 - Tholos 4 de Sainte-Anne, Saint-Vallier (Bouches-du-Rhône), tholos
24 - La Hoguette, (Calvados), arranged pit
25 - Les Barbilloux, Saint-Aquilin (Dordogne), cave
26 - Perpétairi, Mollans (Drôme), hypogeum
27 - La Clauze, Carnas (Gard), combustion area
28 - Macassargue, Corconne (Gard), cave
29 – Rossi, Corconne (Gard), cave
30 - La Coste, Saint-Haon (Haute-Loire), rock shelter
31 - Mont-Jonnet, Coubon (Haute-Loire), rock shelter
32 - Le Rond-du-Lévrier, Salette (Haute-Loire), rock shelter
33 - La Roquette, Saint-Pargoire (Hérault), megalithic tomb
34 - Les Marais de Pouline, Villerable (Loir-et-Cher), pit under stone
35 - Les Canas, Varennes-Changy (Loiret), pits
36 - Maison-Rouge, Montigny (Loiret), pit
37 - Barthes 1, Floirac (Lot), dolmen
38 - Barthes 2, Floirac (Lot), dolmen
39 - Bout de la Combe, Marcilhac (Lot), dolmen
40 - Causse de Montvalent, Montvalent (Lot), dolmen
41 - Les Garroustes, Sérignac (Lot), dolmen
42 - Hauteurs de Lugagnac (Lot), dolmen
43 - Pech Curet, Salviac (Lot), dolmen
44 - Pourquayré, Marcilhac (Lot), dolmen
45 - Dignas, Sainte-Enimie (Lozère), tumulus
46 - Mont-Saint-Michel, Mont-Saint-Michel (Manche), tumulary mound
47 - Le Chemin des Breton, Aulnay-aux-Planches (Marne), passage-tomb
48 - Chiozé, Lignon (Marne), pit
50 - Les Mournouards 3, Mesnil-sur-Oger (Marne), hypogeum
51 - Oyes (Marne), 2 hypogeums
52 - La Pierre-Michelot, Villevenard (Marne), hypogeum
53 - Les Ronces, Villevenard (Marne), cave
54 - Vert-la-Gravelle, Vertus (Marne), hypogeum
55 - Mané-Lud, Locmariaquer (Morbihan), dolmen
56 - Belleville, Vendrest (Oise), passage tomb
57 - Hérouval, Montjavoult (Oise), passage tomb
58 - Layon-Beurrefrais, Peroy-les-Gombries (Oise), pit
59 - Ecuelles, Moret-sur-Loing, near Saint-Mammés, (Seine-et-Marne), passage-tomb
60 - Les Boutards, Saint-Hilaire (Seine-et-Oise), pit under a block of sandstone
61 - Dampont, Us (Seine-et-Oise), passage tomb
62 - Pech 1, Bruniquel (Tarn-et-Garonne), dolmen
63 - Pech 5, Bruniquel (Tarn-et-Garonne), dolmen
64 - Pech 9, Bruniquel (Tarn-et-Garonne), dolmen
65 - Parmain 2, Parmain, Val-d'Oise, pit
66 - Villeneuve-Saint-Georges, Varenne-Saint-Hilaire (Val-de-Marne), pit
67 - La Bouissière, Cabasse (Var), dolmen
68 - La Briande, Ramatuelle (Var), dolmen
69 - Les Cèdres, Le Plan-d'Aups (Var), cave
70 - Gonfaron (Var), cave
71 - Gottaubry, La Londe-les-Maures (Var), dolmen
72 - La Haute-Suane, Grimaud/Sainte-Maxime (Var), dolmen
73 - La Lauve, Salernes (Var), cave
74 - L'Olivier, Les Adrets-de-l'Esterel (Var), cave
75 - Roque-d'Aille, Lorgues (Var), dolmen
76 - Saint-Jeaume, Lorgues (Var), dolmen
77 - San Sébastien 1, Plan de la Tour/Sainte-Maxime (Var), dolmen
78 - San Sébastien 2, Plan de la Tour/Sainte-Maxime (Var), dolmen
79 - La Verrerie-Vieille Saint-Paul-les-Fayence (Var), dolmen
80 - Les Echaffins, Cairanne (Vaucluse), hypogeum
81 - Cheny (Yonne), wooden structure
82 - Monéteau (Yonne), pits
83 - Villeneuve-sur-Yonne (Yonne)
84 - Vinneuf (Yonne), pit with paving on the bottom

abnormalities compared to natural mortality (Sellier, 1996).

At "La grotte du Gardon", young individuals and adults are equally represented and the mortality profile of the immatures indicate a natural mortality (Fig. 3). This is not the case in the dolmen 2 of San Sébastien (Plan-de-la-Tour, Var). There is a great discrepancy between young and adult people (3 under 20 years old among 77 individuals). These two scenarios are frequent in common burials even where cremation is not attested.

Many studies show that cremation is not reserved to particular people, according to sex and age.

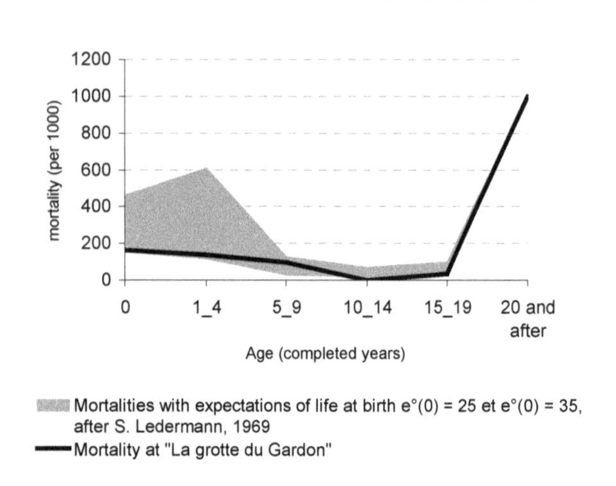

Figure 3 – "La grotte du Gardon" : mortality curve showing the distribution of sixteen immature individuals after « the principe of minizing abnormalities » (Sellier, 1996)

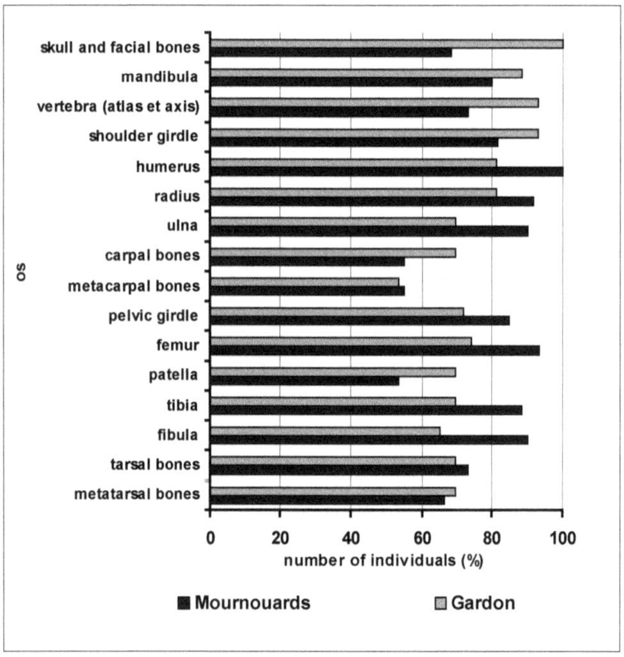

Figure 4 – Osteological diagram illustrating two primary cremation deposits : "La grotte du Gardon" (Ain) and "L'hypogée II des Mournouards" (Marne ; Leroi-Gourhan *et al.*, 1962).

Real values : "La grotte du Gardon", MNI = 43
"L'hypogée II des Mournouards", MNI = 60

7. What parts of the body are cremated?

Did Neolithic people incinerate cadavers, corpses in decay or skeletons? Is it possible to differentiate dry bones, green bones and in-flesh cremations? Several criteria could make the difference: colour, split, cracking, checking... Colour is not a good marker because, in archaeological context, sediment modifies the initial bone colour. From several publications (Binford 1972; Baby 1954), we can assert that:
- deep transverse splitting or checking, curved crack and warping characterize flesh cremation;
- deep longitudinal fracture and superficial checking characterize dry bone cremation.
Few authors (Thurman and Willmore 1981; Buikstra and Swegle 1989) think that it is possible to recognize green bone cremations from in-flesh cremations. The criteria are too subjective to be used.

In fact, when we have to study archaeological burnt bones, it is not easy to identify these differences. Some bones show splits of dry and flesh bone (Legoff 1998), others are not burnt enough to be fractured.

Moreover, some criteria are not clear. What are the differences between deep longitudinal fracturing and straight cracking (Binford 1972)?
We insist on taking these criteria with caution when interpreting cremation process.

8. How cremating?

Reliability of experiments could also be put back into the melting pot. Indeed, corpuses of replications are rare and generally not archaeological. For instance, lots of experiments take place in ovens. As a consequence, combustion is homogeneous and regular. It was probably not the case in the Late Neolithic. Thus, replications do not give a complete idea of possible models.

We propose several replications that could be tested successively on flesh and dry bones:
- first, several modes of combustion should be tested, with high and low temperatures, and with long and short durations. Experiments led by J. Strzalko, J. Piontek and A. Malinovski (1974) show that contraction is limited with low temperature. For example, will the whole organic substance burn without splitting or warping bones during a cremation with slow and limited temperature increase up to 500-600°C during a long period? A flesh bone may not produce warping if dehydration is gradual and slow.
- it would be important to experiment on a sufficient number of bones in order to differentiate reactions between the bones placed in the middle or in the periphery of the burnt area.
- effects of the combustible are also important to study: some wood produces live charcoals and other ashes. Cow-dungs (as in Bali) and ligneous plants are also used as combustible. If the fire is fed during cremation, does it produce particular remains? This phenomenon should be experienced.
- G. Depierre (1995) focused on another phenomenon. If a burnt bone is cremated twice, in which way does the fire process continue? In this case, will a burnt flesh bone show curved cracks? But if it is dry, combustion may produce some deep longitudinal fractures.

- ethnological examples could give us further information on particular cremation. Which kind of remains produces the combustion of mummies (by drying or smoking)? How would the fragments be distributed if the fire were led by a crematist ...?

To test the availability of the whole criteria recorded, it is necessary to make the whole replications on mixed sets of dry and flesh bones.

9. Funerary reconstruction on cremated burials

It is important to study the funerary treatment. Not only the taphonomic movements of the bones should be observed but, also the parts that have been removed following upon the combustion.

The body of a non-constrained cadaver moves during cremation. Due to the dehydration of the tissues (Wahl 1981; Bass 1984), lower and upper limbs bend. As a consequence, a burnt skeleton found in a flexed position, as suggested in "La grotte du Gardon" according to some connected anatomical segments, does not mean that the cadaver was deposited in this position. It is the position at the end of the cremation.

If bones are fragmented and disordered, it is not always due to later altering. In India and Thailand, a shed may be conducted by a crematist (film realised successively by G. Grévin in India and by J.-P. Pautreau and P. Mornais in Thailand).

However, some methodologies give relevant information:
- **observations during excavation:** ashes layer, charcoals, fire *in situ*, burnt sediment...
- **osteological diagrams** (Chambon 1999). First, one should count each anatomical part and then compare the data with a reference set. A diagram for primary deposits tends to be straight: all parts of the body are well represented (Fig. 4). In secondary deposits, small bones are less numerous and sometimes missing (Fig. 5).
- **relative weights**. Weight is currently used in the studies of Bronze and Iron Age incinerations (Taffanel and Janin 1998; Mc Kinley 1994). This tool is particularly relevant when bones are highly fragmented. All fragments are weighed. Weight of each anatomical part is compared to a reference average standard (Lowrance and Latimer 1957). Such a study was realized for "Le dolmen 2 de San Sébastien". It indicates a high representation of the skulls compared to the other parts of the skeleton. However, the osteological diagram tends to show a higher representation of the upper limbs compared to the lower limbs. This impression disappears when looking to the relative weights (Fig. 6).

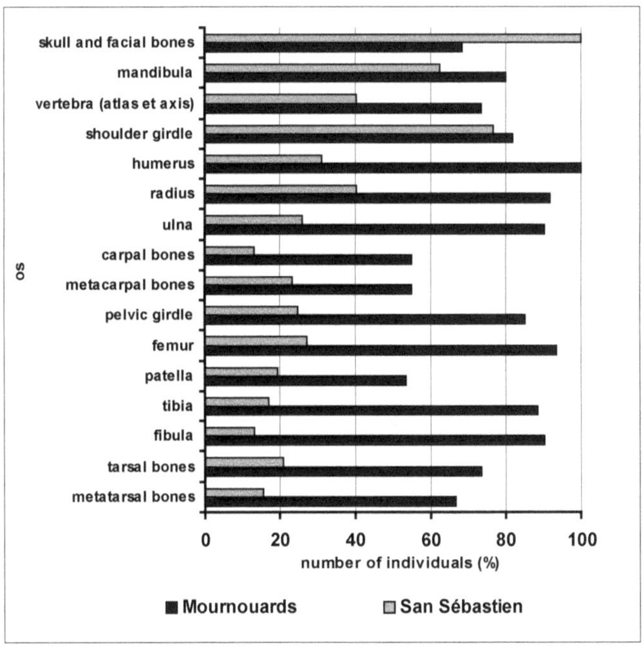

Figure 5 – Osteological diagram comparing a secondary cremation deposit, "Le dolmen 2 de San Sébastien" (Var), and the reference primary cremation deposit, "L'hypogée II des Mournouards" (Marne, Leroi-Gourhan *et al.*, 1962).

Real values : "Le dolmen 2 de San Sébastien", MNI = 77
"L'hypogée II des Mournouards", MNI = 60

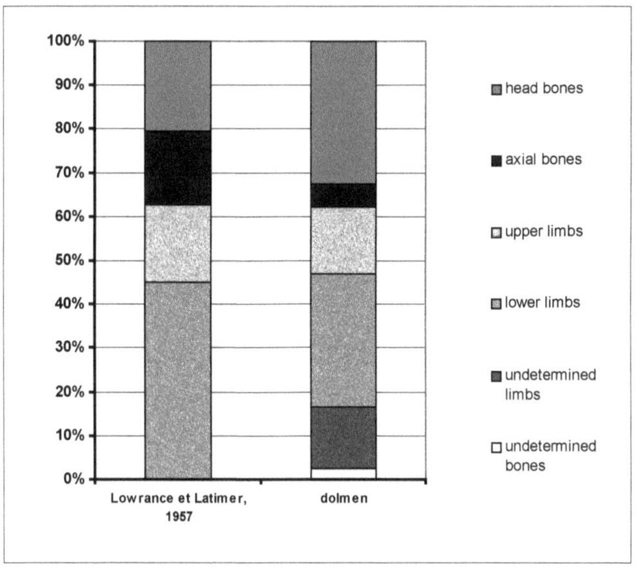

Figure 6 – Relative weights of each anatomical part excavated at "Le dolmen 2 de San Sébastien" (Var) compared to a reference average standards (Lowrance & Latimer, 1957).
Weight of a skeleton = 2881,96 g
Weight of all the bones found in "Le dolmen 2 de San Sébastien" = 41065,94 g

- **spatial analysis**: in common burials, it is possible to work on bone density or anatomical parts of the body on the burial area. This study can also be done after relative weights.

Analysis of bone displacement can be realised from the records or the restoration of osteological bindings (connection, joining, pairing...)
- **vertical analysis**: it is the same process than the one used for spatial analysis. This work enabled to determine successive deposits at "La grotte du Gardon", before finding the thin layers of ashes that divide the burial level in four phases.

Thus, it is possible to distinguish primary burials from secondary ones, and even to identify particular rituals such as collecting some parts of the body, if the secondary deposit is formed by all the collected bones. It is possible to study the amplitude of bone displacements to argue some interpretations:
- cremation has taken place where the body starts decay and it does not seem that fire was led by a crematist, like at "La grotte du Gardon" (Fig. 7);
- secondary burial without particular arrangement, like at the dolmen 2 of San Sébastien;
- the wood-shed fell down. As a consequence, bones are in anatomical position but without strict connections, like at Reichstett-Mundolsheim-Souffelweyersheim (Blaizot 2001).
With vertical analysis, it is possible to find the micro stratigraphy of a burial and hence to have information on the chronology.

Figure 7 – Cremated individual at "La grotte du Gardon" (Ain), lying on the right side, with left upper limb bent.

Conclusion

It seems that fire is associated with the whole types of deposits and shelters known at the end of the Neolithic. Cremation is a funerary treatment contemporary to inhumations. Cremated people are not particularly selected and cremation is not limited to a chronological period and is not associated with unusual burial rituals.

Cremation is not opposed to inhumation, as it is considered in occidental culture and history. However, methodologies to make bones speak are neither numerous nor reliable. Their improvement is crucial if we want to understand cremation.

Acknowledgements

I thank A. Schmitt, P. Murail and Ph. Soler for help with the preparation of the English–language version of the manuscript. I thank P. Sellier for discerning comments.

Bibliography

ALLARD, M.
1984 *Site de Peyrolebade à Espalion (Aveyron). Etude des tertres n° 1, 2 et 3.* Rapport de sauvetage urgent, 45 p.

BABY, R. S.
1954 Hopewell cremation practices. Ohio Historical Society, *Papers in Archaeology*, 1, pp. 1-7.

BASS, W. M.
1984 Is it possible to consume a body completely in a fire? In *Human Identification. Case Studies in Forensic Anthropology,* edited by T.A., Rathbun & J.E., Buikstra, Charles C Thomas publisher, Springfield Illinois, pp. 159-167.

BERARD, G.
1954 Nouvelles découvertes au dolmen de la "Bouissière", Cabasse (Var), *Bulletin de la Société Préhistorique Française*, LI, pp. 666-667.

BILLAND, G., GUILLOT, H. and LE GOFF, I.
1995 Lacroix-Saint-Ouen, Le Prieuré (Oise). *In Allées couvertes et autres monuments funéraires du Néolithique dans la France du nord-ouest, Allées sans retour*, edited by C. Masset & P. Soulier, pp. 169-170.

BINFORD, L.
1972 Analysis of Cremations from three Michigan Sites. *In An archaeological Perspective*, edited by L.R. Binford, New York and London: Seminar Press, pp. 373-382.

BIRKNER, R.
1980 *L'image radiologique typique du squelette*, Paris, Maloine, traduit de l'Allemand par M. Bléry, édition originale: 1977 (Das typische Röntgenbild des Skeletts, München, Urban Scharztenberg).

BLAIZOT, F
2001 Premières données sur le traitement des corps humains à la transition du Néolithique récent et du Néolithique final dans le Bas-Rhin, Dimensions culturelles, *Gallia Préhistoire*, 43, pp. 175-225.

BRUZEK, J.
1991 *Implications à l'étude du dimorphisme sexuel de l'Homme fossile. Fiabilité des procédés de détermination du sexe à partir de l'os coxal,* Thèse de Doctorat, Museum National d'Histoire Naturelle, Paris.

BRUZEK, J.
2002 A Method for Visual Determination of Sex, Using the Human Hip Bone, *American Journal of Physical Anthropology*, 117, pp.157-168.

BUIKSTRA, J. E. and SWEGLE, M.
1989 Bone modification due to burning: Experimental Evidence. *In Bone Modification*, edited by R. Bonnichsen & M.H. Sorg. Center of study of the first American, pp. 247-258.

CHAMBON, P.
1999 *Du cadavre aux ossements. La gestion des sépultures collectives dans la France néolithique.* Thèse de nouveau doctorat d'Ethnologie – Anthropologie – Préhistoire. Université de Paris I. 2 vol.

COURTAUD, P. and DUDAY, H.
1995 Découverte d'une nécropole mésolithique à la Vergne (Charente-Maritime), *Bulletins et mémoires de la Société d'Anthropologie de Paris*, 7, 3-4, pp. 181-184.

COURTIN, J.
1974 *Le Néolithique de la Provence*, Mémoires de la Société Préhistorique Française, 11, 359 p.

DEPIERRE, G.
1995 *Pratiques funéraires gallo-romaines liées à l'incinération. Apport spécifique de l'ostéologie de l'archéologie et de l'ethnologie.* Mémoire de D.E.A., Université de Besançon, UFR des Sciences du langage de l'homme et de la Société, 2 vol., 114 p. 57 fig.

DUDAY, H.
1978 Archéologie funéraire et Anthropologie. Application des relevés et de l'étude ostéologiques à l'interprétation de quelques sépultures pré- et protohistoriques du Midi de la France, *Cahiers d'anthropologie*, 1, pp. 55-101.

DUDAY, H.
1987 Contribution des observations ostéologiques à la chronologie interne des sépultures collectives, *In Anthropologie physique et archéologie, Méthodes d'étude des sépultures*, edited by H. Duday &C. Masset, Paris: CNRS, pp. 169-170.

DUDAY, H and COURTAUD, P.
1998 La nécropole mésolithique de la Vergne. *In Sépultures d'Occident et genèse des mégalithismes (9000-3500 avant notre ère)*, edited by J. Guilaine, Séminaire du Collège de France, pp. 25-37.

DUDAY, H., COURTAUD, P., CRUBEZY, E., SELLIER, P. and TILLIER, A.-M.
1990 L'anthropologie « de terrain »: reconnaissance et interprétation de gestes funéraires, *Bulletins et Mémoires de la Société d'Anthropologie de Paris*, 2, n.s., pp. 29-50.

GATTO, E. and BUQUET, C.
2000 La structure plurielle à crémations de la grotte du Gardon (Ain): une pratique funéraire originale du Néolithique récent ?, *Bulletins et Mémoires de la Société d'Anthropologie de Paris*, 12, 3-4, pp. 303-332.

GATTO, E.
in press La place de la crémation dans le traitement des défunts à la fin du Néolithique en France, *Publication du XIVe Congrès de l'U.I.S.P.P., Liège Belgique/Belgium, 2-8 septembre 2001.*

GISCLON, J.-L.
1983 Les sépultures collectives sous-abri de la Coste 2, Saint-Haon (Haute-Loire), *Bulletin de la Société d'Anthropologie du Sud-Ouest*, XVIII, 3, Bordeaux, pp. 107-116.

GISCLON, J.-L.
1986 Saint-Haon (Haute-Loire). Le gisement funéraire néolithique de la Coste 2, *Revue archéologique du Centre de la France*, 25, 1, pp. 101-102.

LECLERC, J. and TARRÊTE, J.
1994 Incinération. *In Le dictionnaire de la Préhistoire,* A. Leroi-Gourhan, 1st edition 1988, Presses Universitaires de France, Paris, pp. 536-537.

LEDERMANN, S.
1969 *Nouvelles tables-types de mortalité*, Presses Universitaires de France (Travaux et documents; 53).

LE GOFF, I.
1998 *De l'os incinéré aux gestes funéraires, Essai de palethnologie à partir des vestiges de la crémation,* Thèse de doctorat d'Ethnologie - Anthropologie - Préhistoire, Université de Paris I, 507 p.

LE GOFF, I., BILLAND, G., and GUILLOT, H.
1995 Lacroix-Saint-Ouen (Oise), une sépulture collective S.O.M. incendiée?, *Bulletin de la Société Préhistorique Française*, 92, 2, 261-265.

LEROI-GOURHAN, A., BAILLOUD, G., BREZILLON, M., *et al.*
1962 L'hypogée II des Mournouards (Mesnil-sur-Oger, Marne), *Gallia Préhistoire*, V, 1, pp. 23-133.

LOUIS, M. and PEYROLLE, D.
1934 Aire à incinérations de l'époque néolithique, *Congrès Préhistorique de France. Compte-rendu de la XIe session*, Périgueux, pp. 142-144.

LOWRANCE, E. W. and LATIMER, H. B.
1957 Weights and linear Measurements of 105 Human Skeletons from Asia, *American Journal of Anatomy*, 101, 3, pp. 445-459.

Mc KINLEY, J. I.
1994 Bone Fragment Size in British Cremation Burials and its Implications for Pyre Technology and Ritual, *Journal of Archaeological Science*, 21, 339-342

MASSET, C.
1968 Les incinérations du Néolithique ancien de Neuvy-en-Dunois (Eure-et-Loir), II - Etude anthropologique, *Gallia Préhistoire*, I, 1, pp. 205-234.

MASSET, C.
1995 Neuvy-en-Dunois (Eure-et-Loir), Entre 4366 et 3773 avant J.-C. *In Allées couvertes et autres monuments funéraires du Néolithique dans la France du nord-ouest, Allées sans retour*, edited by C. Masset & P. Soulier, Editions Errance, p.192.

MOORREES, C.F.A., FANNING, E.A. and HUNT, E.E.
1963a Formation and resorption of three deciduous teeth in children, *American Journal of Physical Anthropology*, 21, pp. 205-213.

MOORREES, C.F.A., FANNING, E.A. and HUNT, E.E.
1963b Age variation of formation stages for ten permanent teeth, *Journal of Dentist Research*, 42, 6, pp. 1490-1502.

MORARD,
1981 A propos d'une charte inédite de l'évêque Pierre d'Oron : lépreux brûlés à Lausanne en 1321, *Revue d'histoire Ecclésiastique Suisse*, 75, 231-238, cited by A. Susini (1988).

MOREL, P. and PERROT, R.
1976 Anthropobiologie d'un abri-sous-roche préhistorique, Le Rond-du-Lévrier (Haute-Loire), *Travaux et documents de paléoanthropologie et paléopathologie*, 3, Lyon: C.N.R.S., 261 p.

ROUDIL, O. and BERARD, G.
1981 *Les sépultures mégalithiques du Var*, CNRS, Paris.

SAUZADE, G.
1983 *Les sépultures du Vaucluse du Néolithique à l'Âge du Bronze, études Quaternaires. Géologie, Paléontologie, Préhistoire.* Mémoire n° 6, Laboratoire de paléontologie humaine et de préhistoire.

SAUZADE, G.
1987 Le dolmen de Gauttobry, Commune de la Londe-les-Maures (Var), *LAPMO (Travaux du Laboratoire d'anthropologie de préhistoire et d'ethnologie des pays de la Méditerranée occidentale)*, pp. 97-105.

SAUZADE, G.
1988 Le dolmen II de San-Sébastien, Communes de Plan-de-la-Tour et Sainte-Maxime (Var). Réflexion sur les pratiques de l'inhumation dans les sépultures collectives provençales, *Gallia Préhistoire*, 30 (1987-1988), pp. 119-143.

SAUZADE, G., COURTIN, J. and CHABAUD, G.
1975 Le dolmen I de San Sébastien, commune de Plan-de-la-Tour, Sainte-Maxime (Var), *Congrès Préhistorique de France*, Martigues 1974, pp. 581-594.

SAUZADE, G., COURTIN, J. and CHABAUD, G.
1988 Le dolmen dela Haute-Suane (Grimaud-Sainte-Maxime, Var) et la tombe circumaire en bloc de l'Amourié (Grimaud), *Bulletin de la Société Préhistorique Française*, 85, 5, pp. 148-159.

SCHMITT, A.
2001 *Variabilité de la sénescence du squelette humain. Réflexions sur les indicateurs de l'âge au décès : à la recherche d'un outil performant*, Thèse de doctorat en anthropologie, Université Bordeaux I, 350 p.

SELLIER, P.
1996 La mise en évidence d'anomalies démographiques et leur interprétation : population, recrutement et pratiques funéraires du tumulus de Courtesoult. In *Nécropole et société au premier Âge du Fer: le tumulus de Courtesoult (Haute-Saône)*, edited by J.-F. Piningre, Document d'Archéologie Française, 54, Paris, La maison des sciences de l'Home, pp. 188-212.

SCHOUR and MASSLER
1940 Studies in tooth development: the growth pattern of human teeth, part two, *Journal of American Dental Association*, 27, pp. 1918-1931

STRZALKO, J., PIONTEK, J. and MALINOVSKI, A.
1974 Possibilities of burned human bones identifications in the light of experimental investigations. *In Methods, results and consequences of investigations of bones from cremations burials*, edited by H. Gizynska, pp. 31-42.

SUSINI, A.
1988 *Etudes des caractéristiques biophysiques des tissus calcifiés humains (os, émail, dentine) soumis à des traitements thermiques. Applications anthropologiques et médicales.* Thèse de doctorat ès sciences anthropologiques, Université de Genève, Département de Morphologie, Division I, 254 p.

TAFFANEL, O. and J. & JANIN, T.
1998 *La nécropole du Moulin à Maillhac (Aude). Monographie d'Archéologie Méditerranéenne*, 2. Publication de l'UMR 154 du CNRS "Sociétés de la Protohistoire et de l'Antiquité en France Méditerranéenne", Lattes.

THOMAS, L.-V.
1981 *Le cadavre. De la biologie à l'anthropologie*, Edition Complexe, 214 p.

THURMAN, M.D. and WILLMORE, L.J.
1981 A replicative cremation experiment, *North American Archaeologist*, 3, 4, 1980-1981, pp. 275-283.

VALLA, F.
1994 Mungo. *In Le dictionnaire de la Préhistoire,* A. Leroi-Gourhan, 1st edition 1988, Presses Universitaires de France, Paris, p. 756,

VAQUER, J.
1998 Les sépultures du Néolithique moyen en France méditerranéenne. *In Sépultures d'Occident et genèse des mégalithismes (9000-3500 avant notre ère)*, edited by J. Guilaine, Séminaire du Collège de France, pp.167-186

WAHL, J.
1981 Beobachtungen zur Verbrennung Menslicher Leichname, *Archäologisches Korrespondenzblatt*, 11, pp. 271-279.

FIRE AS A MEANS OF RITUAL TRANSFORMATION DURING THE PREHISTORY OF SOUTHERN SCANDINAVIA

Lars Larson

Introduction

Studies of the circumstances surrounding the creation of material culture, as well as of technology and craft organisation, are well represented in research. However, there is an important need to study the causes of the transformation of material culture – deposition in graves, offerings etc., as well. Creating and constructing requires profound practical knowledge. Destruction – or should we say transformation, with fire as one means for this process? – does not require the same physical competence but rather a sense of how to impose and manipulate emotions and impressions among the members of the society on specific occasions, in order to turn an act of destruction into an act of reproduction.

Scandinavian archaeologists have long interpreted collections of artefacts in water or wetlands as results of ritual acts, but they have had a more indifferent attitude to sacral destruction by burning. This has not received the same attention as offerings in wetlands.

These attitudes may mirror our own relation to fire. Fire is seen as something negative – a burning throat in need of some relief, or the importance of adequate insurance coverage against fire damage. Fire is destructive, something final – the true end. Perhaps it is only the fire in the hearth and at the barbecue that evokes positive feelings. The role of fire in crafts may be exemplified by iron forging and associations with soot, dust and sweat; in other words, less positive aspects. This also relates to the difficulties environmentalists have in convincing the general public of the importance of forest fires in maintaining a sound ecological balance.

However, if we move to the Mediterranean area, it is much easier to find evidence for the ritual use of fire. The Old Testament and classical written sources are littered with descriptions of sacrificial burning (Goudsblom 1992). It was the most common means of establishing contact between this world and the metaphysical one. Even today, the use of fire for warmth and food preparation, or for swiddening the soil before sowing, is more common in the Mediterranean area than further north.

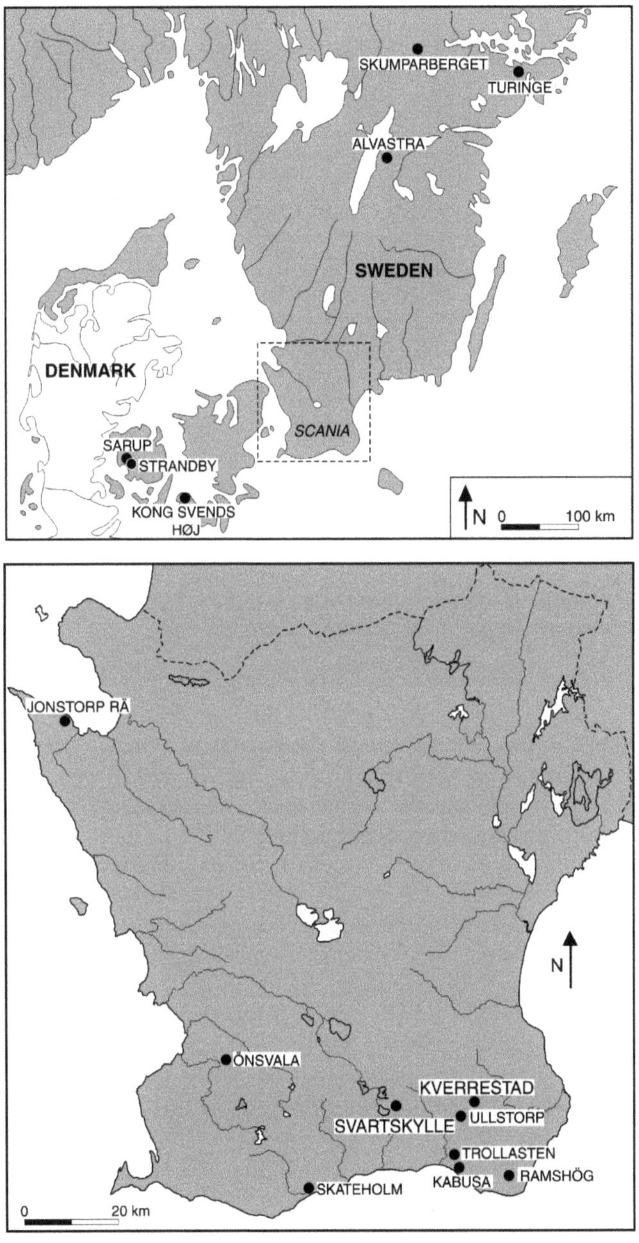

Fig. 1. Southern Scandinavia (top) and Scania – southernmost part of Sweden (bottom) – with the sites mentioned in the text.

For prehistoric people, fire meant warmth, cooking and protection. Fire had much more positive implications, including rebirth and cleansing. I am quite certain that prehistoric ritual burning was much more common than we usually suspect.

The aim of this presentation is to provide some samples of how and where fire was used in Scandinavian prehistory and to concentrate on a few examples in an unlimited temporal perspective.

Massive Demolition by Fire during the Neolithic

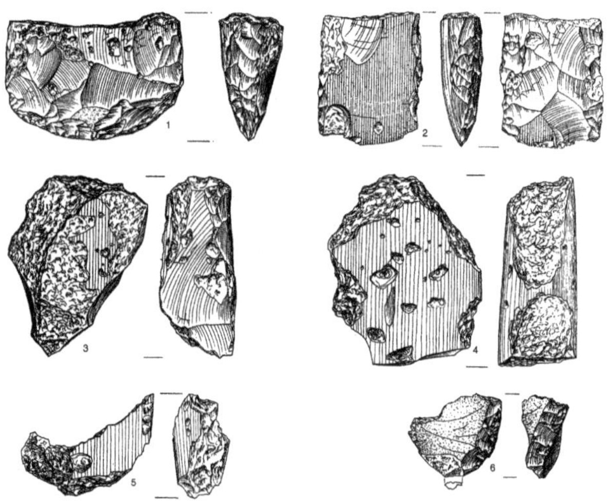

Fig. 2. Artefacts from the Early Neolithic site Svartkylle. 1, 3–5: fragments of thin-butted flint axes, 2: fragment of thin-bladed axe and 6: flake scraper. Scale 2:3.

Svartskylle is situated 12 km from the coast. During the Early Neolithic this area as well as the coastal area bore evidence of well-established settlement (Larsson, M. 1992). The area in between was sparsely settled and probably used as hunting grounds and pastureland (Berglund 1991).

During the extensive revised survey of monuments carried out in Scania during 1990s, fire-damaged flint artefacts were spotted in a field at Kverrestad about 17 km east of the previously mentioned site and at about the same distance — about 15 km — from the sea (Fig. 1). Just like the site at Svartskylle, the other site is situated in a specific setting. A river valley at Kverrestad runs from east to west, delimited to

In south-eastern Scania, in the southernmost part of Sweden, two sites have evidence of massive intentional demolition by fire. During surveys for Neolithic settlement, as part of research involving the cultural landscape in the 1980s, one such site was found at Svartskylle (Fig. 1).

Several thin-butted flint axes (Fig. 2: 1, 3-5) as well as a small number of thin-bladed axes (Fig. 2:2), flake scrapers (Fig. 2:6) and blades, along with some ordinary flakes, had been heavily fragmented by fire (Fig. 2). The axe types date the finds to the later part of the Early Neolithic, c. 3500 cal. BC (Larsson 1989). The burnt flints lay in four concentrations on a hill with a dominant position in the landscape and partly surrounded by wetlands.
the south by a marked ridge. The site is situated on a small plateau on the southern side of the river valley (Fig. 3) with marked slopes on three sides and with a wetland below the site. The bedrock at the site is made up of slate. In the main part of the excavated area only a thin layer of clayish sand, which equals the plough zone, covers the bedrock.

Fire-damaged flints could be found on the plateau within an area of approximately 70 x 70 m (Fig. 4). Excavations revealed a number of pits of varying size and depth in which damaged flint and stone artefacts had been deposited together with a considerable amount of fragmentary pottery (Fig. 5). The largest pit was about 4 m long, the shortest less than 0.5 m. The depth varied from more than a metre to less than ten centimetres below the topsoil. Finds were made throughout the filling, which show that the artefacts were deposited during the entire process of filling in the pits (Fig. 5). A small number of burnt human bones, intentionally cracked into small pieces, were also found.

Fragments from about a hundred thick-butted hollow-ground axes (Fig. 6:1) and chisels (Fig. 6:3) have been found, as well as a small number of thin-bladed axes. There is a variety in the degree of final polishing of the axes. Among the arrowheads, tanged arrowheads have been identified (Fig. 6:5), as well as pressure-flaked projectile points (Fig. 6:4). This type of points has not been identified in southern Sweden before. However, a small number have been found in Denmark (Ebbesen 1980) and interpreted as imports from the Oder area or further to the south. They are present in the Battle Axe culture of the upper Oder area. The finds from Kverrestad make up the largest collection of such points in Scandinavia with about thirty examples. Fragments of a prototype of daggers, known as feeding-knives (Nielsen 1976), have been identified as well (Fig. 6: 6). Flake scrapers and large blades have also been damaged by fire.

Fig. 3. The location of the site at Kverrestad on a on a small plateau in the centre of the picture on the southern side of a river valley.

Non-flint tools such as thick-butted axes and battle-axes (Fig. 6: 2) have been exposed to fire. Among the finds there are battle axes of regional types as well as of a late type from the West Danish Single Grave culture. The stone axes have not only been broken to pieces, like the flint axes they have also been subjected to demolition, such as destruction of the edge.

Clay vessels, mainly of a semicircular shape or rounded with no marked transition between the belly and the neck, decorated with large zigzags on the belly and horizontal impressions just below the rim, also occur.

The dating to the latest part of the Battle Age Culture at about 3900 BP, equal to about 2300 cal. BC, is in good agreement with all the finds (Vandkilde 1996).

Of special interest is the degree of exposure to fire shown by the different tool types. While more than 90% of the axe finds display changes by fire, just one-third of the arrowheads made by pressure-flaking has the same alteration by fire. These marked differences indicate an intentional selection of which type of tool to put in fire and which not to.

That the offerings at Svartskylle and Kverrestad are the result of short-term activities indicates that they should be regarded as completed depositions of mass material. These seem to have been of exceptional size and intended to impress humans as well as metaphysical beings. The destruction of material culture should have been very obvious and the wealth represented by the number of tools and exotics included must have been considerable.

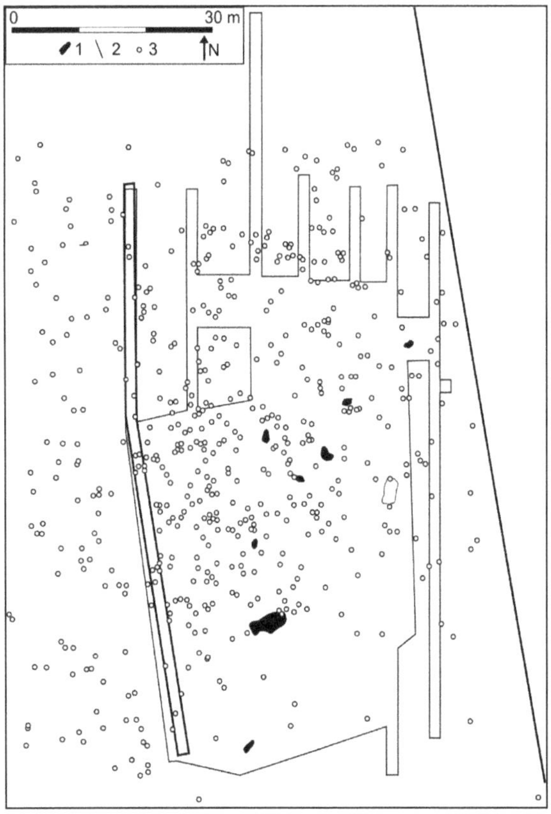

Fig. 4. The distribution of finds and features at Kverrestad. Legend: 1: pits with finds, 2: field boundary, 3: surface finds.

With the exception of Kverrestad and Svartskylle, the finds from just one other site in southern Scandinavia indicate the mass destruction by fire. More than one hundred fire damaged fragments from axes, chisels and scrapers have been found at Strandby, southern Funen (Andersen 2000). The pottery from the site dates the deposition to the earliest part of the Middle Neolithic.

This type of public sacrifice of rare objects and with a direct effect may have been practised on special occasions, probably in combination with external or internal threats. In addition this could be an act, which was primarily meant to legitimate power by impressing representatives of another community.

Fig. 5. Part of a shallow pit at Kverrestad with flint and stone artefacts affected by fire.

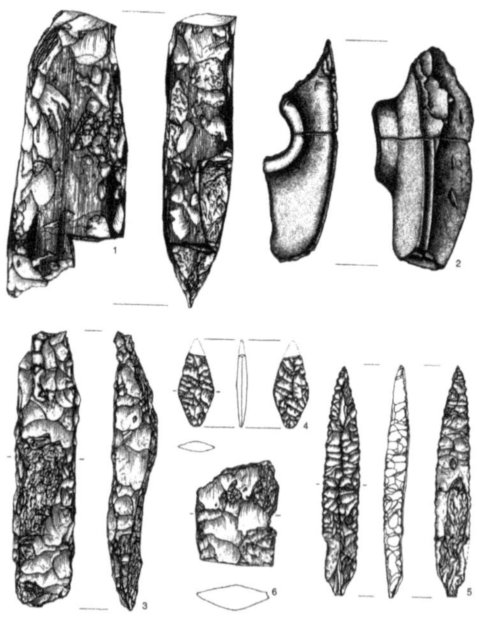

Fig. 6. Artefacts from Kverrestad. 1: fragment of a thick-butted, hollow-ground flint axe, 2: fragment of a battle-axe, 3: fragment of a chisel, 4: pressure-flaked projectile point, 5: tanged arrowhead and 6: fragment of a feeding-knife. Scale 2:3.

Burials and Fire

Although the time difference between Svartskylle and Kverrestad is more than a millennium, there are other finds in the neighbourhood, which might partly bridge the time gap. Outside two megalithic tombs in south-eastern Scania, features have been excavated which included flint tools exposed to fire. Concentrations of burnt human bones, including axes and chisels damaged or destroyed by fire, were found on the stone setting in front of the Trollasten dolmen (Strömberg 1968).

Nine small pits outside the passage of the megalithic tomb of Ramshög contained a filling of soot, cremated human bones and burnt flint artefacts (Strömberg 1971). In some pits fragment of axes destroyed by fire were found. The axes as well as other finds date the features to the late Funnel Beaker culture.

Destruction of tools by fire is relatively frequent at sites throughout the Neolithic (Karsten 1994). The destruction of flint artefacts by fire seems to be less common in burial contexts dated to the Early and Middle Neolithic. However, during a late part of the Funnel Beaker culture axes destroyed by fire are found outside the entrances to megalithic tombs.

Destruction by fire is not common in the context of burials from the Battle Axe culture (Hansen 1917; Christiansson 1953; Malmer 1962; Särlvik and Jonsäter 1974). However, recent excavations have shown that cremation as well as other uses of fire in the mortuary practices are more common than previously thought. At Ullstorp, not more than 5 km to the west of Kverrestad, a grave has been excavated with features related to burning. The grave includes a pit 2.4 m long containing two burials (Larsson 1988). Six post-holes were identified around the pit, which proves that a wooden construction was erected above the grave pit and the wooden coffin in which the interred was placed. The post-holes were the foundation of a rectangular tent-shaped structure. Rites connected with the burial ritual included the deposition or scattering of pottery in the building. Thereafter it was burnt down and a stone frame constructed around the graves, with a layer of stone placed on the coffin. The pit was then refilled. Neither skeletons, parts of skeletons nor grave goods were exposed to heat because of firing. The grave was dated to 3850 BP.

At Turinge in eastern central Sweden a mortuary structure was excavated (Bratt *et al.* 1994; Lindström 1995). Four posts and a shallow trench between the posts delimited an area of 4.6 x 2.8 m. In the trench small pits contained cremated human bones, pottery and flint axes. The artefact had not been exposed to fire. The bones belong to young as well as old individuals. Most bones had been cracked to pieces. However, the buried people must have been cremated elsewhere. Of special interest is that none of the grave goods were exposed to fire. However, a pit nearby contained pieces of burnt flint.

Despite the presence of cremated human bones in Kverrestad, the deposition cannot be regarded as a true grave deposition. No other grave-good depositions show any similarities to the number of tools and vessels found.

Cremated human bones do appear in many other circumstances. One example is the observations made during the excavation of the initial causewayed enclosure at Sarup on south-western Funen, Denmark, dated to an early part of the Funnel Beaker culture. Burnt human bones were found close to the palisade of the first enclosure at the same site (Andersen, 1997). Cremated human bones were also found in two post-holes of a small structure made of altogether four posts in connection with the second stage of the causewayed enclosure (Andersen, 1997: Fig. 117). Flint axes exposed to fire are common in connection to causewayed enclosures in southern Scandinavia as well as henge structures dated to the transition from the Funnel Beaker culture to the Battle Axe culture.

In cases like Sarup and Kverrestad, parts of humans seem to have been an active component of the ritual use of fire as well as the deposition. Parts of the same human body might have been deposited during different activities and at different sites related to the society. Through deposition and different media such as soil, water and fire, the human body was transferred to the supernatural world. In such a way the deposition of some people might have enclosed the entire space in which they played an active part during their lifetime.

Ritual Burning in Different Perspectives

Ritually significant burning is evident in connection with settlement sites from the Battle Axe culture as well (Larsson 1992). At the excavation of a settlement from the late part of the Battle Axe culture at Kabusa, 14 km to the south-west of Kverrestad and situated rather close to the shore of the Baltic Sea (Fig. 1), remains of a house as well as pits were found (Larsson 1992). The sand at one of the pits showed traces of intensive fire, as it was been partly coloured red. Its contents were charcoal and burnt hazelnut shells as well as sherds of pottery.

But traces of intentional fire appear in many other cases during the Middle Neolithic. The pile dwelling from the Middle Neolithic at Alvastra, western Östergötland (Fig. 1), includes remains of ritual activities. The site is dated to about 2800 cal. BC. It consists of two square wooden platforms joined at an oblique angle (Browall 1986). These two square parts are in turn divided into several rectangular "rooms", each with a hearth. Based on dendrochronological studies, the pile dwelling was not used for more than 42 years, but this included long periods when parts of the platform were unused. Large collections of burnt fruits and seeds have been found beside the hearths, such as crab apple, cereals and hazelnuts. Human and animal bones also occur. The structure has primarily been viewed as an assembly place where each family had its own rectangular cell. Such a clear division may have been needed if two social groups, for example, two kin groups, used the same structure for activities of a sacral character. The site includes artefacts characteristic of the Pitted Ware culture as well as the Funnel Beaker culture (Browall 1991).

Fire in ritual acts connected with settlements is by no means restricted to the Middle Neolithic. Intentional burning is thought to be the most plausible interpretation for the remains of a house at Skumparberget from south-eastern central Sweden, dated to the Early Neolithic (Apel *et al.* 1997) (Fig. 1). When houses were destroyed, culture was in a way transformed back into the nature from which the building material originated.

At several Neolithic sites in south-east Europe houses, destruction by fire can be found (Stevanovic 1997). The fire was intentional and a formal way to end the use of a house to show the social and material continuity of Neolithic society. The contents and the arrangement of tools within the houses signify deliberate depositions which were included when the houses were killed by fire (Chapman 1999).

The habit of intentionally destroying structures and clay artefacts by fire can be found as early as the Upper Palaeolithic. In Dolni Vestonice in the Czech Republic, clay figures cracked in pieces when exposed to fire (Soffer *et al.* 1993). These may have exploded when heated. The number of fragmented clay figures is a strong indication of intentional destruction. The same procedures took place in the Funnel Beaker culture and Pitted Ware culture of Sweden (Janzon 1983) as most zoomorphic figures are in fragments. This is also the case for anthropomorphic figurines in the Finnish Neolithic (Kivikoski 1964).

Alteration of flint by fire is practised in other circumstances. Large quantities of unworked flint are often found in connection with megalithic tombs. Burnt flint is found mixed with clay, which covers the orthostats and sometimes the entire chamber. In these cases the flint has been interpreted as constituting a material with a special quality, namely, the ability to absorb moisture in order to keep the chamber dry (Strömberg 1971). Another purpose relates to the aesthetic quality of burnt flint — the white colour. A thick layer of burnt flint covered the surface along the southern side of a passage grave, Kong Svends Høj, on the Danish island of Lolland (Fig. 1). Based on the amount of finds from some test pits, between 4 and 5 tons of burnt flint was used (Dehn *et al.* 1995).

Axes and Fire

That axes represent the majority of finds at Kverrestad as well as Svartskylle is by no means remarkable. Axes seem to be the most used component in many activities of ritual relations during the Neolithic. In the large number of votive depositions in wetland of southern Scandinavia, axes make up the majority of finds (Karsten 1994; Nielsen 1985). Burnt axes rarely appear in wetland depositions.

During the late part of the Middle Neolithic, finds from the Pitted Ware culture contemporaneous with the Battle Age culture provide examples of destruction involving axes. Of the axe fragments from the late Pitted Ware culture site at Jonstorp RÄ in north-western Scania (Fig. 1), about two-thirds had been exposed to fire (Malmer 1969). Of the axe fragments from the excavation, 6% have no polish. But in the surface collection from the same site 21% of the axe fragments are not polished. The interpretation is that offerings of unpolished axes in which fire was involved were performed in the neighbourhood of the settlement, and later spread by ploughing.

In the Funnel Beaker culture cremated human bones as well as axes exposed to fire appear in several different contexts such as settlement sites, graves, causewayed enclosures and wetland offerings. Bones and axes might have acted within a common cosmology but used as agents in different rituals.

In cases like Sarup and Kverrestad, parts of humans seem to have been an active component of the ritual use of fire as well as the deposition in rituals of collective appearances. Parts of the same human body might have been deposited during different activities and at different sites related to the society. Through deposition and different media such as soil, water and fire, the human body was transferred to the supernatural world. In such a way the deposition of some people might have enclosed the entire space in which they played an active part during their lifetime. Axes might have also been regarded as agent for supernatural forces and treated in a way similar to some of the human members of the society. The burning of axes should be understood as a kind of cremation.

Fire – Destroyer and Creator

There is a very marked difference in the attitude towards burning compared with offering in wetlands which was very frequent in southern Scandinavia (Karsten 1994). In wetlands the deposition of artefacts does not change their function. In the former, the destruction of the artefact is easily visible at the point when the practical function of the tool ceases to exist. Fire is the destroyer. Destruction by fire had a direct visual impact. The artefact undergoes remarkable and rapid changes during the act. A colour transformation takes place from natural black or grey to white. Some changes are similar to the cremation of a human body when the colour of the bones changes to white. Ritual burning has a public, direct, evocative and even more magical appearance.

Fire as a medium for transformations connected to *rites de passage* has mainly been applied to the mortuary practice but has also been used in many other circumstances (van Gennep 1909; Douglas 1966; Bloch 1971; Perlès 1977; Huntington and Metcalf 1979). Fire was the destroyer but at the same time the creator.

It is also conceivable that transformation by fire changed character because of purely concrete circumstances. For example, there are major differences as to burning of the forest during the Early and Middle Neolithic. During the former period the trees were cut and burnt – slash-and-burn management. Later, however, parts like the root shoots were protected in order to produce leaf-fodder (Göransson 1996). The significance of fire thereby changed from a medium of total destruction to an element that prepared the way for the budding of new life.

Burning was destructive, but in some cases it had a conserving, preserving result. Posts were singed so that they would last longer in the earth. This was known as early as the Stone Age.

The deposition of artefacts recorded at Kverrestad is a ritual practice, which seems to be a part of a process with several analogies during the Neolithic as well as the Mesolithic.

During the Mesolithic, use of fire in rituals is mainly related to cemeteries from the Late Mesolithic. At the cemeteries at Skateholm I, southernmost part of Scania, a small number of cremation graves were found (Larsson 1993). Cremation graves are forthcoming at Mesolithic sites related to the bog at Vedbæk, eastern Zealand (Kannegaard Nielsen and Brinch Petersen 1993).

However, fire can also be related to ritual without cremation. Traces of wooden structures raised over the grave-pit were found at Skateholm. These had been burnt down prior to the refilling of the grave. One of several dog graves were found just below a fireplace. The fire has not affected the dog.

In Western Europe cemeteries are found in and below shell middens, such as Téviec and Hoïdic on the southern shore of Brittany (Pequart *et al.* 1937; Pequart and Pequart 1954; Schulting 1996). The use of fire in the mortuary practice without effecting the interred is evident.

The earliest Neolithic analogies in southern Scandinavia are probably the destruction by fire of the palisade of posts in front of the grave which was then covered with soil – the first stage in the construction of earthen long barrows during the earliest part of the Neolithic (Larsson 2002). There are examples of firing around some of the earliest stone chambers in long dolmens, which were totally covered by earth (Ebbesen 1990). There are several examples in megalithic tombs where layers of stones covers the deposition of intentionally cracked pottery and other representatives of the material culture some of which combined with fire. After the transitional stage when the artefacts were transferred from this to a supernatural world, they might have been regarded as dangerous or even evil to the existing society. The simplest way to get rid of this material was to cover it order to prevent contact with members of the society. This practice is not much different from the mortuary practice related to dead members of the society. So in a way the artefacts at Kverrestad were actually buried.

What is expressed at Svartskylle and Kverrestad with massive destruction by fire differs from the destruction of single artefacts by fire evidenced at settlement sites and graves. This type of offering including axes is found throughout the entire Neolithic but on certain occasions the act achieves an imposing effect. The cosmology, which dictated burning, just like that related to wetland offerings, was active throughout most of the Neolithic.

The change of colour of the flint artefacts from the natural black or grey to white might be connected to a rite de passage. The colour change could be linked to the process when a human being is cremated. It all ends up with some white bones. In that sense the use of fire on axes could be regarded as the cremation of these flint objects and thereby the mortuary practice as well as rules related to votive depositions might have been interrelated.

Fire in Late Prehistoric Cosmology

"Surt fares northward, lord of the fire-giants;
his sword of flame gleams like the sun;
crashing rocks drag demons to their doom;
men find the way to Hell as the sky splits open"
(*Poems of the Vikings* 1969:10).

and later:

Dark is the sun, the earth sinks below the sea
no bright star now shines from the heavens;
only smoke and menacing fires
leap to the skies in lordly flames

This is the introduction of "The Sibyl's Prophecy" (Völuspá) at Ragnarök – the tale in Norse mythology of how the universe will be destroyed by a combat between giants and gods.
The fire-giants lived in Muspelheim – the world of fire – which was located to the south.
The destruction of the universe, however, results in the creation of a new world – a realm of bliss:

I see rising a second time
out of the waters the Earth, green once more;
an eagle flies over rushing waterfalls,
hunting for fish from the craggy heights.

A rebirth is achieved as the effect of the fire. In this case, as in many others, fire is used as a destroying as well as a purifying element. The new world rises from the sea – one of many examples showing that the dichotomy of fire and water is an important precondition for the cosmic order.

The transition from inhumation to cremation may sometimes be regarded as a drastic change. Yet the change from one form of burial to another, as exemplified, say, in the Scandinavian Bronze Age, was a slow and complex process (Gräslund 1994; Kaliff 1997). Moreover, there are many examples of mixed elements of these two ways of treating the dead. It may be the case that we have concentrated far too much on the treatment of the corpse and devoted far too little consideration to other aspects of the use of fire in conjunction with the burial rite. Even if the deceased was not cremated, this does not mean that fire was not an active element in the rite.

If one believes that the cremation of the body – a destruction of the human matter – was necessary to release the person's soul, then one can view inhumation and cremation as distinct opposites. On the other hand, if one believes that the fire and smoke were sufficient elements for liberating the soul (Kaliff 1997), then the use of fire in mortuary ritual may be seen in a completely different light.

The metamorphosis that took place with cremation, through the release of the body, may have been the aim of the use of fire in inhumation as well, either as one stage in the ritual or as the entire ritual. Traces of burning in graves, in the form of accumulations of ashes and charcoal, indicate ritual burning in conjunction with the filling of the grave.

Fire and Cemeteries

If one looks into the use of fire, cremation is just one of several possible ways to relate fire to mortuary practice. Fire as a medium in mortuary practice extended further in different ceremonials. One such example is a cemetery at Önsvala in the south-west of southernmost Sweden (Fig. 1). It includes some 26 graves deposited during at least three stages between the fourth and eleventh centuries (Larsson 1982). The graves were found because of a rescue excavation resulting in the removal of the plough zone of an area extending far outside the cemetery. Close to the cemetery, twelve strictly rectangular fireplaces were found located in a row fifty metres long (Fig. 7). The radiocarbon dates gave the values 1600±50 and 1590±50 BP (Lu-672 and Lu-671). With a calibrated age of about 375 AD, the fireplaces correlate very well with the first interments of the cemetery. It is of particular interest that there are no cremation graves in the cemetery. The fireplaces might have been used as a kind of ceremonial purification of the site in order to make it acceptable for the settlement of the dead, especially since the area showed traces of settlement during early periods in prehistory.

The row of fireplaces was located about 50 metres from the closest grave. As few cemeteries are excavated in their totality, including a considerable area around, we do not know whether this is a rare example or not. It cannot be ruled out that other rows of hearths, interpreted as examples of a fire cult (Thörn 1996; Björk 1998) may be associated with contemporary graves or cemeteries. But in several graves traces of fire are found in the filling.

The Önsvala cemetery provides more examples of how fire was used in ceremonial contexts to do with mortuary practice.

In a small pit containing just a human skull, the filling was thoroughly mixed with charcoal. The pit, which was marked by a raised stone, is dated to about 675 AD. The skull has not been affected by fire, and although several graves show disturbance by later burials, it is not possible to combine the skull with the postcranial parts of another buried individual. There are a considerable number of graves from the Iron Age where the uncremated body is surrounded by charcoal and ash. In other contexts, horizons of charcoal and ash occur in the filling, showing that a fire burned in the grave before it was completely filled in (Stjernquist 1955; Lamm 1973). This corroborates the use of fire in the burial ceremony, even in graves where the body was not cremated.

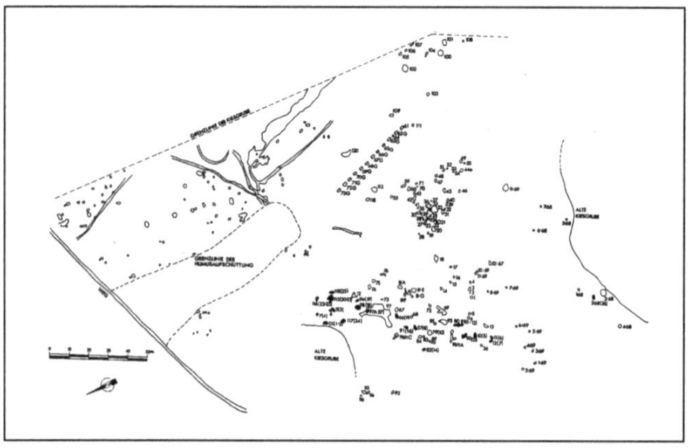

Fig. 7. Önsvala, southernmost part of Sweden. The plan shows different features within the excavation field. The row of rectangular fireplaces (features 61–73G) is situated to the east of the graves marked as filled structures.

Another ritual component in prehistoric society that probably had significant similarities to mortuary practice concerns the deposition of artefacts – a phenomenon that is well attested in Scandinavia since the Early Neolithic (Larsson 1998; 2001). In the Early Iron Age in particular, we find deposits of weapons and equipment associated with an armed group, sometimes on a large scale (Hagberg 1967). These remains may be connected to the testimony of classical authors who describe how a victorious group sacrificed the entire equipment of the defeated group in gratitude for the victory. Some of the objects show destruction by deformation. Sometimes we find deformation by heating, but because of the material in which it is deposited it is at times difficult to detect. In some cases the heavy bending of iron objects could hardly have been achieved without first heating them, but not to a temperature that affected the surface of the objects.

Fire as a Transforming Agent

Another aspect is how the hearth was perceived. The use of the home hearth is a more obvious feature at the transition from the Early to the Late Bronze Age (Skoglund 1999). Plants such as beans, which require more cooking, become more common. At the same time, one can see an increase in the function of fire in ritual contexts, with the change from inhumation to cremation as being another mode of cultural expression.

Burning in ritual contexts has mostly been associated by earlier scholars with cremation. This is related to customs that had a great impact on other aspects of everyday life in prehistory.

The significance of fire as a transforming agent in the production of metals has been well known since Neolithic times. That subject would be far too large for this presentation, but I can cite one example of how patterns of thought concerning transformation can also be expressed in mortuary practice. In Viking Age graves in central Sweden we usually find iron slag. This has often been taken to show that people wanted to emphasise the significance of iron production. Another explanation could be that people saw obvious similarities between the transformation of the human body by cremation and the purification of ore into iron by the use of fire (Burström 1990). The occurrence of flint in Iron Age graves could be explained in the same way. The flints have often been perceived as irrelevant elements, with the exception of the finds that have been interpreted as parts of a fire production process and hence also a possible metaphor for the presence of fire in the burial ritual. Flint has also been used as fluxing agent in the final process of obtaining high-quality iron: the addition of flint raises the temperature and gives a better finished product. Flint may therefore be regarded as a symbol of increasing the ability of the fire to purify and "refine" the body in the cremation process.

To study both the concrete uses of fire and its symbolic function, it will be extremely important in future to carry out careful documentation both of the circumstances reflecting the treatment of the actual body and of the filling of the grave.

References

ANDERSEN, N. H.

1997 The Sarup Enclosures. The Funnel Beaker Culture of the Sarup site including two causewayed camps compared to the contemporary settlements in the area and other European enclosures. *Jutland Archaeological Society Publications* XXXIII: 1. Aarhus.

2000 Kult og ritualer i den ældre bondestenalder. *Kuml* 2000, 13–57.

APEL, J., HADEVIK, C. and SUNDSTRÖM, L.

1997 Burning down the house. The transformational use of fire and other aspects of an Early Neolithic TRB site in eastern central Sweden. *Tor* 29, 5–47.

BERGLUND, B. E.

1991 The Early Neolithic landscape. Landscape, land-use, and vegetation. In Berglund, B. E. (ed.), *The cultural landscape during 6000 years in southern Sweden — the Ystad Project.* Ecological Bulletins 41, 69–70. Copenhagen.

BJÖRK, T.

1998 Härdar på rad. Om spåren efter en kultplats från bronsåldern. *Fornvännen* 1998, no. 2, 73–79.

BLOCH, M.
1971 Placing the Dead. Tombs, Ancestral Villages, and Kinship Organization in Madagascar. London.
BRATT, P., LINDSTRÖM, J. and LUNDSTRÖM, Å.
1994 *Dödshuset från stridsyxetid. Stockholms läns museum*. Stockholm.
BRINCH PETERSEN, E., ALEXANDERSEN, V. and MEIKLEJOHN, C.
1993 Vedbæk, graven midt i byen. *Nationalmuseets Arbejdsmark* 1993, 61–69.
BROWALL, H.
1986 *Alvastra pålbyggnad: social och ekonomisk bas. Theses and Papers in North-European Archaeology* 8. Stockholm.
1991 Om förhållandet mellan trattbägarkultur och gropkeramisk kultur. In Browall, H., Persson, P. and Sjögren, K-G. (eds.), *Västsvenska stenåldersstudier*. Gotarc Serie C 8, 111–142. Göteborg.
BURSTRÖM, M.
1990 Järnframställning och gravritual. En strukturalistisk tolkning av järnslagg i vikingatida gravar i Gästrikland. *Fornvännen* 1990, no. 4, 261–271.
CHAPMAN, J.
1999 Deliberate house-burning in the prehistory of Central and Eastern Europe. Gustafsson, A. and Karlsson, H. (eds.), *Glyfer och arkeologiska rum – en vänbok till Jarl Nordbladh*. Gotarc Series A 3, 113–126. Göteborg.
CHRISTIANSSON, H.
1953 Hyllingefyndet och BF 13–16 i minatyrutförande. *Fornvännen* 1953, No. 2-3, 65–83. Stockholm.
DEHN, T., HANSEN, S. and KAUL, F.
1995 *Kong Svends Høj. Restaureringer og undersøgelser på Lolland 1991. Stenaldergrave i Danmark* 1. København.
DOUGLAS, M.
1966 *Purity and danger. An analysis of the concepts of pollution and taboo*. London.
EBBESEN, K.
1980 En Oldtidshøj ved Systofte. *Lolland-Falsters Stiftsmuseums årsskrift* 1980, 13–33.
1990 The Long Dolmen at Grøfte, South-West Zealand. *Journal of Danish Archaeology* 7, 53–69.
GOUDSBLOM, J.
1992 *Fire and civilization*. London.
GRÄSLUND, B.
1994 Prehistoric Soul Beliefs in Northern Europe. *Proceedings of the Prehistoric Society* 60, 15–36.
GÖRANSSON, H.
1996 Om skottskogsbruk och utfodring med kvistar under mellanneolitisk tid och om skogutnyttjandet under mesolitisk tid. In Slotte, H. and Göransson, H. (eds.), *Lövtäkt och stubbskottsbruk. Människans förändring av landskapet – boskapsskötsel och åkerbruk med hjälp av skog*, 409–452. Stockholm.
HAGBERG, U. E.
1967 *The Archaeologogy of Skedemosse*, I–II. Stockholm.
HANSEN, F.
1917 Några enmansgravar från stenålder. *Fornvännen* 1917, 67–89.
HUNTINGTON, R. and METCALF, P.
1979 *Celebration of Death. The anthropology of mortuary ritual*. Cambridge.
JANZON, G.
1983 Zoomorphic clay figurines and beads from Ire, Hangvar parish, Gotland. *Fornvännen* 1983, 1–20.
KALIFF. A.
1997 *Grav och kultplats. Eskatologiska föreställningar under yngre bronsålder och äldre järnålder i Östergötland*. Aun 24. Uppsala.
KANNEGAARD NIELSEN,E. and BRINCH PETERSEN,E.
1993 Burials, People and Dogs. In Hvass, S. and Storgaard, B. (eds.), *Digging into the Past*, 76–80. København.
KARSTEN, P.
1994 Att kasta yxan i sjön. En studie över rituell tradition och förändring utifrån skånska neolitiska offerfynd. *Acta Archaeologica Lundensia* 8: 23. Lund.
KIVIKOSKI, E.
1964 *Finlands förhistoria*. Stockholm.
LAMM, J. P.
1973 En folkvandringstida kammargrav vid Torsätra. *Fornvännen* 1973, No. 2, 81–89.
LARSSON, L.
1982 Gräbern und Siedlungsreste der jüngeren Eisenzeit bei Önsvala in südwestlichen Schonen, *Schweden. Acta Archaeologica* 52: 129-195.
1988 Mortuary Building above Stone Age Grave. A Grave from the Battle Axe Culture at Ullstorp, Southern Scania, Sweden. *Papers of the Archaeological Institute University of Lund* 7 (1987-1988), 81–98.
1989 Brandopfer. Der frühneolithische Fundplatz Svartskylle im südlichen Schonen, Schweden. *Acta Archaeologica* 59: 143-153.
1992 Settlement and environment during the Middle Neolithic and Late Neolithic. In Larsson, L., Callmer, J. and Stjernquist, B. (eds.), *The Archaeology of the Cultural Landscape. Field Work and Research in a South Swedish Rural Region*, 91–159. Acta Archaeologica Lundensia 4: 19.
1993 The Skateholm Project: Late Mesolithic Coastal Settlement in Southern Sweden. In Bogucki, P. (ed.) *Case Studies in European Prehistory*, 31–62 Ann Arbor.
1998 Prehistoric Wetland Sites in Sweden. In Bernick, K. (ed.), *Hidden Dimensions: The Cultural Significance of Wetland Archaeology*, 64–82. Vancover.
2001 South Scandinavian wetland sites and finds from the Mesolithic and the Neolithic. Purdy, B. A.

(ed.), The Environmental and Cultural Heritage of Wetlands. In press.

LARSSON, L. (ed.)
2002 *Långhögar. De äldsta monumentala gravmonumenten i sydsvenskt neolitikum.* In press.

LARSSON, M.
1992 The Early and Middle Neolithic Funnel Beaker Culture in the Ystad area (Southern Scania). Economic and social change, 3100–2300 BC. In Larsson, L., Callmer, J. and Stjernquist, B. (eds.), *The Archaeology of the Cultural Landscape. Field Work and Research in a South Swedish Rural Region.* Acta Archaeologica Lundensia 4:19, 17–90. Lund.

LINDSTRÖM, J.
1995 Gläntan — dödshuset från stridsyxetid. *Sörmlandsbygden* 1995, 59–70.

MALMER. M. P.
1962 Jungneolithische Studien. *Acta Archaeologica Lundensia* 8: 2.
1969 Gropkeramikboplatsen Jonstorp RÄ. *Antikvariskt Arkiv* 36. Stockholm.

NIELSEN, P. O.
1976 De såkaldte 'madknive' af flint. Aarbøger for nordisk Oldkyndighed og Historie 1974, 104–118. København.

NIELSEN, P. O.
1985 Neolithic Hoards from Denmark. In Kristiansen, K. (ed.), *Archaeological Formation Processes. The representativity of archaeological remains from Danish prehistory,* 102–109. København.

PEQUART, M., and PEQUART, S.-J.
1954 *Hoïdic. Deuxime station-nécropole du mésolithique cotier Armoricain.* Anvers.

PEQUART, M., PEQUART, S-J., BOULE, M. and VALLOIS, H.
1937 Téviec. Station-nécropole mésolithique du Morbihan. *Archives de l'Institut de Paléontologie Humaine* 18. Paris.

PERLÈS, C.
1977 *Préhistoire du feu.* Paris.

Poems of the Vikings 1969. "The Sibyl's Prophecy" translated by Patricia Terry. Indianapolis and New York.

SCHULTING, R.J.
1996 Antlers, Bone Pins and Flintblades: the Mesolithic Cemeteries of Téviec and HoÎdic, Brittany. *Antiquity* 70: 268, 335–350.

SKOGLUND, P.
1999 Diet, Cooking and Cosmology. Interptreting the Evidence from Bronze Age Plats Microfossils. *Current Swedish Archaeology* 7, 149–160.

SOFFER, O., VANDIVER,P., KLIMA,B. and SVOBODA,J.
1993 The Pyrotechnology of Performance Art: Moravian Venuses and Wolverines. In Knecht, H., Pike-Tay, A. and White, R. (eds.), *Before Lascaux. The Complex Record of the Early Upper Paleolithic,* 259–275. Boca Raton.

STEVANOVIC, M.
1997 The Age of Clay: The Social Dynamics of House Destruction. *Journal of Anthropological Archaeology* 16, 334–395.

STJERNQUIST, B.
1955 Simris. On Cultural Connections of Scania in the Roman Iron Age. *Acta Archaeologica Lundensia.* Series in 4o. No. 2. Lund.

STRÖMBERG, M.
1968 *Der Dolmen Trollasten in St. Köpinge, Schonen.* Acta Archaeologica Lundensia 8:7. Lund.
1971 *Die Megalithgräber von Hagestad. Zur Problematik von Grabbauten und Grabriten.* Acta Archaeologica Lundensia 8:9.
1975 *Studien zu einem Gräberfeldt in Löderup. Grabsitte–Kontinuitet–Sozialstruktur.* Acta Archaeologica Lundensia 8:10. Lund.

SÄRLVIK, I. and JONSÄTER, M.
1974 Prästgårdskulle i Veddinge. Undersökning av en förhistorisk gravhög. *Varbergs Museums årsbok* 1974, 85–94.

THÖRN, R.
1996 Rituella eldar. Linjära, konkava och konvexa spår efter ritualer inom nord- och centraleuropeisks brons- och järnålderskulturer. In Engdahl. K. and Kaliff, A. (eds.), *Religion från stenålder till medeltid,* 135–148. Riksantikvarieämbetet. Arkeologiska undersökningar Skrifter nr. 19. Stockholm.

VANDKILDE, H.
1996 *From Stone to Bronze. The metalwork of the Late Neolithic and earliest Bronze Age in Denmark.* Jutland Archaeological Society Publications XXXII. Aarhus.

VAN GENNEP, A.
1909 *Les Rites de Passage.* Paris.

ASPECTS OF THE USE AND FIRE MANAGEMENT OF BRACKEN FERN (Pteridium esculentum)

Bruce Veitch

Plant and ecosystems management in Australia

In 1969 the late Professor Rhys Jones published a landmark paper entitled 'Fire Stick Farming'. He synthesised ethnographic accounts of fire use by Aboriginal people and its ecological impacts, predominantly for the island of Tasmania. Subsequently, his ethnographic research in northern Australia outlined the extensive use of fire as a tool of ecosystems and food plant management, and also conveyed the notion that Aboriginal people practised complex management of ecosystems through the use of fire. His work brought widespread attention to the possibility that the structured use of fire by Aboriginal people may have determined the nature of much of Australia's floral associations by the time of European arrival. Jones further suggested that vegetation associations in Tasmania (and elsewhere in Australia) were changed and maintained to such an extent by the structured use of fire by Aboriginal people that the altered vegetation associations "must be regarded as human artifacts" (Jones 1973: 26).

Jones's thesis drew heavily on a range of ethnographic material, evidence for the use of fire by Aboriginal people across continental Australia, dating from the first visits to the area (e.g. Flinders 1814; Hunter 1793; King 1827 Vol 1: 323; Labilladierre 1800; Menzies 1791; Nind 1831; [but see Pelsart 1629; Stokes 1846: II; Tasman 1644; Vlamingh 1696; Volkersen 1658; all cited in Hallam 1975: 17, 29-33, 16-17, 33]).

Subsequent explorers began to note more detailed associations between the regulated use of fire by Aboriginal people. The historical sources connecting the use of fire and the beneficial effects are so abundant that only a small sample of them will be discussed here[1]. For example, the often quoted passage (below) from Major Thomas Mitchell's explorations in eastern New South Wales reflect a growing awareness of the impact of the regulated use of fire by Aboriginal people.

> Fire, grass, kangaroos and human inhabitants seem all to depend on each other for existence in Australia .... Fire is necessary to burn the grass and from those open forests in which we find the large open forest kangaroo; the native applies that fire to the grass at certain seasons, in order that a young green crop may subsequently spring up and so attract and enable him to kill or take the kangaroo with nets. In summer the burning of the long grass also discloses vermin, birds nests etc., on which the females and the children who chiefly burn the grass feed. But for this simple process, the Australian woods had probably contained as thick a jungle as those of New Zealand or America instead of open forests (Mitchell 1848 cited in Jones 1969: 225; and Mulvaney 1975: 242-243; see also Gott 1982; Gould 1971; Kimber 1976, 1983; Leichhardt 1847: 354; Mitchell 1848: 306; Nind 1831: 28; Stokes 1846 II: 228).

In recent decades it has become increasingly apparent that the use of fire to manipulate floral and faunal associations entailed detailed and complex ecological knowledge, which was articulated through an overarching cultural lense (see footnote 3) (Gould 1971: 23; Hallam 1975; Haynes and Chase 1982; Jones 1969, 1973; Kimber 1976, 1983; Lewis 1982, 1985, 1989). For example, Grey's comment on the management of *Typha* sp. reeds:

> The natives must be acknowledged to bestow a sort of cultivation on this root as they frequently burn the leaves of the plant in the dry season in order to improve it (Grey 1841: 229; see also Gott 1982),

suggests both intricate ecological knowledge and a sophisticated use of it. The burning of areas containing cycads by Aboriginal people is seen as providing a predictable, synchronised returns of increased returns from otherwise staggered yields (Beaton 1977, 1982).

The use of fire to protect productive ecosystems that are fire sensitive is one feature of fire use that has emerged during more recent decades. It has become known that Aboriginal people in several parts of northern Australia, managed monsoon vine forests (a form of rain forest) by either:

[1] The reader is directed to sources such as Jones (1969) and Hallam (1975) for excellent syntheses and discussions of the use of fire as noted from historical texts, and to Jones (1973), Kimber (1976, 1983) and Lewis (1982, 1985, 1989) for excellent examples of contemporary use of fire by Aboriginal peoples in Australia.

- Burning fire breaks around them;
- Intentionally leaving them unburnt;
- Ensuring that they were subject to very occasional and very low intensity fires before the leaf litter had fully cured. This enabled the removal of leaf litter that inhibited the growth of tubers such as yams, without damaging fire sensitive flora (e.g. Lewis 1985: 25, 1989: 949; Haynes and Chase 1982; Jones 1973: 25; Mangglamarra, Burbidge and Fuller 1991; Stocker 1966).

The attention paid to monsoon vine forests was to protect the rich floral and faunal associations within them. Interestingly, in some areas there is evidence to suggest that Aboriginal people may not have protected monsoon vine forests at all (Russell-Smith and Dunlop 1984). Thus there appears to be evidence for nuanced and varied management practices in similar environmental areas of northern Australia (e.g. Lewis 1985, 1989; cf. Head 1994).

The Challenge

Horton (1982) challenged Jones' (1969) "Fire Stick Farming" hypothesis by also drawing upon a broad range of ecological evidence. Horton's basic premise is that the Aboriginal people in Australia "observed and made use of existing natural fire regimes, and they did not attempt to develop a new one". He further states that they were "observers relying on detailed knowledge of their surroundings" and could not make a deliberate attempt to change existing fire regimes (1982: 237, 248-9).

Horton (1982) sees the role of human firing as supplementing and maintaining natural fire regimes. Horton argues that firing by Aboriginal people was confined by natural fire potentials, which are governed by the time taken for particular ecosystem to reach maturity (as outlined by Luke and McArthur 1978, and Walker 1981). "They (the Aboriginal people) could do no more than this, they could not burn beyond the fire potential" (Horton 1982: 247). He argues that to burn beyond this potential would have caused massive extinctions of both flora and fauna (Horton 1982: 241-3). Horton concludes that the ethnography does not conclusively demonstrate a close connection between changes in vegetation and Aboriginal burning (Horton 1982, see also McBryde and Nicholson 1978).

A number of challenges could be made against Horton's argument. These include the fact that the fire potentials of Luke and McArthur (1978) and Walker (1981) only consider major fires and not smaller annual burns (Walker 1981: 117). Also, Horton appears to rely implicitly on climax theory, or classical succession theory. Climax theory has been shown to be of little relevance to the study of Australian ecosystems as it regards fire as an aberration and not a natural part of the ecosystem (Noble and Slayter 1981). The Australian flora has been associated with increasing sources of stress in general, and specifically with fire, as evident from a cursory glance at the available literature. Indeed, much of Australia's flora has evolved to become fire dependent (for example Kemp 1981; Gill 1975, 1977, 1981; Newsome, McIlroy and Catling 1975; Catling and Newsome 1981). Therefore fire is a regular phenomenon in the context of Australian ecosystems and more importantly is a constant factor of natural selection, aspects of which can be manipulated (such as frequency, season and intensity). It seems illogical to assume that sufficient fuel loads to support any kind of fire will not occur before a climax association has reached biological maturity (see also Bowman and Brown 1986).

In response to Horton's challenge, Bowman and Brown systematically assessed Horton's thesis (Horton 1982; cf. Bowman and Brown 1986). Bowman and Brown concluded that his arguments were mostly based on simplistic assessments of ecosystems dynamics. In particular, they note that:

> "The likelihood of a [*vegetation*] community to burn is not only related to fuel mass but to its distribution, moisture content and available ignition sources. In contrast to the views of Mount and Horton the open conditions of regeneration communities allow drying and hence increase the probability of fire rather than the converse..." (Bowman and Brown 1986: 167).

Horton's assertion that reliance on the ethnographic record is problematic is generally upheld, with the exception of Tasmania. However, Bowman and Brown (1986) noted the need for further anthropological work regarding the use of fire as well as studies on autecology or single species ecology.

Discussion

Generally, Horton's premise has received little attention in the published literature since Bowman and Brown's 1986 paper. The common view today in Australia is that fire was and is still used by Aboriginal people to manage their country, to the extent that many areas of Australia are viewed as "landscapes socialised by fire" (Head 1994). Horton's challenge, while addressed in detail by Bowman and Brown (1986), remains valid in the sense that it is incumbent on researchers dealing with issues of plant and ecosystems management to establish the measurable extent to which specific ecosystems and organisms can be impacted by the regulated use of fire (Bowman and Brown 1986). In this context, the examination of Bracken Fern may yield some useful insights into the extent to which a single plant could be managed via the regulate use of fire. Although Bracken Fern is commonly regarded as a fireweed (e.g. Rymer 1976; Watt 1976), Bracken's rhizomes were used as food in a number of places around the world, and in at least Australia and New Zealand, it was managed by fire (e.g. Cremer and Mount 1965: 315; Gott 1982: 64; Jones 1969, 1973; Lourandos 1980a, 1980b). Gott (1982) has noted that Bracken will regenerate quickly from rhizomes and dominate an area for several years until shaded out by competing shrub growth. She also speculates that cyclic burning could ensure a constant supply of Bracken and that the combined effects of burning and gathering would enable dense stands of Bracken to be maintained.

Ethnographic evidence for the use of bracken in Australia

Bracken rhizomes were used as food from Shoalwater Bay on the Central Queensland coast through to Tasmania (Table 1). Bracken has been described as a carbohydrate staple of the Aboriginal people of southeast Australia. It was known to the Aboriginal people of Tasmania as Tara (Backhouse 1843, Appendix D; Maiden 1889: 54), Kuwan to the Brabirrawulung (Mathews 1902: 103), Guange to the Kurnaj of southern Victoria, (Hope Coutts 1971: 107) and Mukine to the Aboriginal people of western Victoria (Smyth 1978: 209; Lourandos 1980a, 1980b, 1983) (Table 1).

Table 1. Ethnographic sources attesting to the use of Bracken rhizomes by Aboriginal people in eastern Australia

Location in Australia	Source
Shoalwater Bay, Queensland	Flinders 1814 (noted in the year 1802)
Sydney area, New South Wales	Bradley 1786-92:107, Collins (1798 cited in Lawrence 1968:198) Hunter 1973:65
Southeast New South Wales	Dawson 1881:20
Victoria	Backhouse 1843, Dawson 1881:20, Lourandos 1980a, 1980b, Smyth 1978:210,214,
Tasmania	Labillardiere 1800; Ling Roth 1890:96, Plomley 1983:133, Robinson 24-6-1831, 1841.

Bracken rhizomes contain carbohydrates and a range of trace elements. The food value of bracken rhizomes is high and contains more protein per unit weight than carrots and pumpkin, and more carbohydrate, fat and calories than potato, carrots, spinach and pumpkin (Cane *et al.* 1979). However, Bracken's reproductive mechanisms and general physiology make it difficult to manipulate as a cultigen. Its rhizomes are intolerant to disturbance, making transplanting difficult (Braid 1959).

Bracken fern: some ecological considerations

It is necessary to provide some basic principles of plant biology before discussing Bracken response to fire. Carbohydrates in plants are necessary for respiration, energy production and growth and are produced by photosynthesis. Plant growth is a function of the amount of leaf tissue exposed to sunlight (Hyder 1972: 304). Stored carbohydrates in root and rhizome reserves decrease in direct proportion to the severity and frequency of defoliation. Carbohydrates are translocated from root and rhizome reserves to areas of foliage growth after defoliation to replace the lost photosynthetic potential. The more severe the defoliation, the longer the time required to reach the stage of stability in energy supply (Youngener 1972). Deterioration may be caused in some plant types through excessive defoliation causing exhaustion (Trlica 1977: 89). Areas of understorey not defoliated for some time, however, may conversely suffer reduced photosynthesis owing to competition for light (Youngener 1972: 300).

Relative to many other understorey species, Bracken can be favoured by fire. A great proportion of the mass of the rhizomatous plant is comprised by the rhizome which is well protected underground, and carries both developing and dormant frond primordia (O'Brien 1963). Consequently, Bracken has the advantage of being able to re-establish fronds without having to undergo extensive germination, re-establishment or structural repair after fire. This bestows the double advantage on Bracken of:

- Requiring less time required to re-establish foliage projective cover than species which regenerate from seed and spore only;
- Requiring a lower energy investment to reach and surpass pre-fire cover levels compared to seed germinating plants.

Bracken and its life cycle

Bracken requires well-drained soil to survive (Jones and Clemesha 1976: 190; cf. Hope and Coutts 1971: 107). Bracken has two means of reproduction, that of spore propagation and by vegetative separation of younger rhizomes from older ones (Crookes and Dobbie 1963: 171; Jones and Clemesha 1976: 32). The details of spore propagation will not be discussed in detail here, suffice to say that for spore propagation to succeed moist conditions are required[2].

The process of vegetative propagation occurs after the sporophyte has established itself. Bracken rhizomes creep and branch as growth and age progress (Braid 1959). Rhizomes will continually branch unevenly so that a single plant can cover a large area in a short time (Crookes and Dobbie 1963: 171). As the rhizome axis grows and branches at one end, it dies at the other (Watt 1976: 218). Consequently, areas of older rhizome between younger areas of rhizome rot and die leaving several individuals (Crookes and Dobbie 1963).

Freshly broken ground provides easy access for Bracken's creeping rhizome. It is quick to occupy and dominate areas where the ground has been disturbed such as burnt out or uprooted tree stumps and recently graded roadside ditches. Woodlands can provide shelter from frosts, which can kill Bracken fronds and apices. Bracken is also protected from the effects of excessive sunlight, wind and localised drought

[2] Further details of the nature of spore propagation in Bracken can be found in more detailed biological literature (eg. Heath and Chinnock 1974:10; Jones and Clemesha 1976; Sporne 1962: 10-11).

which individually or combined can cause the stomata to close and reduce rates of photosynthesis and plant vigour (Watt 1976: 230, Braid 1959: 24).

Excessive litter accumulation can lead to a decline in Bracken vigour. For example, conditions in woodland are usually good for litter decomposition. Where litter accumulation exceeds decay, Bracken will decline, often choked by its own litter. In such cases, young fronds have less access to sunlight, leading to difficulty in breaking through the litter layers and buds may not be able to sprout (Watt 1976: 234). Despite the advantages of wooded areas, bracken's propensity to readily invade open space and to occupy recently cleared areas is a direct response to increased amounts of sunlight reaching the understorey. Not only does woodland shelter Bracken, but it also controls its spread by limiting sunlight penetration (Braid 1959: 29; Cremer and Mount 1965: 315-6; Rymer 1976).

The removal of competing understorey and inhibiting litter cover serves to increase the stimulus of solar radiation to ground level, effectively encouraging more frond growth. In tall grasslands grass cover largely controls bracken frond numbers and dimensions. While frond numbers fluctuate but the average remains fairly constant roughly following a pattern of one new frond growing for every old one dying. In heathland and short grasslands bracken is subject to a state of fluctuating equilibrium. Watt (1976: 223) has described five stages of bracken habitation of heath/grasslands, which are as follows:

1. Grassland and Pioneer phase
2. Building
3. Mature
4. Degenerate
5. Grassland and Pioneer phase.

The Rise and Fall of a Frond

New fronds sprout each spring from buds attached to the rhizome that have remained dormant for some time (Wakefield 1975; Watt 1940, 1947, 1976). As the new frond grows, carbohydrates are translocated from rhizome reserves to the new frond (Hendrick 1921: 161). As the frond continues to grow, the reliance on rhizome reserves decreases until later in the spring when the frond returns carbohydrates to the rhizome and a period of rhizome growth ensues from late spring through autumn and winter (Hendrick 1921: 161; Williams and Foley 1976).

Bracken and Fire

Bracken fern has often been observed to be selectively favoured by burning in a range of ecosystems (Cremer and Mount 1965: 315-6; Gimingham 1972; Gott 1982; Jones and Clemesha 1976: 90; Rymer 1976; Wakefield 1975: 42; Watt 1971, 1976). The dominance of Bracken in recently burnt areas is accomplished by the stimulation of dormant buds through the removal of insulating litter (Wakefield 1975; Watt 1940,1947: 234). In heath and grasslands the cycle of fluctuating equilibrium described above can be arrested at the mature phase, if burnt, by the removal of competing understorey, giving regenerating fronds a temporary monopoly on available sunlight. Repeated burning at the threshold of the "degenerate phase" may maintain the dominance of bracken, especially if the fire regime were to be governed by a perceived threshold of litter accumulation[3].

Conversely, very frequent firing of bracken, especially just after seasonal sprouting in spring, may have the same effect as excessive defoliation. Rhizome reserves become depleted with excessive stimulation of frond growth and plant mortality may result in the long term (Jones 1969; Preest and Cranswick 1978; Watt 1971: 983). Indeed, annual spring-summer cutting of bracken rhizomes is a prescribed and very effective method of Bracken control and removal (Braid 1959; Jones 1969).

The Experimental Fire and its Results

The conduct of the experimental fire is described here in basic detail only with a fuller description is provided in Veitch (1990). An area of open wet sclerophyll forest containing bracken adjoining unimproved pastureland in northeast New South Wales was selected for the experiment. An area of this forest was burnt in the late spring (November)[4]. The fire was of low intensity and patchy distribution. Regrowth in the burnt and unburnt (control) areas was then monitored for 10 months, being sampled in November, March, July and October.

The overall doubling of frond numbers in the burnt area compared with a reduction of approximately 75% in the unburnt area, suggests a stimulation of bracken growth and apparent vigour resulting from fire. This increased frond growth is despite greater exposure of fronds in the burnt area owing to removal of litter cover. The stimulation of frond growth seems to have had some effects on the rhizomes as well.

The slightly greater amounts of gross energy in the rhizomes of the burnt sample from March to July suggest an increased production and movement of carbohydrates from a greater number of fronds. The inversion of burnt and unburnt gross energy levels in September is thought to represent the effect of greater relative photosynthesis occurring in the unburnt area compared with the earlier commencement of frond sprouting in the burnt area. This may indicate that the effects of the fire extend into the next growth flush.

---

[3] Such perceptions of tidiness are commonplace in ethnographic accounts of the use of fire from northern Australia (eg. Jones 1973; Kimber 1976, 1983; Lewis 1982, 1985, 1989).

[4] With the permission and supervision of the local Volunteer Bush Fire Brigade

Table 2. Major features of wet-sclerophyll forest community where the effect of fire on bracken was investigated

Variable	Treatment	November Sampling	March Sampling	July Sampling	September Sampling
Frond Numbers	Burnt	73	300	170	140
	Unburnt	202	285	104	54
Rhizome Analyses					
Gross Energy Content (Kcal/100g)	Burnt		4190	4030	3952
	Unburnt [a]		4100	3970	4050
	Unburnt [b]		3952	3822	3962
Potassium content (ppm)	Burnt		7963	14000	8750
	Unburnt [a]		9119	9850	12500
	Unburnt [b]		14531	9400	13500
Phosphorous content	Burnt		1358	1225	1050
	Unburnt [a]		1517	525	1200
	Unburnt [b]		1450	725	1160

a, b Independent replicate samples

The uniformly lower levels of potassium and phosphorus in the burnt rhizome samples in March may represent a residual effect of increased investment of these nutrients into earlier-accelerated frond growth. This is not the case with the March gross energy levels by virtue of the higher intercellular (apoplastic) compared with intracellular (symplastic) mobility of carbohydrates in the plant. By July there is a uniform increase in potassium and phosphorus concentrations in the burnt sample, compared with lower concentrations of these elements in the unburnt samples. This, together with the inversion of the relative concentration of these elements in the September sampling may reflect an acceleration of the seasonal growth cycle of frond and rhizome in the burnt area. Such an acceleration is opposed to relatively stable element concentrations and the trend toward decreasing frond numbers in the unburnt area. In the burnt area, the elimination of ground litter and most of the competing understorey allowed more light to penetrate to ground level. A fire regime with a frequency of 3-6 years would arrest any trend toward declining frond numbers observed in the unburnt area and maintain the dominance and rhizome production of bracken without exhausting the plant (Veitch 1990). Such a periodicity would be consistent with the buildup of dominance of bracken over time after burning of sclerophyll forest (cf. Cremer and Mount; Jones 1969; Tolhurst 1990).

The Use of Bracken in Australia and New Zealand; a Brief Comparison

Australia

Australian Aboriginal society is generally understood not to be socially stratified in the sense that Maori and other Polynesian societies are (cf. Hiatt, L. 1968; Sahlins 1972). Aboriginal concepts of land ownership are complex and foreign to the European eye, so much so that they are still being defined in Australia's court systems. However, residential patterns are generally thought to be characterised as relatively mobile, to varying degrees, within defined territories (e.g. Lourandos 1977; 1980a; 1980b; Veth 1993).

The gathering activities of women in mainland Aboriginal society provided most of the food used by any one group, particularly vegetable food, such as bracken rhizomes (Bowdler 1981; Hamilton 1981; Hiatt, B. 1976; Hiatt, L. 1968). The implement most likely used for gathering bracken rhizomes was the digging stick, a piece of fashioned timber of variable length (Bowdler 1981). Unlike Maori society, there are no records for the systematic storage of vegetable foods in Aboriginal Australia. In Australia, socially stratified societies that produced large-scale architecture did not arise (e.g. Attenbrow 1987: Chapter 1). Access to land is not usually restricted (Lourandos 1977: 214; Pilling 1968:157), and gathering rights on areas of land are often given in return for reciprocal gathering rights (e.g. Lourandos 1977; Rose 1968:201; Tonkinson 1978: 43-54; Veth 1993). These reciprocal gathering rights appear to hold even where areas of transplanted yams, for example, have been marked by the traditional owners (e.g. Haynes and Chase 1982).

Family units were generally associated in fluid groups including kin relations and friends (Allen 1972: 94-8; Hallam 1975: 40; Tonkinson 1978: 43-54; Veth 1993). These groups often gathered in times of periods of resource abundance for ceremonies (Thompson 1949; Lourandos 1983).

The frequency with which areas of land containing bracken were burnt cannot be accurately estimated, given the general and indirect nature of much of the Australian ethnography (cf. Bowman and Brown 1986; Horton 1982). There is, however, sufficient testimony to support the notion that areas often used by Aboriginal people received varied and regular firing which would have included management of bracken fern (e.g. Gould 1971: 23; Hallam 1975; Haynes and Chase 1982; Jones 1969, 1973; Kimber 1976, 1983; Lewis 1982, 1985, 1989).

New Zealand

Maori society is internally stratified (Sahlins 1972: 139-40). Status within the society ranged from hereditary chiefs to enslaved enemies. The basic unit of Maori society was the extended family which was grouped in traditional lines of descent called *Hapuu*. *Hapuu* could constitute a village and combined to form the largest unit, the *Iwi* or tribe (Keesing 1975: 94-96).

Bracken fern was highly valued in Maori society and is considered to have served as a primary food staple in some areas of New Zealand. It was regarded as more reliable than the cultivated sweet potato or *Kumara* in some areas (Best 1902; Shawcross 1967). Bracken rhizomes were gathered by groups of men who would combine to turn large amounts of earth. This was done using a *Ko* (a 3-6ft length of timber fashioned for the cultivation of *kumara* gardens) or a *Kaheru* (a 2-3ft length of timber, rounded at the handle, flattened and sharpened at the digging end) (Best 1925: 71). The rhizomes were dug in large quantities and carried back to the villages. There they were stacked on stages (*titora aruhe*) and exposed until dry. They were then packed into layers and stored in cooking huts or food stores. From these stores they were removed for roasting, then pounded by women with wooden clubs 10-11 inches long. This pounding was repeated to remove any remaining fibrous material and finally to produce a meal of starchy paste called *Komeke*, which was either eaten raw or roasted into cakes (Best 1902: 51-2. Colenso 1880: 21-2; Cook 1951: 60).

The placement of food items into stores restricted access to them. Visits to storehouses were undertaken in a formal manner and only under the authority of the chief (Colenso 1880:10). Although chiefly status did not convey the same degree of ritual and economic power as among the high chiefs of Tonga and Hawaii (Keesing 1975: 95-6), the Maori chief was able to govern the distribution of wealth, especially food (Best 1902: 103; Sahlins1972: 139-40, 242). Firth (in Sahlins 1972: 139-40, 242) describes the Maori chief as "a kind of channel through which wealth flowed, concentrating it only to pour it out freely again.". Sahlins (1972) sees this as a case of leadership generating domestic surplus, with not all of it redistributed to the people.

The removal of the gathered and processed food from general access to the ritual control of the chief represents a political organization of reciprocities (Sahlins 1972: 209). Fernfields were regularly visited and good fernlands were considered absolute property. Trespass onto the fernfields of others often resulted in harsh retribution even death (Best 1902: 52). Colenso (1880: 21) records that several battles had been fought over the occupancy of an isolated hill, the loose rich soil of which had been famous for the fine fern roots it produced.

Fernfields were burnt triennially, usually in August. This was done in part to control litter levels and old fern growth, but primarily to encourage fresh fern regrowth, which reportedly improved the edible properties of the rhizomes. Newly grown fronds had accessible and very succulent rhizomes compared with the more mature mealier rhizomes (Colenso 1880: 21; Best 1902: 51; Shawcross 1967: 342). The practice of early spring burning of areas with considerable Bracken resources synchronizes the stimulation of new frond growth with the period of natural frond growth.

Discussion

Any given area of Bracken, properly managed by fire could arguably produce more food than a similar area cultivated with *kumara* (Shawcross 1967: 339). The way in which labour was organised to gather and process Bracken rhizomes and to use the gathered food in New Zealand was in marked contrast to the methods used in Australia. The role of Bracken in New Zealand as a key staple seems to be a consequence of the inability to grow most of the tropical Polynesian cultivars in the temperate climate (Shawcross 1967; Jones 1973: 30-1). Such restrictions may have played a lesser role in Australia.

The division of labour among the Maori is clearly different from that within Australian Aboriginal society. In Maori society, half of the labour force (men) gathered rhizomes, while the other half (women) processed them. The storage of Bracken rhizomes and the generation of surplus comprised elements of a redistributive economy geared to a level of production beyond levels of subsistence observed in Australia. Food was not stored to any great extent and it does not seem to be considered as property in Australia Aboriginal society in the way it is in New Zealand.

The material technology used to gather rhizomes in New Zealand was not greatly different from the Aboriginal digging stick (for a discussion of the digging stick see Bowdler 1981). Without detailed comparison it is impossible to assess whether one set of equipment was more efficient than the other. The methods of processing were also similar, except that in New Zealand the Bracken was pounded with wooden clubs, while in Australia stone was used. The construction of drying racks and elaborate storage pits in New Zealand represents a significant departure from Aboriginal use of bracken. These aspects of material culture reflect social and economic relations not found in Aboriginal society.

Generally, it can be suggested that in areas where productive resource zones meet, regular firing associated with repeated visitation may extend Bracken's favoured habitat. As well, the amount of Bracken in any one area may be increased and Bracken levels may be stabilised at a relatively high level of productivity and floral dominance. Bracken exploitation may only constitute a low-key resource to begin with, but with repetitive visitation and firing of favoured food resource areas, Bracken growth and dominance may result.

Also, the period over which large numbers of people could be supported at any particular place may be extended. Examples of this tendency may be found in areas where grasslands lie between swamps, rivers and coastal margins, and open and closed woodlands. The initial attraction of aquatic and nearby terrestrial resources will lead to an increased fire frequency. As Bracken will rarely dominate without human intervention, such areas may be used as an indicator of recent management by fire (Watt 1976: 218).

Conclusions

Bracken can provide a reliable source of food and can be managed by fire. The limits of management are determined by the plant's biological characteristics. Specifically, Bracken cannot be readily transplanted, and its reproductive strategies do not lend it to cultivation. The results of the firing experiment, when combined with extant biological and ecological knowledge about bracken fern, suggests that bracken can be managed by the use of fire to produce enhanced rhizome yields (see also Jones 1969; 1973; Preest and Cranswick 1978; Rymer 1976; Tolhurst 1990; Veitch 1990; Watt 1971; 1976). There are limits, however, to the extent to which Bracken can be managed by fire. Too frequent burning may exhaust Bracken rhizomes, especially if the timing of the burn is after the spring sprouting (say in summer). Also, infrequent burning may allow litter and competing vegetation to impede bracken growth. It appears, however, that the optimum periodicity for enhancing the growth of Bracken Fern may lie between 3-6 years (cf. Tolhurst 1990; Veitch 1990). Any less than three years and rhizomes would be exhausted, any more than six years and competing understorey may outcompete Bracken.

No specific conclusions can be drawn about the periodicity of firing Bracken fields in Australia, given the relative vagaries of the ethnographic record (Bowman and Brown 1986; Horton 1982). Nevertheless, Aboriginal people do appear to have managed the plant with the regular use of fire. Similarly, in New Zealand the Maori people managed Bracken with regular firing. Indeed, fernfields were burnt triennially in the spring (Colenso 1880: 21; Best 1902: 51; Shawcross 1967: 342). Consequently, it appears that both Australian Aboriginal society and Maori society of New Zealand may have either taken the fire management of Bracken Fern to its biological limits, or at least very close to them. In this sense, Jones' proposition that there are real similarities between the economies of Aboriginal people in southeast Australia and Maori people in New Zealand is well founded in a purely ecological sense (Jones 1973).

While there are similarities between the way Aboriginal people of Australia and Maori people of New Zealand managed bracken fern, the similarities seem defined more by the biological limits of bracken, rather than inherent similarities of the two societies. For example, Maori communities owned the land on which Bracken grew in a way that appears different from that in Australia. In Maori society, wars were fought over productive fernfields. Also, in New Zealand, the rhizomes that were gathered were owned and stored. By contrast access to land in Australia appears more governed by principles of reciprocal access to the resources on it. Bracken rhizomes in Australia were gathered and cooked and processed by women, whereas in New Zealand men gathered the rhizomes while women processed them. The presence of elaborate storage facilities associated with centralised control represents a different kind of social organization than in Australia. These differences suggest greatly different cultural ecologies and highlight the sense of complexity of plant management and the use of fire by Aboriginal people in Australia.

Bibliography

ALLEN, H.

1972 Where the Crow flies backwards. Unpublished Ph.D. Thesis, Australian National University, Canberra.

ATTENBROW, V.

1987 The Upper Mangrove Creek Catchment. A study of quantitative changes in the archaeological record. Unpublished Ph.D. Thesis, Sydney University.

BACKHOUSE, J.

1843 *A narrative of a visit to the Australian colonies.* London: Hamilton.

BEATON, J. M.

1977 The dangerous harvest: investigations in the late prehistoric occupation of upland south east and central Queensland. Unpublished Ph.D. Thesis. Australian National University, Canberra.

1982 Fire and Water. Aspects of Aboriginal management of Cycads. *Archaeology in Oceania* 17: 51-8.

BEST, E.

1902 Food plants of' Tuhoeland: being notes on the food supplies of a non-agricultural tribe of the Natives of New Zealand: together with some account of various customs, superstitions, pertaining to foods. *Transactions and Proceedings of the N.S. Institute* (35) 45-111.

1925 *Maori Agriculture.* Wellington: A.R. Shearer.

BOWDLER, S.
1981 Hunters in the highlands: Aboriginal adaptations in the eastern Australian uplands. *Archaeology in Oceania* 16: 99-111.

BOWMAN, D. M. J. S. and BROWN, M. J.
1986 Bushfires in Tasmania: a botanical approach to anthropological questions. *Archaeology in Oceania* 21: 166-171.

BRADLEY, W.
1876-92 *A voyage to New South Wales*. Sydney: Trustees of the Public Library, 1969.

BRAID, K. W.
1959 *Bracken. A review of the literature*. Berkshire: Commonwealth Bureau of Pastures and Field Crops.

CANE, S., STOCKTON, J. and VALLANCE, A.
1979 A note on the diet of the Tasmanian Aborigines. *Australian Archaeology* (9): 77-81.

CATLING, P.C. and NEWSOME, A. E.
1981 Responses of the Australian Vertebrate Fauna to fire: an Evolutionary Perspective. In Gill, A. M., Groves, R. H. and Noble, I. R. *Fire and the Australian Biota*. Canberra: Australian Academy of Science.

COLENSO, W.
1880 On the vegetable food of the ancient New Zealanders. *Transactions and Proceedings of the New Zealand Institute* (13): 3-38.

COOK, J.
1951 *Captain Cook in New Zealand: Extracts from the journals of Captain James Cook*. Wellington: A. H. and A. W. Reed.

CREMER, K. W. and MOUNT, A. B.
1965 Early Stages of plant succession following the complete felling and burning of Eucalyptus regions forest in the Florentine Valley, Tasmania. *Australian Journal of Botany* (13): 303-22.

CROOKES, M. A. and DOBBIE, H. B.
1963 *New Zealand Ferns* (6th ed.). Wellington: Whitcome and Tombs Ltd.

DAWSON, J.
1881 *Australian Aborigines: Languages and Customs of several Tribes of Aborigines in the Western district of Victoria, Australia*. Canberra: Australian Institute of Aboriginal Studies.

FLINDERS, M.
1814 *A Voyage to Terra Australis undertaken for the purpose of completing the discovery of that vast country, and prosecuted in the years 1801-3*. London: G. & W. Nicol.

GILL, A. M.
1975 Fire and the Australian Flora: a review. *Australian Forester* (38) 4-25.
1977 Management of fire-prone vegetation for plant species conservation in Australia. *Search* (8): 20-25.
1981 Adoptive responses of Australian vascular plant species to fire. In Gill, A. M., Groves, R. H. and Noble, I. R. (eds) *Fire and the Australian Biota*. Canberra: Australian Academy of Science.

GOTT, B.
1982 The ecology of root use by the Aborigines of South Australia. *Archaeology in Oceania* 17: 59-65.

GOULD, R.
1971 Uses of fire amongst the Western Desert Aborigines of Australia. *Man* 8: 14-24.

GREY, G.
1841 Journals of two expeditions of discovery in North West and Western Australia, during the years 1837, 38 and 39. London: T. & W. Boone. 2v.

HALLAM, S.
1975 *Fire and Hearth*. Canberra: Australian Institute of Aboriginal Studies.

HAMILTON, A.
1981 Dual social systems. technology, labour, and womens secret rites in the eastern Western Desert of Australia. *Oceania* (51): 4-19.

HAYNES, R. A. and CHASE, A. K.
1982 Aboriginal influence on plant communities in Cape York Peninsula. *Archaeology in Oceania* 17(1): 38-50.

HEAD, L.
1994 Landscapes Socialised by Fire: post-contact changes in Aboriginal fire use in northern Australia, and implications for prehistory. *Archaeology in Oceania* 29(3): 172-181.

HEATH, E. and CHINNOCK, R. J.
1974 *Ferns and Fern Allies of New Zealand*. Wellington: A. H. and A. W. Reed.

HENDRICK, J.
1921 The composition of the rhizomes of Bracken and its variations. *Kew Bulletin* 157-66.

HIATT, B.
1976 Woman the gatherer in Gale, F. (ed.). *Women's role in Aboriginal Society*. Canberra: Australian Institute of Aboriginal Studies Canberra (36).

HIATT, L.
1968 Ownership and use of land among the Australian Aborigines. In Lee, R.B. and Devore, I. (eds) *Man the Hunter*. Chicago: Aldine.

HOPE, G. S. and COUTES, P. J. F.
1971 Past and present Aboriginal food resources at Wilsons Promontory, Victoria. *Mankind* (8): 104-14.

HORTON, D.
1982 The Burning Question, fire and the Australian Ecosystem. *Mankind* 13(3): 237-51.

HUNTER, J.
1793 *Journal of the Transactions at Port Jackson and Norfolk Island*. J. London; Stockdale.

HYDER, D. N.
1972 Defoliation in relation to Vegetative Growth. In Youngener, Y.B. and McKell, C.M. (eds) *The Biology and Utilization of Grasses*. New York: Academic Press.

JONES, D. L. and CLEMESHA, S. C.
1976 *Australian Ferns and Fern Allies*. Wellington: A. H. and A. W. Reed.

JONES, R.
1969 Fire Stick Farming. *Australian Natural History* 224-28.
1973 The Neolithic Palaeolithic and the hunting gardeners - man and land in the antipodes. An address to the *International Union for Quaternary Research* (INQUA), *IX International Conference*: 21-35.

KEESING, R. M.
1975 *Kin Groups and Social Structure.* New York: Holt Rinehart and Winston.

KING, P. P.
1827 *Narrative of a survey of the intertropical and western coasts of Australia, performed between the years 1818 and 1822.* London: John Murray.

KIMBER, R.
1976 Beginnings of farming? Some man plant relationships in central Australia. *Mankind* (10): 142-150.
1983 Black lightning: Aborigines and fire in Central Australia and the Western Desert. *Archaeology in Oceania* (18): 38-45.

KEMP, E. M.
1981 Pre Quaternary fire in Australia. In A.M. Gill et al (eds) *Fire and the Australian Biota.* Canberra: Australian Academy of Science.

LABILLARDIERE, W.
1800 *Voyage in search of la Perouse performed by the order of the Constituent Assembly during the years 1791, 1792, 1793 and 1794.* Drawn up by M. Labillardiere. London: John Stockdale.

LAWRENCE, M. A.
1968 *Aboriginal habitat and economy.* Occasional paper No. 6 Dept. Geography M. A. thesis. Canberra: Australian National University.

LEICHHARDT, L.
1847 *Journal of an overland expedition in Australia, from Moreton Bay to Port Essington...during the years 1844-1845.* London: T. and W. Boone.

LEWIS, H. T.
1982 Fire technology and resource management in Aboriginal North America and Australia In N. M. Williams and E. S. Hunn (eds) *Resource Managers. North American and Australian Hunter-Gatherers,* pp. 45-68. Canberra: Australian Institute of Aboriginal Studies.
1985 Burning the 'Top End': kangaroos and cattle. Australian Heritage Publication Series No. 7 (1). In J. R. Ford, (ed.) *Fire Ecology and Management in Western Australian Ecosystems*, pages 21-32. Perth: Western Australian Institute of Technology.
1989 Ecological and technological knowledge of fire: Aborigines versus park rangers in northern Australia. *American Anthropologist* 91: 940-961

LING-ROTH, H.
1890 *Port Mackay, Queensland.* Halifax: F. King and Sons.

LOURANDOS, H.
1977 Aboriginal spatial organization and population: South Western Victoria reconsidered. *Archaeology and Physical Anthropology in Oceania* 12(3): 202-25.
1980a Forces of change. Unpublished Ph.D. Thesis, Sydney University.
1980b Change or stability?: hydraulics hunter gatherers and population in temperate Australia. *World Archaeology* 11(3): 345-64.
1983 Intensification: a late Pleistocene-Holocene archaeological sequence from S.W. Victoria. *Archaeology in Oceania* (18): 81-94.

LUKE, R. H. and McARTHUR, A. G.
1978 *Bushfires in Australia.* Dept of Primary Industry. Forestry Timber Bureau. Canberra: Commonwealth Scientific and Industrial Research Oganisation Division of Forest Research.

MAIDEN, J. H.
1889 *The Useful Native Plants Of Australia.* Melbourne: Compendium 1975.

MATHEWS, R. H.
1902 The Aboriginal languages of Victoria. *Journal and Proceedings of the Royal Society of New South Wales* (36): 71-106.

McBRYDE, I. and NICHOLSON, P.
1978 Aboriginal man and the land in South Western Australia. *Studies In West Australian History* (3): 38-42.

MANGGLAMARRA, G., BURBIDGE, A. A. and FULLER, P. J.
1991 Wunambal words for rainforest and other Kimberley plants. In N.L. McKenzie, R.B. Johnston and P.G. Kendrick (eds) Kimberley Rainforests of Australia, pp. 413-421. Chipping Norton, NSW: Surrey Beatty and Sons.

MITCHELL, T. L.
1848 *Journal of an expedition into the interior of Tropical Australia.* London: Longman, Brown. Green and Longmans.

MULVANEY, D. J.
1975 *The prehistory of Australia.* Melbourne: Pelican.

NEWSOME, A.E., McILROY, J. and CATLING, P.
1975 The effects of an extensive wildfire on populations of 20 ground vertebrates in S.E. Australia. *Proceedings of the Ecological Society of Australia* (9): 107-23.

NIND, S.
1831 Description of the natives of King George's Sound (Swan River Colony) and adjoining country. *Journal of the Royal Geographical Society* Vol. 1: 21-51.

NOBLE, I. R. and SLAYTER, R. O.
1981 Concepts and Models of succession in Vascular Plant communities subject to recurrent fire In Gill, A. M., Groves, R. H. and Noble, I. R. (eds) *Fire and the Australian Biota.* Canberra: Australian Academy of Science.

O'BRIEN, T. P.
1963 The morphology and growth of *Pteridium aquilinium* var. *esculentum* (Forst.) Kuhn. *Annual Bot.* NS 27: 253-267.

PILLING, A. R.
1968 South Eastern Australia: level of social organisation. In Lee, R. B. and Devore, I. (eds) *Man the Hunter*. Chicago: Aldine.
PLOMLEY, N. I. B.
1983 *The Baudin expedition and the Tasmanian Aborigines 1802*. Hobart: Rubber Head.
PREEST, D. S. and CRANSWICK, A. M.
1978 Burn Timing and Bracken Vigour. *Proceedings of the 31st New Zealand Weed and Pest Control Conference,* pages 69-73, Auckland.
ROBINSON, G. A.
1829-34 *Friendly Mission. The Tasmanian journals and papers of G. A. Robinson*. N. J. B. Plomley (ed.) 1966.
1841 Journals of G.A. Robinson March - May 1841. *Records of the Victorian Archaeological Survey* (6) 1977.
ROSE, F. G. G.
1968 Australian marriage, land owning groups, and initiations. In Lee, R. B. and Devore, I. (eds) *Man the Hunter*. Chicago: Aldine.
RUSSELL-SMITH, J. and DUNLOP, K.
1984 The status and condition of monsoon vine-forests in the Kakadu region: a management report. Unpublished report to the N.P.W.S. Australian National University, Canberra.
RYMER, L.
1976 The History and Ethnobotany of bracken. *Botanical Journal of the Linnean Society* 73: 151-76.
SAHLINS, M. D.
1972 *Stone age economics*. Chicago: Aldine.
SHAWCROSS, K.
1967 Fern Root and the total scheme of C18th Maori food production in Agricultural areas. *Journal of the Polynesian Society:* 330-52.
SMYTH, R. B.
1978 *The Aborigines of Victoria*. Melbourne: Government Printer.
STOCKER, G. C.
1966 Effects of fires on vegetation in the Northern Territory. *Australian Forester* 30: 223-30.
STOKES, J. L.
1846 *Discoveries in Australia: with an account of the coasts and rivers explored and surveyed during the voyage of H.M.S. Beagle in the years 1837-38-39-40-41-42-43*. London: T. and W. Boone.
TOLHURST, K. G.
1990 Response of bracken to low intensity prescribed fire in open eucalypt forest in west-central Victoria. In Thompson, J. A. and Smith, R. T. (eds). *Bracken Biology and Management,* pages 53-62. Sydney: Australian Institute of Agricultural Science Occasional Publication Number 40.
TONKINSON, R.
1978 *The Mardudjara Aborigines. Living the Dream in Australia's Desert*. New York: Holt, Rinehart and Winston.
TRLICA, M. J.
1977 Distribution and utilization of Carbohydrate Reserves in Range Plants. In Sosebee, R. E. (ed.) *Rangeland Plant Physiology*. Colorado: Society for Rangeland Plant Management Number 4.
VEITCH, B.
1990 Aspect of Aboriginal use and manipulation of bracken fern. In Thompson, J. A. and Smith, R. T. (eds). *Bracken Biology and Management,* pages 215-226. Sydney: Australian Institute of Agricultural Science Occasional Publication Number 40.
VETH, P. M.
1993 *Islands in the Interior: The dynamics of prehistoric adaptations within the arid zone of Australia*. International Monographs in Prehistory 3. Michigan: Ann Arbor.
WAKEFIELD, N. A.
1975 *Ferns of Victoria and Tasmania*. Field Naturalists Club of Victoria.
WALKER, J.
1981 Fuel dynamics in Australian vegetation. In Gill, A. M. Groves, R. H. and Noble, I. R. (eds) . Fire and the Australian Biota. Canberra: Australian Academy of Science.
WATT, A. S.
1940 Contributions to the Ecology of Bracken (*Pteridium aquilinium*). The Rhizome. *New Phytologist* 39: 401-22.
1947 Contributions to the Ecology of Bracken (*Pteridium aquilinium*) IV. The Structure of the Community. *New Phytologist* 46: 97-121.
1971 Contributions to the Ecology of Bracken (*Pteridium aquilinium*) VIII The Marginal and the Hinterland Plant: a Study in Senescence. *New Phytologist* 70: 967-86.
1976 The ecological status of bracken. *Botanical Journal of the Linnean Society* 73: 217-39.
WILLIAMS, G. H. and FOLEY, A.
1976 Seasonal variations in the carbohydrate content of Bracken. *Botanical Journal of the Linnean Society* 73: 87-93.
YOUNGENER, Y. B.
1972 Physiology of Defoliation and Regrowth. In Youngener, Y.B. and McKell, C.M. (eds) *The Biology and Utilization of Grasses*. New York: Academic Press.

BURNED HOUSES IN THE NEOLITHIC OF SOUTHEAST EUROPE

Mirjana Stevanovic

Introduction

This article summarizes the results of extensive research on burned houses of the Neolithic of Southeast Europe, which were extensively published (Stevanovic 1996, 1997). The research comprised of excavation, recording, sampling, and the analyses of Neolithic Vinca culture houses at the sites of Vinca Belo Brdo, Gomolava, Opovo, and Selevac. The research was designed with the intention to collect and analyze the evidence for building construction and for the causes and mechanism of house burning and their destruction. The aspects of Neolithic architecture in the region have been largely understated, and at the same time, they potentially hold a wealth of information on behaviour of the Neolithic people.

Past and present architecture has been analyzed from a variety of perspectives and Neolithic houses have been perceived as the defining characteristic of cultural behavior. Houses have been considered as central to Neolithic society but viewed as simple, functional features of the cultural landscape. Therefore, in Neolithic archaeology of Southeast Europe, the excavators have been concerned mostly with the house interiors and very superficially with the house appearance and their structural elements. There are difficulties in working with the Neolithic burned house remains, which can account for this limited view being applied to the houses in Southeast Europe. One is that architectural remains are difficult to record because of their burned state, another, that analyzing the burned remains is a completely new view that has not been considered before. In addition, the explanations of this material culture have mostly regarded Neolithic houses as containing human action and processes and not so much as the expression of those. This meant that the excavators of the houses have focused on recording and analyzing the activity areas and the mobile artifacts in the house, not on the house itself. My study of Neolithic houses begins with the view that the house is a process and that it has a life-history, and thus, a beginning and an end. I believe that a house construction materials and its multiple transformations should be studied as a complex artifact, which can express physical, social, and symbolic aspects of the lives of the people who dwell in it.

In Southeast Europe house building as well as house burning are the characteristics of the settlements during the Neolithic period. I believe that Neolithic house construction and destruction were closely related practices, and through my research I have been trying to examine their relationship. The practice of house construction and deliberate house burning may seem to be on opposite sides of the spectrum, but they can also be seen as complementary practices. I suggest that a deliberate social strategy of house building with the intent to destroy them at the end of their use-life might have existed during the Neolithic. The conclusions reached through my research are that these house fires were deliberate and most likely a part of a social practice to end, in a ceremonial way, the use-life of a house and to give the house remains a highly symbolic value. Consequently, the houses were destroyed by burning, which was initiated and facilitated by the very people who built and inhabited them.

Time and Place Framework

The destruction of houses by conflagration is characteristic of the Middle to Late Neolithic/Early Eneolithic of Southeast Europe (5300-3500 B.C). This time period begins with early farmers, continues with mature farmers, and reaches the climax of farming societies, and it ends with the transition to the Copper Age (see Chapman 2000: 3). Burned house remains have been observed in the prehistoric periods before and after the Neolithic to Copper Age period but not nearly on the same scale. The similarity among the archaeological remains of houses in the Southeast European Neolithic is quite remarkable. The likeness in architecture of this large area can be seen not only in the house inventories but also in the volume of building materials and especially in the burned state of the building materials that is universal at the Neolithic settlements of Serbia, Macedonia, Thessaly, Bulgaria, Romania, and Southern Hungary (see Chapman 1981, 1994, 2000; Garasanin 1973, 1979, 1984; Jovanovic 1965, 1994, Stevanovic 1996, 1997; Tringham and Stevanovic 1991; Whittle 1985; and others). This type of building remains occurs at both types of Neolithic sites, the tell settlements and the flat settlements, which existed throughout the region.

If examined within a wider context - that is, as a part of the Balkan, Aegean, and Anatolian Neolithic processes, the phenomenon of house burning is unique. The Southeast Europe is considered to have been the crossroads between Anatolia, the Aegean, and Europe at the time and, thus, the area where many cultural influences were felt and many cultural links were maintained. The building traditions of Anatolia and the Aegean share some similarities with the central Balkan area – for example that clay is the dominant building material and wood is the second dominant - but there are also important differences. In Anatolia it is mud-brick that dominates house construction, in the Aegean it is a mixture of mud-brick and wattle-and-daub architecture including stone as a building material, while in the Balkans wattle-and-daub was the only method used in house construction. These differences in house construction were also followed by differences in settlement organization. Whereas in Anatolia the Neolithic villages comprise attached houses, which form a honeycomb pattern of buildings; in the Aegean we come across both attached and detached houses; in the Balkans the villages were comprised of individual houses that were built in relative proximity to each other and were never attached.

It is also important to consider the tradition of burned houses in its time context. The main characteristics of the Neolithic of Southeast Europe are uniformity of material culture, the intensification of production and consumption and the increased complexity of organization of physical and social life, the importance of the domestic domain, and the apparent lack of monumental and ceremonial material culture (Tringham and Stevanovic 1991). Sedentary life in village communities along with increased population density and housing density brought about new complexities in the use of space and other resources in the villages. As we have seen from numerous villages, sedentary life in the Neolithic also caused an increase in the density of features and artifacts within the houses, such as storage, ovens and hearths, and food preparation areas. It seems that Neolithic house life and village life in Southeast Europe, with its intensive domestic activities provided conditions for an increase in either accidental or deliberate fires in any stage of a house life, during its occupation, maintenance, or eventual abandonment. Both the storage of plant materials and their accumulation as surplus goods and prestige items and the use of fire in ovens and hearths inside houses could have been the sources of house fires. On the other hand, increased demands on a village space, and its other resources and assets that come with it, must have led to an increase of complexity in social relations. We could imagine that in these circumstances Neolithic people needed to establish and maintain property rights on their houses, a situation which could have necessitate expressive symbolic behaviour in regard to houses. This could have given a new significance to the Neolithic house, to its presence and its durability in the settlement. Thus, house burning as a way of preserving a house place in a village, by leaving the built material remains in place, could have been a way to cope with the increasing complexity of social life.

House Burning

Neolithic houses of Southeast Europe were built in the wattle-and-daub technique. This method uses a framework of wood (including main upright posts and wattling, planks, half-timbers, or reeds which were woven or packed horizontally or placed vertically between them) and layers of varying thickness of clay, which were daubed over the frame on both surfaces (interior and exterior). In the archaeological record these structures are comprised of areas of compacted, burned and collapsed superstructure – rubble made of the clay covering of the wood frame that is transformed under high temperatures into a durable ceramic-like material. The burning has preserved the Neolithic houses in a surprising way, so that the floor with its fixed features as well as the mobile artifacts were kept intact by the baked and crumbled walls on top. The most commonly found features are different floor dividers and borders such as a clay-lip or low interior walls, small size floor pits, ovens and hearths, fixed or mobile ceramic vessels, and an array of complete or fragmented clay, stone, bone, and shell artifacts. The fired remains that can be archaeologically analyzed are the house floor, its artifactual content, and fragmented remains of house superstructure. The standard archaeological analysis of burned house remains is applied to the artifactual content but not to the remains of the house floors or its superstructures. Therefore, the size, shape, and construction of the house walls and the roof, as well as the causes for their burning, remain elusive. Because the walls and roof are partially lost in the conflagration and their survived remains are crumbled and collapsed on and around the house floor, they are very difficult to analyze and can only be reconstructed by careful mapping and lifting the fragments. At the Neolithic sites of Selevac, Opovo, Gomolava, and Vinca we have excavated and analyzed house remains with the intention to reconstruct the houses and to discuss the conditions and causes for house burning. For this task we needed to design a specific set of analyses since a standard set of procedures and analyses for these materials had not existed. The procedure that we developed includes mapping all of the burned remains, their selective recording, and analyses. The results that are briefly described here illustrate that, though the reconstruction and the analysis of the burnt building materials are difficult, they can be effectively used in the archaeological examination (see also Stevanovic 1997).

Even though burned house remains are a very noticeable part of the archaeological record, they have not been systematically studied until now. There are few examples of authors who have based, on their limited research, suggested explanations for house burning (e.g., Semenov 1968; Shaffer 1984, 1993). In addition, experimental studies (e.g., Bankoff and Winter 1979; Hansen 1961) have tried to re-enact the process of house burning on modern examples of Neolithic-like houses and they offered more insights into the problem. It has been suggested by Semenov (1968) that houses were selectively fired to provide a rigid and durable skeleton to the dwelling. I would agree with Semenov that, in some instances and to a limited extent, firing of structural materials was practiced during the Neolithic. However, the completeness of house destruction that we find in the Neolithic material to which I am referring cannot be explained in this way. Shaffer (1984, 1993) suggested that houses in the Italian Neolithic were burned to provide re-usable clay for future house building. This argument, again, cannot account either for the completeness of house destruction found in the Southeast Europe Neolithic, nor is there evidence in the buildings for re-used burned clay. There have been instances of archaeologists who commented on Neolithic burned houses by using ethnographic examples[1].

A widely shared assumption within the archaeological community has been that the Neolithic houses of Southeast Europe were destroyed in accidental fires. As already mentioned, this belief is based on a vague and unexamined notion that the risk of house fire during the Neolithic was great. Many prehistorians in Southeast Europe have expressed the assumption that domestic activities, such as hearth or oven mismanagement caused fire ignition within a house, which later turned into full-fledged and uncontrollable fire that destroyed all the houses in a settlement. Some doubts about the likelihood of the above scenario were affirmed by the experimental studies (e.g., Bankoff and Winter 1979; Hansen 1961), which demonstrated that it is very challenging if not unworkable to fire the houses built of wood and clay to the extent that we find it in the archaeological record, unless the burning was helped by people. The help could have consisted of providing additional fuel and draft for the fire or of not acting to contain the fire. The examples from recent history illustrate this point (Gordon 1953)[2].

The experimental studies conducted by Bankoff and Winter (1979), as well as Shaffer (1993) and earlier Hansen (1961), comprised of igniting fire from within a house and letting the fire develop on its own with the expectation that it would burn down the house. However, in neither case the final result was as expected. The experiments have shown that even if wood is a highly flammable building material, if it is coated with a thick layer of clay, it is hard to burn. That is, if fire is ignited from within the house it tends to stay on the outer surface of the walls and does not penetrate inside the walls where it could reach the wood construction. The experiments also exposed that the hardest part of house burning was to sustain the fire long enough to have it create any visible effect on the house walls. The authors concluded that even if initially a fire was easily ignited by highly flammable materials stored inside the house, this fire would have not lasted long enough to get inside the inner wall layer that is made of wood, which if caught on fire could burn the walls down. The experiments concluded that to achieve an effect of burning the walls through it would have been necessary to help the initial fire by adding fuel to it. This would create conditions for fire to reach high temperatures, and then the temperatures would need to be sustained until they could thoroughly burn the house wood and clay construction as found in the Neolithic remains.

There is other evidence that further contradicts the assumption that the Neolithic houses burned in an accidental village conflagration. One of the most difficult aspects of analyses of Neolithic houses is their chronology - which two or more houses in a settlement were inhabited at the same time. Consequently, there is a problem of finding verification that all the houses in one village burned simultaneously or that they burned at separate times. At the site of Opovo, we concluded that the burned houses, which may have been built and occupied concurrently, were most likely burned at separate times and not in one blaze since burned remains were not found in between the houses, which would indicate the spread of fire from one house to another (Tringham and Stevanovic 1990a, 1990b; Tringham, et. al. 1992). Furthermore, many questions need to be posed regarding the subject, and hopefully some evidence will be provided by future research. For instance, could the indoor ovens be a source of ignition for house fire? Even if the ovens could start a fire, was there enough of fuel in the house construction or/and in the storage to sustain a fire of such intensity that it would burn down the walls covered with a thick layer of clay? In other words, what is the flammability potential of a Neolithic house? Also, what about the people, the residents of the burned houses - where were the people during the house fire? Could they have done something to try to stop the fire or to prevent it from occurring in the first place? It is puzzling to know that there are no instances of human remains associated with burned houses except for two Koros culture houses (Raczky 1983). Is it likely that there were no occasional attempts on the part of Neolithic people to rescue some objects from inside burning house, in which case there would have been instances of people caught in fires who died in the house. What did Neolithic people do to prevent house fires? In order to prevent future fires, Neolithic people could have changed the way their houses were constructed. The changes could have been introduced in the building materials to minimize the use of wood and to maximize use of clay. Another preventive solution could have been to introduce larger spacing between the houses in a settlement to prevent "fire-jumping" from one building to the next. Presently, our capability to affirm any of these possibilities is extremely low. Future excavations and analyses of the building materials must keep these questions in mind as research is carried out.

My study of the burned house remains at Opovo comprised of detailed mapping of the wall fragments and some likely roof fragments that entailed recording their size, shape, orientation, and exact location. All of the mapped and unmapped fragments (small sized fragments which could not

1 In the ethnographic record, house fire of limited extent and burning a house down have been reported from different parts of the world. This record has been selectively used by archaeologists. For instance, fumigation of houses infested with pests has been reported in many ethnographic accounts and cited by some archaeologists as a possible source of the destructive house fire. In the case of the Southeast European Neolithic, the infestation of dwellings is very unlikely considering that oak was the most common type of wood used as the primary construction material and, according to Bakels (1978), this species is very resistant to infestation. Also, fire smoke is used as an agent for fumigation - understandably without it resulting in the complete house destruction.

2 An example of house destruction that was conducted in an organized manner was the raiding of villages in Afghanistan during the Afghan-British war in the beginning of the twentieth century, as presented by Colonel Gordon (1953). A house of mud and rubble walls and a flat mud covered roof had to be prepared for burning or it would not burn at all. The two essentials were extra fuel and good draft. These houses would not burn by the simple application of a torch to their framework made of wood. Even when houses had been thoroughly burned, Colonel Gordon claims, they could be, and in fact were, made habitable after a few months' work.

provide more than the basic information for the study were not mapped) were taken through the first level of analysis, which meant that they were classified within my three-type classification, and then were weighed. The mapped fragments were taken into the second level of analysis, when they were drawn, photographed, and described in detail so that they could be subjected to statistical analysis. The selected samples then went through the third level of analysis, consisting of geomorphological and x-ray tests conducted in order to establish the mineralogical content and thermal properties of these materials (Stevanovic 1996, 1997). The final step in the procedure was to conduct an experiment, which consists of firing different raw materials from the vicinity of the archaeological sites in order to make comparisons to the originally fired clays from the Neolithic settlements. The analyses and the experimental study provided a fairly good understanding of the composition of the raw materials and also of the thermal transformations that they had undergone. The next challenge was to assess how the different raw materials would behave in a fire. Taking into consideration the experimental results and the ethnographic information regarding how difficult it is to set on fire house walls that are primarily built of clay, I have looked at the ratio between the volume of construction wood and the volume of clay, as well as at the types of wood used in construction and its flammability potential.

A series of analyses by which I examined the temperatures of burning, the ignition points in the house, and the fire paths was conducted with the idea to try to follow the ways by which fire moved through the building and how fire might have consumed it. This series of analyses is based on the principles of modern arson investigations, which seek to discover sources and causes of destructive fires. I believe that some of the principles of arson investigation could be applicable to the investigation of Neolithic house fires. These are determining the origin of the fire in space, its causative agent, i.e., the nature of the initial fuel, and the nature of the ignition. If we can provide evidence for these fire-related elements, we should be able to get an informed idea about the nature of Neolithic house fires, and then we could begin the discussion on the intentional or accidental house burning. Fire begins with ignition, and in my investigation I proposed that a point of vitrification that occurs on the lowest level within a house is to be considered the ignition point. The spread of temperature away from this point from the highest towards lowest temperature should be considered a fire path. Based on the arson principles, I created maps of fire paths for the Opovo houses. According to the same principles I have proposed that if there is more than one vitrification point, which I consider to be a fire ignition point within one house, this will be considered an act of a deliberately set fire (Stevanovic 1996, 1997).

The analyses of the Opovo houses showed more then one vitrification point on the floor of each house, and the following conclusions were drawn for the Opovo houses: (i) the temperatures of house burning were much too high to have been achieved only by the fire of the house construction wood; (ii) the fire path in each Opovo house indicates several ignition points within a house; (iii) the fire path shows that the ignition most likely did not start in the house roof but at floor level; (iv) the ovens were fired in temperatures of lower intensity than those that burned the house walls. The result of the analyses did not support the suggestion that the indoor ovens were ignition points for house fires because they were not the most burned features of the house, which they should have been had they been the source of the house fires. And furthermore, the wood use in the construction of the analyzed houses would have been insufficient as the sole fuel for the house burnings.

Closure of the House

The results of the analyses I have interpreted are not supportive of the assumption that houses at Opovo were burned in accidental fires. The completeness of the house burnings and their subsequent destruction, along with the evidence that the houses were not used secondarily, were not even "scavenged" for utilitarian or symbolic objects which even though burnt were afterwards still functional, indicate the intention of the occupants to preserve things as they were. The absence of human or animal skeletons inside or around the burned houses could be interpreted that the residents had the complete control of the house fires. This behavior could further indicate that the house residents wanted to bring their houses to a closure rather than try to save their houses from the fires. A much more plausible interpretation of burned houses in the Southeast European Neolithic, in my view, is that the house conflagrations were the result of deliberate and symbolic social action. The house burnings and their orderly collapse could have been an organized and strategic effort of "sealing it off" from future use. House construction, their use, and their destruction were the segments of a continuous process. At the endpoint of their use-lives the houses cease to be residential features, they were brought to a closure by being burned down, and at that point their symbolic role began.

It has been suggested by Tringham (Tringham and Stevanovic 1990a) that at Late Vinca culture tell settlements where the land available or preferred for residence was more restricted, the houses would have symbolized the continuity of land use and possibly land ownership. Building on this idea, I contend that Neolithic houses, by being burnt at the end of their use-lives could have been used to secure the visibility and continuity of the social unit in a settlement. For Neolithic people the burned houses could have embodied a number of important roles. For instance, the role to make a continuous claim on a land property by a household in a village. The most immediate role that the burned house could have had was the linking of the spiritual and living worlds of a household residents. It may seem paradoxical to suggest that house destruction can assure continuity in a society in which houses seem to have been the largest asset for an individual or a family. However, if a house for structural and/or symbolic reasons, such as- the death of the household head or an important elder, the cases that are often mentioned in ethnography- had to be replaced at the end of its use-life, the firing and the resulting solidification would have been a good strategy to take. The process of burning and material results could preserve the house in the memory of the individuals and the community. Also, it would not be surprising to me that the Vinca people would have chosen fire to carry out the closure of their houses in a ritualized action. It was during the Neolithic that domestication of fire had been achieved by bringing it within the realm of the houses (in hearths and ovens). Fire as an agent of transformation of raw-food into cooked food was a crucial part of the process that focused on bringing the wild within the house and into the cultured/domesticated space of the humans (Hodder 1990). We have evidence that in the later phases of Vinca culture Neolithic people used sophisticated pyrotechnology, the knowledge of which they could have used in relation to house burning, one way or the other[3]

For all of the reasons mentioned, it seems highly improbable to me that all of the Neolithic villages in the Southeast European Neolithic during this long period of prehistory (ca 1500 years) excavated so far would have been burned down by coercive activity of the kind described by Vitruvius[4]. This is especially hard to accept if we believe that Neolithic people could have prevented such a catastrophic outcome by introducing changes in their method of house construction, in the choice of materials used, or in the intra-site organization of the dwellings. The technological knowledge of the Vinca culture people, for instance, has proved to be substantial and I believe it included the possibility to introduce some changes to their architecture, had they wanted it. The possibility of altering the method of house construction to include mud-brick and thereby reducing the risk of destructive fire would have been feasible since this method must have been known to the Vinca culture people from time of somewhat earlier settlements in the bordering region to the South, such as at the site of Anzabegovo. Also, by simply increasing the distance between houses in one village, they could have prevented an accidental fire that occurred in one house from spreading to the next. It is also feasible to imagine that Neolithic people could have developed other methods of fighting the fires so that they did not consume entire settlements.

On the other hand, if we persist in assuming the coercive causes for the village fires, we have to rethink our firmly embedded belief that the Neolithic was a period of great stability, peacefulness, gradual change, and development in which no large movements of population, such as migrations, that would involve territorial fights were taking place. If all Neolithic settlements of Southeast Europe were the targets of coercive attacks and all their houses were destroyed in conflagration caused by raiders, the question arises: who were the people that undertook these attacks?

House-life and Death

From the ethnographic sources we can see that houses can serve to their people as residences and as burials. In the ethnographic record we can see that the planned or unplanned abandonment and replacement of houses and sites for structural or social reasons is a worldwide practice known from a variety of time periods. The phenomenon of abandonment can be episodic, seasonal, or permanent. They can occur on the settlement or individual household level. The social reason for abandonment on the village level is known from the historic Pueblos at Orabi where the partial abandonment of the site was due to a social dispute (Wilshusen 1984, 1988, 1989). According to Brooks (1993), the Plains societies of North America feature several types of abandonment, from planned abandonment that was related to the life-cycle of the house and a need for its replacement to the less planned or very sudden abandonment in the case of a sudden death. The long-term planned house and/or site abandonment could take two years of preparation in which help was provided by the kin. In these instances, the houses that were abandoned show a pattern of "scavenging" for high-reproduction-cost goods or other valued items, including removing and transporting of the old wood beams in the areas where wood is scarce, and exhibit burning of the dwellings as a ritual associated with the abandonment of a residence. House destruction or abandonment for fear of ancestral ghosts, if a person dies in a house for reasons other than old age, is known among the Athapaskans in the Southwestern United States (Jett and Spencer 1981). Among

---

[3] The ability of the Vinca culture potters to reach high temperatures in conjunction with reducing atmospheres on a routine basis and prior to the first appearance of copper ore processing in late Vinca sites has been now demonstrated (Kaiser 1984, Kaiser *et al.*, 1985). Moreover, it is suggested that high temperature firings had become "culturally conventionalized" in the Vinca culture. The Vinca people's pyrotechnological experience included plaster-lime production, as well as copper ore processing. The pyrotechnological practice of the Vinca culture people could have a twofold consequence for this study. Firstly, the knowledge of fire and skill in handling it may mean that they knew of ways to create, control, and prevent fires. For instance, in the case of house fires the control could have been achieved by introducing changes in the building materials, building methods such as mud brick architecture, or in a village-wide change of house proximity. Taking into consideration the lengthy time period during which the practice of house burning occurred it seems realistic to assume that such changes were possible and likely if all other necessary conditions were met. Secondly, the pyrotechnological knowledge of the Vinca culture people was such that it could have allowed for a deliberate setting up of a successful house fire and, hence, house destruction. Knowing that complete burning of a wattle-and-daub Neolithic house is not necessarily an easy task (Bankoff and Winter 1979) it is being claimed in this study that the knowledge of pyrotechnology was desirable for achieving it.

[4] A rare historical account of house fires is presented by the Roman historian Vitruvius who believed that the risk of fire was great and the results catastrophic, as demonstrated by Verulamium when the city was fired by Boudicca in 61 AD. Vitruvius (as referenced in Davey 1961: 41) writes that the part of the city of Verulamium that was built in wattle work burned down during this attack. Vitruvius' account describes wattle- and-daub houses as an easy target for the flames. I would suggest, however, that there was a difference between the Neolithic and Roman wattle and daub houses that are demonstrated by their different behavior in fire. The Roman houses of Verulamium ignited more easily for a number of reasons. The Roman cities, in which certain residential quarters were built in wattle-and-daub according to Vitruvius' account were famous for being overcrowded with multi-story buildings that were tightly packed, very poorly built, with very thin walls and large thatched roofs. What is most important is that the clay cover of these buildings seems to have been quite thin. Equally important for understanding the Roman Verulamium city fire of 61 AD is the nature of the fire itself. The city was burned in an organized fashion and houses were set on fire at the same time. This would have caused a large concentration of the gases that occur in fires, which could then have carried the fire as long as there was suitable fuel available.

the Navajo a house in which a person died may be burned; if not, a hole is torn through the north wall and the roof beams are allowed to fall in, indicating that the place should be avoided (Reichard 1963). These ethnographic examples are not only important to illustrate the practice of house abandonment but also to draw attention to a related practice of partial or complete house destruction.

In the European Neolithic, discussion of site abandonment and replacement has been limited[5]. Explanations such as the one given for LBK villages of Central Europe assume that complete villages were abandoned as a single event and later inhabited again in the similar fashion. I believe that even though a complete village abandonment and replacement was a possible strategy in the Neolithic, another strategy, that of abandonment and replacement on individual or household level, must not be ruled out. I would argue that such a strategy can be village-wide practice, even a cultural convention during the Vinca culture, but still executed at the household level.

We know that the house in addition to its domestic role, was a place of burial. This is exhibited in the actual intramural burials, a tradition well known in the Neolithic of Anatolia but not present in the Southeast European Neolithic. There is so far only one example of house and burial relationship in Southeast Europe, published by Raczky (1983) who has interpreted the two Koros culture skeletons as deliberately burned together with the house as a form of human burial. Recently, Ozdogan and Ozdogan (1998) have suggested buildings intentionally destroyed by fire at Çayönü and interpreted it as the symbolic burial of houses. They also give the example of "intentional burning of building before their burying at Beidha" (Ozdogan and Ozdogan 1998: 590). Apel et al. (1997) have suggested house burning at an Early Neolithc TRB site in Sweden. Other ethnographic accounts indicate how the house life cycle and the human life cycle can be interwoven (Trojanovic 1990; Djordjevic 1984; Filipovic 1982, Halpern 1958; and others). The example of a Toraja chief's burial ceremony shows that a special house is built, which is then burned down as a part of the ritual (Waterson 1996).

The tell settlements of Southeast Europe and the Near East are prime examples of the importance of a village location, but also of the household location within the village, which is expressed by numerous generations of people building their houses in the same place. At Opovo there are 3 building horizons which, on the basis of 14C dates, are believed to have lasted for nearly 200 years. In my mind the places of Neolithic people that were 'kept alive' and visible for generations to come indicate a continuity of social structure, such as lineage, through the mechanics of house construction, their use and destruction being the segments of a continuous process.

Even though a burned and collapsed house becomes invisible by being covered by humus and/or by another house on top, it retains its visibility and its mnemonic potential. Its existence would have been known to the people who built a new house on top of it. There are multiple examples from the site of Opovo, as well as from many other Vinca culture sites of the practice of house construction by incorporating older outdoor hearths and ovens, as well as older burned houses, into the foundation of new houses. This was achieved by erecting the new house not directly on top of the older features, but resting on one part of the older house or feature (in some instances to incorporate more than one outdoor oven and/or house). This positioning of a house may have given it a structural advantage in the sense that the stable, solid ground of the older features was used as a foundation for the new house. However, the fact that they do not always lie directly on top of the older feature I interpret as a need for symbolic support rather then as the mechanical support and part of the house foundation. The Neolithic house could have marked social and material continuity, which were important to Neolithic societies. As a part of the domesticating process, the Neolithic house was conceptually central and the visible symbol of the Neolithic social unit. It could have been objectifying the successful social reproduction, and its enduring visibility could have been crucial in establishing and maintaining the social order.

Bibliography

APEL, J., C. HADEVIK, and L. SUNDSTROM
1997 *Burning down the house. The transformational use of fire and other aspects of an Early Neolithic TRB site in eastern central Sweden.* Upsala. Tor 29: 5-47.

BANKOFF, A. and F. WINTER
1979 A House-burning in Serbia. *Archaeology* September: 8-14.

BAKELS, C. C.
1978 Four Linearbandkeramik Settlements and their Environment: a paleoecological study of Sittard, Stein, Elsloo and Hienheim. *Analecta Praehistorica Leidensia* 11. Leiden: University Press.

BROOKS, R. L.
1993 Household abandonment among sedentary Plain societies: behavioral sequence and consequences in the interpretation of the archaeological record. pp. 178-191. In Cameron, C. and S. Tomka (eds.) *Abandonment of settlements and regions.*

[5] Some causes for village abandonment and later replacement during the Neolithic of Central Europe have been hinted at by archaeologists that worked on the LBK culture. Childe (1957), Soudsky′ and Pavlu (1972) analyzed the LBK horizons of houses in terms of interrupted occupation. Their argument suggests that between the two phases of houses the same people lived at another site. They locate the cause of such alternating of the settlements, in the type of agricultural practices conducted by the LBK populations. Soudsky′ talked about a restricted territory within which a group of related people would alternate settlements staying at each location for 10-15 years at one time or until the fields were exhausted by over cropping. Moderman (1973; n.d.) argued against a cyclical habitation pattern of the LBK settlements, remarking that this explanatory model, developed for the tropical regions characterized by the slash and burn agriculture, was inappropriately imposed on the Neolithic settlements of Europe, which were located on different types of soil.

Ethnoarchaeological and archaeological approaches. Cambridge: Cambridge University Press.

CHAPMAN, J.
1981 *The Vinca Culture*. Oxford: BAR International Series 117.
1994 The early Balkan village. pp. 33-55. In *Neolithic of Southeastern Europe and its Near Eastern Connections*. ed.. Budapest.
2000 *Fragmentation in Archaeology. People, places and broken objects in the prehistory of South Eastern Europe.* London and New York: Routledge.

CHILDE, V. G.
1957 *The Dawn of European Civilization*. London: Routledge & Kegan Paul.

DAVEY, N.
1961 *A History of Building Materials*. London : Phoenix House.

DJORDJEVIC, T.
1984 *Nas narodni zivot*, 1. Beograd: Prosveta.

FILIPOVIC, M.
1982 *Among the People: Native Yugoslav Ethnography.* (Edited by) Hammel, R. Ehrich, R. Fabijanic-Filipovic, J. Helpern, and A. Lord. Ann Arbor: The University of Michigen.

GARASANIN, M.
1973 *Praistorija Srbije (Prehistory of Serbia)*. Belgrade: Srpska Knjizevna Zadruga.
1979 Centralnobalkanska Zona. pp. 79-213. In *Praistorija Jugoslovenskih Zemalja*. II, Basler, D., A. Benac, S. Gabrovec, M. Garasanin, N. Tasic and K. Vinski-Gasparini, (eds.). Sarajevo: Akademija Nauka i Umetnosti Bosne i Hercegovine.
1984 Vinca u Mladje Kameno Doba. pp. 5-10. In *Vinca and its world.* Srejovic. *et. al*., (eds.). Beograd: Galerija Srpske Akademije Nauka i Umetnosti.

GORDON, D. H. Colonel
1953 Fire and the Sword: the Technique of Destruction. *Antiquity* 27: 149-152.

HALPERN, J. M.
1958 *A Serbian Village*. New York : Columbia University Press.

HANSEN, H. O.
1961 Mudhouses. *KUML*. Årbog for jysk arkaeoloisk selskab.

HODDER, I.
1990 *The Domestication of Europe*. Oxford: Basil Blackwell.

JETT, S., and V. SPENCER
1981 *Navajo architecture: forms, history, distributions*. Tucson: University of Arizona Press.

JOVANOVIC, B.
1965 Starija Vincanska grupa u juznom Banatu. *Radovi Vojvodjanskih Muzeja* 14: 15-42.
1994 Gradac Phase in the Relative Chronology of Late Vinca Culture. *Starinar* XLIII-XLIV: 1-11.

KAISER, T.
1984 *Vinca Ceramics: Economic and Technological Aspects of Late Neolithic Potery Production in Southeast Europe*. Ph.D. thesis, Department of anthropology, University of California, Berkeley.

KAISER, T., U. M. FRANKLIN, and V. VITALI
1984 Pyrotechnology and Pottery in the Late Neolithic of the Balkans. 24th *International Archeometry Symposium*, pp. 85-93.

MODDERMAN, P. J. R.
(Unpublished) Some Remarks on the Structure of Linearbandceramic Settlements.
1973 Bespiegeling ovre de constructie van een bandkeramisch huis. pp. 131-140. In van Es W. A. *et al.* (eds.) *Archeologie en Historie*.

OZDOGAN, M., and A. OZDOGAN
1998 Buildings of Cult and the Cult of Buildings. pp. 581-603. In Arsebuk, G., M. Mellink, and W. Schirmer (eds.), Light on Top of the Black Hill. Studies presented to Halet Cambel. Istanbul: Ege Yayinlari.

RACZKY, P.
1983 Origins of the custom of burying the dead inside houses in South-East Europe. *Szolnok Megyei Muzeumi Evkonyv* 1982-83: 5-10.

REICHARD, G. A.
1963 *Navaho Religion. A study of symbolism.* New York: Princeton University Press.

SEMENOV, S.A.
1968 *Razvitie tehniki v kamenom veke.* Institut Arkheologii: Academia Nauk SSSR.

SHAFFER, G.
1984 Neolithic Building Technology in Calabria, Italy. Unpublished Ph.D. thesis, State University of New York, Binghamton.
1993 An Archaeomagnetic Study of a Wattle and Daub Building Collapse. *Journal of Field Archaeology* 20: 59-75.

SOUDSKY, B. and I. PAVLU
1972 The Linear Pottery Culture settlement patterns of central Europe. In Tringham R., G. Dimbleby and P. Ucko (eds.) *Man, Settlement, and Urbanism*, London: Duckworth.

STEVANOVIC, M.
1996 The Age of Clay: The Social Dynamics of House Destruction, Ph.D. thesis, University of California, Berkeley.
1997 The Age of Clay: The Social Dynamics of House Destruction. *Journal of Anthropological Archaeology* 16: 334-395.

TRINGHAM, R. and M. STEVANOVIC
1990a The nonceramic uses of clay. pp. 323-397. In Tringham, R. and D. Krstic (eds.) *Selevac: A Neolithic Village in Yugoslavia*. Los Angeles: UCLA Institute of Archaeology.
1990b Field Research. pp. 57-156. In Tringham, R. and D. Krstic (eds.) *Selevac: A Neolithic village in Yugoslavia.* Los Angeles: UCLA Institute of Archaeology.

1991 Vinca Domestic Architecture as Arenas of Household and Gender Tensions. 56th SAA Meetings, New Orleans, April 23-28, 1991.

TRINGHAM, R. and D. KRSTIC

1990 Conclusion: Selevac in the Wider Context of European Prehistory. pp. 567-617. In Tringham R. and D. Krstic (eds.) *Selevac: A Neolithic Village in Yugoslavia* Los Angeles: UCLA Institute of Archaeology Press.

TRINGHAM, R., B. BRUKNER, T. KAISER, K. BOROJEVIC, N. RUSSELL, P. STELI, M. STEVANOVIC, and B. VOYTEK

1992 Excavations at Opovo, 1985-1987: Socioeconomic change in the Balkan Neolithic. *Journal of Field Archaeology* 19(3): 351-386.

TROJANOVIC, S.

1990 *Vatra u obicajima i zivotu srpskog naroda.* Beograd: Prosveta.

WATERSON, R.

1996 House, Place and Memory in Tana Toraja (Indonesia). Paper presented in the session "Opening up the house: A dialogue across the discipline" of the 95th Annual Meeting of the American Anthropological Association.

WILSHUSEN, R.

1984 *The Abondonment of Structures.* Dolores Archaeological Program Technical Reports. NTIS, DAP-279.

1988 Architectural Trends in Prehistoric Anasazi Sites During A.D. 600 to 1200. pp. 593-708 In Blinman, E., C. Phagan and R. Wilshusen (eds.) *Dolores Archaeological Program: Supporting Studies: Additives and Reductive Tehnologies.* Denver: US Department of Interior, Bureau of Reclmation, Engineering and Research Center.

1989 Unstuffing the Estufa: Ritual Floor Features in Anasazi Pit Structures and Pueblo Kivas. pp. 89-111. In Lipe W. D. and M. Hegmon (eds.) *The Architecture of Social Integration in Prehistoric Pueblos* Cortez, Colorado: Crow Canyon Archaeological Center.

WHITTLE, A.

1985 *Neolithic Europe: A Survey.* Cambridge: Cambridge University Press.

SOME COMMENTS ON THE TECHNOLOGY OF PREHISTORIC POTTERY IN THE WESTERN BALKANS IN THE LIGHT OF ETHNOARCHAEOLOGICAL RESEARCH

Richard Carlton

1.1 Introduction

Fire, the visible heat energy released during the rapid oxidation of a fuel, is the principal transformational agent of the pottery-making cycle. Successful firing, when clay paste becomes ceramic fabric, is achieved by directing and controlling the exothermal release of heat from burning fuel to alter the physical state of clay molecules constituting a formed object, thereby transforming earth to stone. In order to avoid cracking, warping or melting, however, the temperature, duration and atmosphere of firing must be controlled within strict boundaries according to the fabric and form of objects being fired.

The application of the maximum amount of available heat energy within the constraints permitted by mechanical considerations is a precarious business; the process is difficult to control and results may be variable. It is precarious because it is an attempt to harness a force of nature which has been familiar to humans from the earliest times as a potential extreme hazard. There is a tension here, however between the knowledge of fire as a destructive hazard and the appreciation, also from an early period, of its various positive qualities; as a source of warmth, light and cooking heat, for example, and as an agent for hardening wood, splitting rock and searing wounds. Arguably, because of this ancient, precarious tension between creation and destruction, the phenomenon of fire remains enigmatic and continues to stir the human psyche. It retains a magical, metaphysical quality in the sense that its essence can not be fully comprehended nor tamed. The use in pottery firing of violent fire deliberately created to generate temperatures unbearable to humans still arouses a certain awe and charges the event with added significance. Furthermore, the creation useful tools which a raging fire can not consume or fracture stands as a metaphor for human interaction with, or power over, nature. Certainly, no other stage in the pottery-making sequence is accompanied by such intense, transcendental emotions as firing, its defining act.

The purpose of the present paper is not to explore further the essence of fire, but to discuss the ways in which it is used as the primary agent of transformation in the context of West Balkan pottery-making complexes, where, it is argued, the constraints imposed by material choices make the control of heat energy particularly critical. Further, it will be argued that the interface between the latter stages of the firing process - including strategies developed to ameliorate the effects of over-firing - and the entry of pottery into the arena of use, is crucial to the understanding of the technological process of pottery-making as a whole and of its survival in the face of apparent competition from other technological processes and ceramic products. Ample reason will be found to resist the assumption that firing methods lie beyond the active choice of traditional potters who, fixed into a socially-embedded evolutionary state, have neither the knowledge or inclination to change, in favour of an argument accentuating the positive benefits of open firing. Thus, rather than assuming that the use of coarse pastes is conditioned entirely by the available 'primitive' firing technology, it is suggested that the reverse may be the case. Technological relationships between different facets of modern pottery-making, material and non-material, will also be used to shed light on the nature of prehistoric pottery-making in the same location and to determine the extent of continuity between the two.

1.2 Context of the Study

The south Slav-speaking lands of the Western Balkans, incorporating much of the former Yugoslavia, probably exhibit, and certainly exhibited until very recently, one of the most diverse pottery-making cultures on earth, employing a wide range of production technologies to produce a diverse range of pottery types, some of which appear throughout the region whilst others are more localised (Carlton 1999b). The archaeological record supports the view that such diversity has prevailed since at least the medieval period and may extend into antiquity. Within the area with which this discussion is mainly concerned, the highland and coastal zone, centred upon Bosnia-Herzegovina (Fig. 1), pottery-making since at least the medieval period has been characterised by the use of hand-wheels and bonfire-firing technology (Fig. 2) in association with calcareous pastes (Kalmeta 1954; Popović 1956, 1957, 1959; Randić-Barlek 1990). Within this group, however, a considerable degree of variation is exhibited in all aspects of pottery-making,

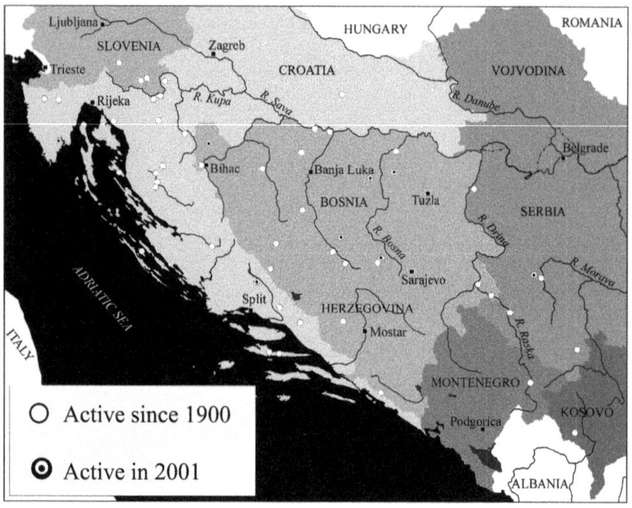

Fig. 1. Locations of potters using the hand-wheel in the Western Balkans

including firing, for which kilns are used by a minority of potters, suggesting that the use of open firing by the majority is a matter of deliberate choice.

Fig. 2. Fehim. A hand-wheel in use at Demiševci, Western Bosnia

Beyond the description of the firing process as part of a detailed ethnoarchaeological enquiry into the production of calcite-tempered pottery on the island of Iž, the broader aim of the present work is to cast light upon the character of Iron Age, or Romano-Illyrean, pottery-making on the island by invoking the results of that enquiry and its implications. This is carried out by focusing on the firing process as the core of the ceramic technological sequence. Although micro-regional in scope, this localised study benefits from more extensive research into the origins, range and diverse character of traditional pottery-making in the western Balkans region (Carlton 1988a, 1988b, 1989, 1998, 1999).

1.3 Theoretical Background

Clays and non-plastics can be mixed to achieve an extremely wide range of mineralogically and texturally different pastes, from which can be formed an equally wide range of sizes and shapes of vessels. Hardness, porosity, colour, decoration and surface texture are also immensely variable qualities over which the potter can exercise a measure of control. Given this potential for infinite variety in pottery products, the appearance of similarity between different vessels is indeed significant, since it implies purposeful replication by a potter or group of potters, contemporaneously or over time. Indeed, such behaviour is the norm for a given potter or group of potters, and the concept of replication, with its material correlate 'similarity', is used to identify potting 'traditions' using archaeological material.[1] The perception and consideration of continuity and change in such traditions, revealed by material similarities and differences over time, enables inferences to be made about a range of social and economic phenomena which pottery is held to reflect. Indeed, in addition to its ubiquitous availability, the uses of pottery as an instrument of dating and indicator of social change best account for its popularity in archaeological studies.

As noted above, by virtue of its durability and relatively good survival in archaeological contexts, the perceived importance of pottery and consequent expectation placed upon its analysis are often out of all proportion to its original role and status. Indeed, it is increasingly apparent through a multitude of 'cautionary tales' (Kramer 1985) that in many cases it is actually rather a poor barometer of social status and change (*c.f.* Adams 1979; Rice 1987a). Whereas pottery is often taken to closely reflect various aspects of human social systems, including identity groups, social status and population levels, a number of studies have shown this not always to be the case. Saraswati and Behura (1966: vii), for example, conclude that there is a lack of association between zones defined by pottery technologies and the main linguistic and cultural divisions of India, while Courtney cites excavated and documentary evidence in concluding that ceramics are a poor indicator of status in both medieval English and nineteenth-century North American society (Courtney 1997: 101).

Recent approaches to categorisation, therefore, whilst accepting that a level of correspondence between pottery-making behaviour and defined pottery types must be assumed (Van der Leeuw 1984: 711), do not accept the common assumption that types defined by the analyst reflect the ideas

[1] The term 'tradition', though much used in pottery studies, is the source of some confusion. In general terms a tradition of pottery making, or 'technological tradition', refers to long-term stability in the process of pottery making, the knowledge of which is passed down directly from one generation to the next. Technological tradition is not synonymous with 'typological tradition', which is defined by similarity of form and decoration (Rye 1981, 5). Thus, pots that are identified as coming from different typological traditions may in fact be the products of the same technological tradition.

of those who originally made and used the pots. Since archaeological analysts are bound to interpret remains by drawing inferences from them on the basis of observations made in the present, the way forward is suggested by Watson's comment that, "*ethnoarchaeolgy, formally or informally applied, is the source of all observations that enable archaeological interpretation*" (Gould and Watson 1982: 356). Watson argues that such inferences are drawn by analogical reasoning, therefore, that explanatory mechanisms are most likely to be derived from the rigorous use of ethographic analogy, notably by constructing theories and broad law-like generalisations about relationships between human behaviour and material culture resulting from that behaviour (Gould and Watson 1982: 356).

The increased and more rigorous archaeological use of ethnography, and the development of ethnoarchaeology and pottery ethnoarchaeology as discrete fields of study (Kramer 1985; Deal 1998), followed the surge of interest in the roots of archaeological knowledge that accompanied the New Arcaheology movement. Ethnoarchaeology, combining archaeological and ethnographic techniques and approaches, has been used to build models of the interaction between behaviour and material culture, and to identify principles that condition these actions. An underlying assumption of interpretations based on ethnoarchaeology is that various aspects of human behaviour have material correlates, and that in the archaeological record such materials appear as patterned deposits whose behavioural correlates may be inferred by the use of some kind of analogy (Gould and Watson 1982: 356-363; Kramer 1979) - much theoretical debate in ethnoarchaeology centres on the validity or otherwise of using ethnographic analogy for archaeological interpretation. Two of the most important theoretical approaches derived from ethnoarchaeology and espoused as unifying strategies are ceramic ecology (Matson 1965; Arnold 1985) and organisational modelling (Peacock 1981: 1982; Van der Leeuw 1976 and 1984), which attempt to interpret pottery within general frameworks of context, production and use, providing a much-needed conceptual link between the material remains of pottery-making and the activities, organisational states and economic modes of production responsible for them. Importantly in the present context, the generalised models proposed by these authors present a means for comparing pottery-making systems in different locations and from different periods.

1.4 Approach to the Present Study

The analysis of European prehistoric or native Roman pottery using such models remains relatively rare, although the Assendelver Polders study of pre-Roman and Roman Iron Age pottery in North Holland provides a mechanism for such work (Van der Leeuw *et al.* 1987). In that study, the authors set off with the premise that the technology used to manufacture ancient pottery reflects the overall organisational complexity of its production system, which in turn "*is related to that of the society in which it occurs*" (Van der Leeuw *et. al.* 1987: 226). An understanding of the way in which pottery is made, based upon the technological analysis of ancient pottery remains may, therefore, provide the best or only indication of the organisational and economic context of its production. In order to do this it is essential first to understand pottery-making technology and be able successfully to interpret the material traces of production processes using fragmentary remains. Based upon this evidence, an attempt must then be made to assess the decision-making processes of potters in the light of constraints imposed by the social and natural environment, including the properties of the available raw materials.

The present study is concerned with the nature of late prehistoric pottery-making interpreted through its material remains in the light of findings derived from an investigation of modern pottery-making in the same location, the Dalmatian island of Iž. The prehistoric pottery assemblage was recovered when an Iron Age or Romano-Illyrean settlement was encountered on Košljin hill during archaeological field survey of an area close to the modern pottery-making village of Veli Iž. Pottery fragments recovered from surface scatters there displayed certain characteristics in common with modern pottery made on the island, along with some apparent differences. Anlaysis of the Košljin pottery assemblage was carried out, mainly from a technological perspective, to illuminate not only the character of pottery-making, but the anthropological context of its manufacture and use. Indeed, as is so often the case in archaeological interpretive studies, the paucity of other classes of artifactual remains from Košljin means that the analysis of recovered pottery offers virtually the only source of potential information on the social organisation, status, economy and trade contacts of the group presumed to be responsible for its manufacture, certainly for its use and discard.

In the present case we are fortunate to have information on a modern pottery-making tradition which shares the same physical environment and at least some of the technological characteristics of the ancient pottery under analysis. The modern tradition can be used, therefore, as a model against which to compare the ancient material, both with respect to individual attributes, such as forming or firing technology, and within frameworks of organisational and economic complexity suggested by Van der Leeuw (*ibid.*) and others (*cf.* Balfet 1965; Peacock 1982). The particular importance of the study of fire in the present study is that firing lies at the crux of the technological complex, while something of its nature can be ascertained from surviving remains.

As a first step it is proposed to examine the physical environment of pottery-making and its known and likely history on the island. The nature of modern pottery-making is then examined, and the implications of the use of certain classes of inclusions emphasised. A conceptual basis for modern pottery-making on the island is then be suggested and its main material correlates identified, thereby providing a basis from which to conduct a comparative analysis of the Iron Age material.

1.5 Location and History

The island of Iž (Fig. 3) lies at the centre of the Zadar archipelago, a group of islands in the north part of the Adriatic Sea associated with the main port city of the North Dalmatian littoral. The islands of the archipelago are orientated north-west to south-east, parallel with the high limestone mountains of the coastal Velebit and other Dinaric mountain ranges, and contiguous with the low foothills of the Velebit in the Ravni Kotari region inland of Zadar.

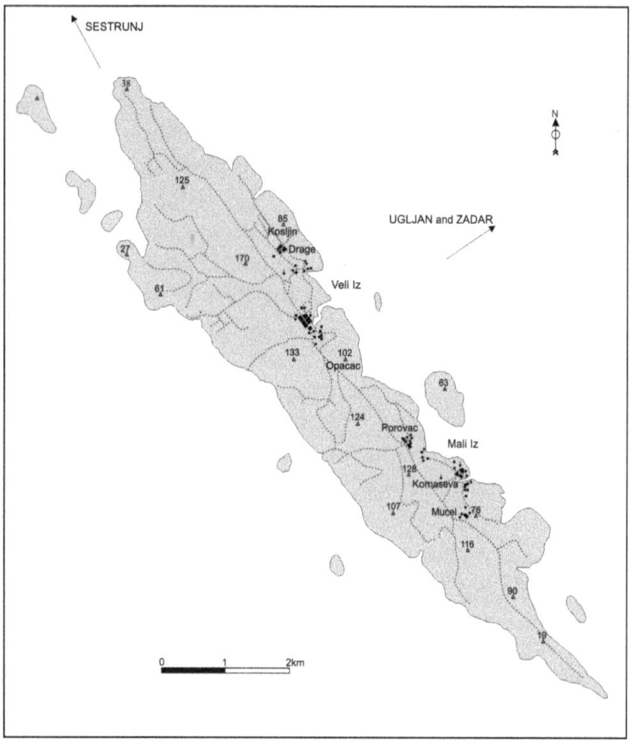

Fig. 3. The island of Iž, North Dalmatia

Although primarily consisting of steep ridges, many of the Adriatic islands, including Iž and its larger neighbour, Ugljan, contain sheltered valleys and coastal lowlands which add significantly to their agricultural potential. Although a small island, 12km long and averaging about 1.5km wide, with a total land area of some $17.5km^2$, Iž displays a wide range of the features and landforms commonly associated with limestone geology, typical of *karstic* landscapes elsewhere. Thus, it is hilly and rather rocky, with an absence of surface water for much of the year and, with the exception of a freshwater spring at sea-level, no running streams. Natural soils are rather poor, but due to widespread stone clearance and terracing, soil cover is generally fairly good on all but the steepest slopes and higher hill-tops. The result of this intensive soil management and agricultural activity, largely post-medieval in origin, is a constructed landscape of small, terraced fields enclosed by massive dry-stone walls and clearance cairns. Parts of the island considered too steep, remote or barren for agricultural exploitation by terracing have in the relatively recent past been exploited for timber, fuel and rough grazing, leaving few areas that may in any sense be considered truly 'natural'. Although a range of land-use classes may be identified similar to those recently described on the adjacent mainland (Chapman *et. al* 1996: 28-31), an ever expanding area is now covered by a dense scrub, called *macchia*, consisting of the principal evergreen species (notably oak, but excluding pine), in stunted form, with a plethora of mainly unpalatable, thorny bushes such as juniper which are often taken as 'anthropogenic indicators', associated with degradation (or, in this case, re-establishment) of the natural forest vegetation (Chapman *et al.* 1996: 41-43).

The island is known with certainty to have been populated since at least the Iron Age (Batović 1974). Indeed, the Iron Age is the only period prior to the post-medieval era which is associated with significant, surviving remains of occupation and land-use, and two centres of Iron Age activity have been identified around Veli Iž (Fig. 4), with a third at Mali Iž (Batović 1974).

Fig. 4. Košljin wall. Iron Age enclosure wall at Košljin, Veli Iž

Though somewhat smaller, in general character the structural remains of the Iron Age population on Iž are broadly similar to a group of enclosures, the Ravni Kotari group, recently investigated in the Zadar hinterland and regarded as a discrete North Dalmatian tradition (Chapman *et. al.* 1986: 154). Batović maintains that the hill-forts on the north Dalmatian coast, including this group, are exclusively Iron Age (Batović 1977), although similar settlement sites are claimed as early the third millenium BC in central Bosnia (Čović 1983), and in most areas probably begin in the middle to late bronze age (Benac 1985). The emergence or greater prominence of monuments in the landscape during the Iron Age can be regarded as a product of the increased complexity of political organisation in society and the associated expansion of commercially-oriented production and regional trade.

It is often stated that the 'Celts' influenced the fourth century Illyrians before being assimilated (Singleton 1985: 10; Fine 1991: 12), and the same ill-defined grouping is often credited with introducing the pottery wheel and other innovations to the region (Stipčević 1980: 125). Certainly, the pottery wheel seems to have been introduced sometime in the mid-first millenium BC, although the direction of its introduction is disputed - north of the study area it is documented on the Hungarian plain by the sixth century BC, where it is regarded as a product of Scythian incursions from the Black Sea region, ultimately derived from Greek populations there (Almassy, *pers. com.*); locally wheel-made pottery in association with pottery kilns is also found in early Iron Age contexts in Albania (*op. cit.* 125-6), notably at Gryke e Ujkut near Little Prespa Lake, and from *La Tene*, or Venitic, mid-Iron Age contexts in Slovenia and northern Croatia. Wilkes (1992: 230) notes that in Dalmatia pottery continued to be made by hand into the Roman period except where contacts with Italy and Greece led to the use of the potter's wheel. Although direct evidence for sites of pre-Roman pottery making in the Dinaric region is lacking (Griffiths 1988: 77), most of what is assumed to be locally-produced pottery recovered from Iron Age domestic contexts is hand-made (Mason (1988: 242) specifies "*coil-built*") and rather coarse, variously tempered with sand, limestone, calcite and organic materials.

Although the first documentary reference to the island is by the 10th century Byzantine emperor, Constantine Porphyrogenitus (Cvitanović 1964: 69), and there are references in Venetian documents to specific people and places from the later thirteenth century (Cvitanović 1964: 72), it is not until the beginning of an extended period of intermittent Venetian-Ottoman conflict in the mid-sixteenth century that archival records relating to Iž begin to be abundant. Ottoman raiding from the sea began around 1570, and it is clear that the activities of pirates, both Ottoman and Venetian, during this period impacted upon the settlement pattern of the islands, forcing settlement away from the coast onto hill-tops and increasing the population as a result of immigration from the Ravni Kotari region in the Zadar hinterland. It appears that the population continued to rise by internal growth, peaking around the late nineteenth and early twentieth centuries at a figure way above the sustainable carrying capacity of the island,[2] since when it has fallen by over 90%.

[2] The following selected population figures are given by Cvitanović (1954: 70): 1527 - V. Iž 174, M. Iž 135; 1598 - V. Iž 220; 1754 - V. Iž 397, M. Iž 189; 1879 - V. Iž 1152, M. Iž 596; 1937 - V. Iž 1521, M. Iž 1398; 1954 - V. Iž 1185, M. Iž 777.

1.6 Settlement, Economy and Land-use in the Modern Period

Living conditions for the majority of households in the highly nucleated island settlements were until fairly recently extremely cramped, with large families eating and sleeping together in a single room, most domestic activities centred upon the hearth. Naturally, in a settlement of this size, there were considerable variations in wealth and living conditions within the population. The ability to create wealth depended largely upon land-holdings, but also on access to capital which allowed boats to be acquired for the purpose of trade.

The acute shortage of available land upon which to expand or intensify agricultural production led the growing population in the post-medieval period to intensify the production of goods to exchange for the basic food staples produced in surplus elsewhere. Thus, a number of activities usually oriented towards subsistence and based in the domestic sphere have also been practiced by elements of the population at various times in order to produce a surplus for exchange or sale. Changing the economic orientation of an activity in this way has the effect of raising its status from a household task to specialist trade, with concomitant increases in time and other investments devoted to it. In such agriculturally marginal environments, these activities tend to be the principle means by which peasant farmer(-fishers) procure additional resources in the form of goods, services or, in recent times, money.

The range of activities recently practiced at both domestic-subsistence and professional levels on the island includes fishing, building and shoe-making. In addition, a number of occupations came to be practiced by restricted segments of the population which demanded special skills not known to (or forgotten by) the wider population. Among these may be listed barrel-making, boat-building, stone-carving and pottery-making. Although most such activities, which may be regarded as specialist trades, are still carried out by individuals working alone, some have become collective concerns, with division of labour and shared profits built into their organisation. Boat-building, which seems to have been established on Iž earlier than most other islands (Petricioli 1988: 17), is the most obvious example of this, although here again it should be cautioned that individuals may practice boat-building alone whilst selling their labour to collective enterprises for larger-scale projects. The presence of boat-builders on Iž enabled some islanders to take advantage of trade possibilities which began to open with a return to regional peace and, most importantly, the decreasing influence of pirates, from around the end of the seventeenth century. Several families from the island, such as Banić, Strgačić and Vojvodić, are known to have traded between Venice and Zadar throughout the eighteenth century (*ibid.*), building large houses on the island as a result. Pottery-making seems to have been one of the first craft activities on Iž to have achieved group organisation, possibly pre-dating boat-building in this respect, though perhaps following the example provided by communal fishing.

2 Ceramic Ethnoarchaeological Study of Modern Pottery-making on Iž

2.1 Pottery-making at Veli Iž

Pottery-making on the island of Iž as restricted to parts of the main, highly nucleated part of Veli Iž, principally the neighborhood of Galija, a district set well above the shoreline which is recognised as the oldest inhabited part of the village. Despite unprovenanced suggestions by Marijan (1990) and Ehrlich (1985: 2), there is no hard evidence supporting pottery manufacture at Veli Iž earlier than 1743, when the potter, Ive Ivićin-Sutlović, is mentioned (Cvitanović 1964: 99). By the later nineteenth century, and probably much earlier, pottery-making had become the main specialist activity at Veli Iž, occupying more individuals seasonally than any other non-subsistence activity. Indeed, Iž conforms well to Stark's definition of community specialisation as, "*the production of goods by a large group of households for a broad consumer population, often on a full-time basis in order to earn a livelihood*" (Stark 1991: 63), with interdependent village communities in symbiotic relationships within a regional exchange system (Stark 1991). Close parallels clearly exist here with Peacock's 'nucleated industry' mode of production (Peacock 1982: 9; Van der Leeuw 1984: 754-7).

2.2 Approaches to Study

The fieldwork upon which the present study is based was carried out over two summer seasons totalling four months fieldwork, with four follow-up visits of one week's duration over the next four years. Fieldwork focused on the ethnographic observation of pottery-making activities and their material consequences, the latter focussing on firing processes, but various other strategies were adopted as required to pursue certain areas of interest in more depth. Thus, oral history, field survey, limited excavation and laboratory analyses were all used to consolidate and expand the results of basic ethnography. Two approaches, ethnoarchaeology and participant observation, were of particular importance to the study. Ethnoarchaeology provided a conceptual focus, giving the study potential significance beyond that of 'rescue ethnography'. Participant observation provided information and insights which may not otherwise have presented themselves, since active participation by the observer changes not only the observer's conception of potting, but also the potters' perceptions of the observer, and is likely to lead to more involved discussion on the finer points of technique. Being subliminal in nature, the insights provided by participant observation, or observer participation, tend to lie outside the scope of normal data recording strategies, but should be regarded as permeating the entire study.

The focus of the present study is firing technology, but the implications of firing on other aspects of technology necessitate a brief description of the entire pottery-making sequence.

2.3 The Production Sequence[3]

The principle raw materials of pottery-making, clay and calcite, are known to have been taken from various sources. Traditionally, clay was quarried from sources close to the bay of Soline, 2.5km from the village, but since around 1960 clay a more accessible source at Privlaka on the mainland has been used, while other sources are sometimes exploited when they become available. The quarried clay is prepared by drying in the sun on flat roofs or in enclosed yards, whereupon it is crushed, cleaned of visible roots and stones, and soaked by stirring slowly into a barrel of water.

Calcite was formerly quarried from rock outcrops close to the village, using pick axes to prize it free from fissures in the limestone.[4] When local sources began to be exhausted, it was brought from other islands such as Sestrunj and Kornati, but modern road cutting activities on the island have exposed new seams close to the village. The quarried blocks are prepared by beating with a metal hammer, then sieved to remove the finer fraction for use. The coarser fraction - up to c.0.7mm fragment size - may then be ground to reduce preparation time if the potter owns or has access to a hand-mill (Fig. 5).[5] In 1989 only one of the five potters then active owned a mill, but he allowed access to another; the others, working as two 'firms', used only hammers and sieves.

Fig. 5. Iž mill. Grinding calcite at Veli Iž

Clay and calcite are mixed in a proportion of approximately 3:1 by treading on sacking on the floor of the workshop (Fig. 6), producing a soft, coarse paste which, to avoid over-stiffening before use, is formed into a rounded heap and left in a cool place, usually covered with wet sacking.

[3] Details of the pottery production sequence used at Veli Iž have previously been given elsewhere, most comprehensively by Ehrlich (1985) and Carlton (1988a), with lesser accounts by Carlton (1988b, 1989 and 1990), Stojanović (1965), Oštrić (1978), Lulić (1989) and Marijan (1991).

[4] Gavazzi filmed this operation in 1930.

[5] Grinding stones, are said to have been produced on neighbouring islands, most adjacently on Sestrunj and at Sali on Dugi Otok (Lulić 1988a: 39).

Fig. 6. Iž treading. Treading the clay and calcite mixture at Veli Iž

The forming of vessels takes place on a composite-type hand-wheel, or *kotor*, of variable size and generally rather poor rotational efficiency.[6] Typically, such a wheel consists of a wooden disc attached by short connecting poles to a cross-shaped, centrally-perforated sub-structure; the entire appliance rotating upon a pointed hard-wood pivot (*tulac)*, set into a perforated, dressed limestone block (*trupica)*. Since the use-life of a wheel may extend over several generations,[7] potters often own several of different ages, sizes and individual functional characteristics. In addition to the wheel, wooden tools are used in shaping, cutting and piercing during forming.

With regard to vessel forming, the technique of ring-building is used, along with secondary scraping, to build the vessel walls to the required height from a flat base (Fig. 7 and Fig. 8).

Fig. 7. Iž forming base. Forming a flat base on a hand-mill

Fig. 8. Iž coiling Ring building the vessel walls

The walls are then shaped using clockwise rotation of the wheel and lateral force applied with a wooden shaping tool. Most rims, being straight or slightly inverted, are formed by compressing downwards between finger and thumb, but reverted rims are bent laboriously in segments, using the digits of both hands and slow, rhythmically-interrupted rotation of the wheel, before finishing in the usual way. The vessel may then be decorated simply with lines or dots quickly applied with the tip of a forming tool, before being freed from the wheel by cutting vertically though the lower wall of the vessel, in order to remove the outer part of the base – the remaining part of the base, having been dusted outset with a separating material, does not adhere. It is important to note that the forming process is a single-stage process, with no gaps for drying except when handles or legs need to be added to the base.

Firing, which follows a period of drying, is arguably the definitive stage in the pottery-making process, when the shaped vessels are physically transformed by the application of extreme heat. This is also the most stressful and demanding stage in the production cycle, both for the pots which are liable to crack or shatter, and for the potters required to make critical judgements in conditions of extreme discomfort and potential danger.[8] In the final decade or so of pottery-making at Veli Iž, potters fired irregularly up to four times each season, but Mate Sutlović, last of the full-time (though still seasonal) potters, until the mid-1980s fired approximately monthly, with 100 pots each time split into batches of three. Formerly, levels of production were such that some potters may have fired even more frequently as part of a fixed schedule.

Firing takes place on level ground upon wasteland close to the harbour which is used variously for waste incineration, vegetable growing, chicken rearing and rough pasture. The remains of firing cover up to $50m^2$, but less than $10m^2$ is used intensively at any given time. This site has been used by potters since around 1945, when it replaced a number of sites that had previously been used for firing around the north-west side of the harbour, principally at Slanac. Informants also suggested that dried pots were on occasion taken to neighbouring islands by boat to be fired and sold, taking

[6] The term 'rotational efficiency' is used here as a relative term to describe the ease with which a wheel may be turned and its rotational movement maintained, but not as a judgement on its efficiency as a pottery-making tool.

[7] For example, in 1994 the Sutlović brothers were still using wheels known to be between 50-100 years old, passed down to them by relatives and colleagues.

[8] The Veli Iž potter Ive Vlahov died soon after firing his last batch of pots in July 1992, and at Ularica in northern Bosnia, Mate Gavran died whilst firing in May 1988.

advantage of fuel sources and markets at the host location. The communal use of a single firing place, necessitated by an acute shortage of space, is rare, though not unique amongst western Balkan potters,[9] and does not lead potters to co-operate in firing, except to the extent of tacitly arranging schedules to avoid clashes.

Fuel for firing mainly consists of dried olive clippings, sometimes from winter pruning activities, but usually derived from the practice of using small branches and twigs as fodder for sheep and goats. Having been stripped of their leaves by sheep and goats they are stored in heaps around the stock compounds, but have little domestic fuel value and tend to be burnt as bonfires by their owners if not removed by potters. In earlier times fuel supply was a problem and potters were sometimes driven to purchasing wood from neighbouring islands.

Fig. 9. Iž staking pots. Staking pots on the firing site at Veli Iž

Potters attempt to exercise some advance control over the firing process in a number of ways, notably by waiting for the optimum ground and weather conditions, warming the pots in the sun in advance of firing and pre-heating the firing site by burning grass, domestic rubbish and brush-wood. The sun-warmed pots are arranged in rows, facing downwards at a slight angle, using stones and sherds as supports where necessary (Fig. 9), then a covering of olive clippings is laid on top and lit on the upwind side, or at several places if there is little wind. When it has burned completely, another batch of fuel is thrown onto the upturned vessels (Fig. 10).

Fig.10. Iž firing blaze. Firing with olive clippings

This is repeated several times during the course of a firing, with the amount of fuel increasing after the first two or three cycles, producing intense bursts of heat as the fire burns fiercely for a minute or two, rapidly consuming the tinder-dry twigs and leaves. Potters are able to use changes in the colour of the pots - from buff-orange, to black, to orange-red - to gauge the end of the process, or to assess how much more fuel to add to which areas of the stack.

Firing temperatures were recorded at Veli Iž on three occasions using six thermocouples - heat sensitive probes located at the tip of two metre long flexible steel tubes - attached to a single, central control box and digital monitor. Based on the measured results of the three firings it was seen that firings varied in duration and maximum temperature. However, firing is always of short duration (well under an hour) and achieves temperatures between 650 and 850°C, the latter being well above the suggested upper limit for firing pottery with a significant calcareous component (Rye 1981: 107; see also Brakespear 1986). Temperature variation within the stack, in both vertical and horizontal dimensions, caused by the relatively sheltered conditions beneath the upturned pots and by localised draughts, is very considerable.

Upon completion of the firing, the heap of vessels is swept clean of remaining ash and embers and the pots removed with iron hooks attached to long poles, known as *kuka*, and stacked some distance away to cool (Fig. 11 and Fig. 12). As well as providing a pre-heated firing site upon which to warm and fire another batch of pots, the swift cooling of the fired pots probably helps to diminish the risk of over-firing (see below). It is interesting to note in this respect that the practice of quenching, almost universal amongst potters elsewhere in the western Balkans (Carlton 1999a), was formerly carried out at Veli Iž, perhaps as late as the 1980s.[10]

[9] Communal firing sites are also noted at Demisevci (Kalmeta 1954) and Baljivac (Mićević 1933).

[10] Ive Vlahov stated that dipping to blacken pots was formerly practised, although no evidence of it has been seen on any pots collected at Iž from 1930 onwards. However, some pots made by the Ivan Sutlović 'Gazda' in 1984 have a grey accretion on the lower exterior walls, apparently from dipping in a solution of ash from the firing site.

Pottery broken in the course of firing tends to be removed from the area and dumped elsewhere, often in the sea. Deposits of ash from the surface of the firing area are also removed for use in the vineyards and vegetable gardens as a fertiliser.

Fig. 11. Potrav removing pot. Removing a pot from the bonfire at Potravlje, Dalmatia

Fig. 12. Iž removing pots. Removing pots from the firing area at Veli Iž

3 Experiments with Calcareous Pastes and Fire

3.1 Introduction

A notable and important feature of pottery-making at Veli Iž is the use of calcite, a form of calcium carbonate ($CaCO_3$), as a component of clay pastes. Participant observation has highlighted the great effort and expense involved, particularly in comparison with other potential raw materials, in procuring and preparing calcite for use, as well as the risks involved during firing, suggesting that calcite is regarded as superior materials by potters. Exactly which qualities make it superior and how, and at which stage in the pottery-making cycle, these qualities become manifest, do not appear to be matters of conscious concern to the potters, however. Indeed, although a common component of pottery derived from archaeological contexts in many parts of the world and from all periods, the specific properties of calcite as a component of fired clay pastes remain poorly explored, though it seems likely that a combination of factors are at work. Amongst this range of factors, the role of quenching, almost universally practiced by potters using calcareous pastes in the western Balkans - including, until 1990, at Kaluderovac in the Lika region bordering North Dalmatia (Gavazzi 1978) - has never previously been explored but clearly merits investigation.

3.2 Calcium Carbonate Inclusions

The advantages and limitations of including calcium carbonate ($CaCO_3$) in a clay paste are very well known and frequently cited in studies of ceramic technology (c.f. Rice 1987, Woods 1985, Rye 1981). Briefly, its most significant properties are that it is easily crushed to angular particles and displays a similar Coefficient of Thermal Expansion (CTE) to that of fired clay. In practical terms these properties are advantageous in a number of ways. During the forming stage the addition of angular calcite adds to the structure and mechanical strength of a soft clay paste, making it more manipulable. During firing it is liable to expand and - just as importantly - contract at a similar rate to the clay matrix, thereby causing minimal stress within the fabric and improving the chances of successful firing. The angularity of the particles also enhances their value in arresting incipient cracks within the fabric, thereby further increasing the survival chances of the pottery. These qualities are in addition to the reduced shrinkage during drying caused by the net decrease in relative clay content of the paste, which is a quality induced by the inclusion of any non-plastic, whatever its nature. The most significant drawback to calcite as an inclusion in clay pastes is that it is chemically transformed above about 750°C to Calcium Oxide (CaO), the subsequent rehydration of which in air leads to pressures within the ceramic fabric, often leading to spalling, cracking or disintegration (Fig. 13).

Fig. 13. Flaking pot. Vessel made from a calcareous paste showing the delayed effects of over-firing

Previous experiments have shown that the damage likely to result from over-firing -tempered pottery increases with temperature and duration of firing (Rye 1981), so that extended firing at, or even a little below, 800°C will result in greater damage than a firing briefly peaking at 800°C or a little above. Rather more predictably it has also been shown that the risk of damage increases in correlation with the proportion of $CaCO_3$ in the matrix and to some extent with increasing particle sizes of $CaCO_3$/CaO inclusions.

Therefore, the risk of damage to pottery made from calcium carbonate-rich fabrics may be reduced by controlling firing temperatures and lowering the $CaCO_3$ content of the original clay paste to a minimum. Throughout most of the Dinaric region calcium carbonate in various forms - principally calcite, sometimes limestone and shelly river sand - is frequently the only material considered suitable for use by potters as a tempering agent. The potters' dilemma is that the proportion of non-plastic in the clay paste can not be lowered without endangering vessels made from it during drying, firing and subsequent use; yet the inclusion of high proportions of calcium carbonate means that any over-firing is more likely to lead to damage by lime spalling.

Quenching, by submerging the pots in a soupy organic solution immediately after firing, traditionally has been the most universally-practiced of several post-firing treatments used by potters in the western Balkans to improve the saleability of their products. Flour is usually used in this role - wheat,[11] maize,[12] barley,[13] rye[14], and millet[15] flours are all mentioned in one or more contexts - but alternatives such as soot or wood-smoke resin,[16] charred and crushed fern root,[17] wallnut casing,[18] and oak or ash bark,[19] are occasionally preferred (Kalmeta 1954). Rice notes a range of alternative organic substances, many of them food substances, used in this procedure in non-European contexts (Rice 1987: 163-4).

The methods used to quench pots vary little within the study region. Some potters deliberately heat the prepared solution to boiling point prior to use, while others prepare it a day or more in advance. Where materials other than flour or bran are used, the same mixture can be used repeatedly and continuously, being topped up when necessary. Most potters take pots straight from the fire and dip them while red-hot in a large metal container of the mixture. Immersion of the vessel for a few seconds causes vigorous effervescence as the solution around the vessel boils. Such effervescence is universally held to be an essential part of the process and indicates that the desired effects (see below) have been achieved. Upon removal from the solution, the vessel is seen to change colour rapidly as the solid component of the solution carbonises. In a few cases the quenching mixture is poured or splattered onto the vessel, leading to incomplete blackening (Fig. 14).

Fig. 14 Potrav quench. Quenching using the pouring method at Potravlje

In the few cases where information has been collected on the reasons for the practice of quenching, potters concentrate on its decorative and apparent strengthening, or hardening, effects.[20] It may also be the case that potters or their customers prefer pots without the colour variations created by differential conditions in open firing, and can also be noted that darker colours more effectively absorb the energy of (cooking) fires, but this is probably of marginal, if any consequence, particularly since pots are rapidly blackened by use on the open hearth. Another plausible reason, though not one given by potters, is that the creation of a thin layer of carbonised deposit on the surface of the vessel may cover cracks or other defects formed during firing. Similarly, it may also be suggested that carbonisation seals pores in the surface of vessels, thereby making them more water-tight, although this may be achieved just as effectively by other means, notably by sealing with oil or starch during cooking.

It was in order to explore the properties of calcite as an inclusion in low-fired pottery, particularly in association with the practice of quenching, that a series of experiments was carried out using calcareous pastes. Two of these are briefly described below.

[11] Recorded at Vrkašic, Brezova Kosa, Potravlje, Demiševci, Dobrinja, Malesići, Prebidoli and Ivanjska.
[12] Recorded at Rajovići, Zlakusa, Kučin/Dzurovo, Vrkašić, Demiševci, Liješevo, Dobrinja, Malešići and Prebidoli. Also noted as the worst kind of flour for this purpose at Ivanjska.
[13] Recorded at Lončari (Crikvenica), Kaluderovac, Ervenik, Vrkasić, Krasinac and Donji Rujani.
[14] Recorded at Lončari (Crikvenica), Kučin/Dzurovo, Kaluderovac, Ervenik, Vrkašić and Ivanjska.
[15] Recorded at Liplje and Ivanjska.
[16] Soot is recorded at Liplje, Veselići, Baljevac and Golo Brdo; tar only at Podklanec.
[17] Recorded at Grič, Ularica and Dobrinja.
[18] Recorded at Grič and Lončari (Crikvenica).
[19] Recorded at Grič and Lončari (Crikvenica).

[20] Potters themselves variously refer to the process as kaljenje or kaliti meaning 'tempering', or 'to temper'; alternatively as obara, obaru, obarisi or obariti, meaning 'stewed' or 'cooked'; and in one case as kalajsali, or 'tinning'.

3.3 Methodology

3.3.1 ***Experiment One*** used a mixture of clay samples from the three pottery making locations, Kaluderovac (using clay from Otes) in the Lika region, Potravlje in Dalmatinsko Zagorje and Veli Iž on the coast (using clay from Privlaka on the mainland). The calcite used in the procedures was derived from potters at Veli Iž, Kaluderovac and Potravlje.

The clay samples were crushed and rehydrated in fresh water then thoroughly mixed. The hydrated clay was then divided into two equal parts, one of which was dried and rehydrated with sea water. In this way two uniform masses of hydrated clay were produced, one containing fresh water, the other sea water. Each mass was then further divided into six equal parts to be mixed in varying proportions with calcite, the latter being prepared by crushing with a hammer and sieving through a 2mm mesh. Each of the 12 clay masses was then accurately weighed in grammes to two decimal places and a calculation made to quantify the exact amount of calcite required in order to make up the following pastes (expressed in % clay: % calcite): 100:0, 95:5, 90:10, 85:15, 80:20, 75:25.

Each of the 12 prepared pastes was then used to produce four briquettes of consistent dimensions which were allowed to dry at room temperature. Batches of 12 briquettes (i.e. containing 12 different paste mixtures) were then fired at 50°C intervals ranging from 750-900°C. In order to minimise exposure to high temperatures and replicate as far as possible open firing conditions, the contents were fired quickly and removed from the kiln immediately upon reaching the required temperature. The briquettes were allowed to cool at room temperature before being arranged on white card. They remained undisturbed for three weeks before being photographed, and for a further three months before final photographs were taken and observations made.

3.3.2 ***Experiment two*** used clay paste samples mixed by potters at Kaluderovac, Potravlje and Veli Iž. Each sample was rehydrated in fresh water, except half of that from Veli Iž which was rehydrated in sea water. Eighteen briquettes were manufactured from each of the resultant four pastes. Six batches of dried briquettes, comprising three of each paste mixture, were then fired at 50°C intervals ranging from 700-950°C. One briquette in each of the 24 groups of three was put to one side and left as a control. Another was allowed to cool until warm then immersed completely in water for 24 hours. The third was quenched when red-hot by immersing for five seconds in a mixture of flour and water (200g flour per litre of water), then withdrawn and left to cool. The samples were then arranged on card and observed daily for two weeks before being photographed and left for a further two months to allow any longer term changes to emerge (Fig. 15).

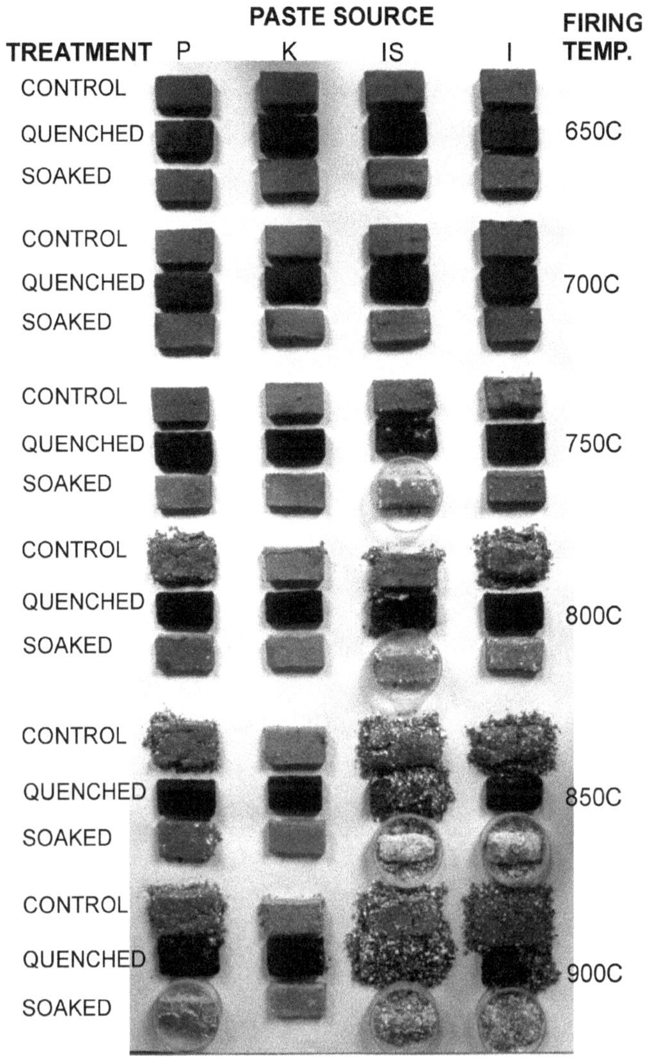

Fig.15 Clay briquettes two weeks after firing

3.4 Results

The results of experiment one were apparent within two weeks and generally comply with those expected following Rye (1976), though the use of pure calcite predictably caused transformations to occur at generally lower temperatures. As may be seen from figure 15, the most profound physical changes occurred to briquettes containing high proportions of calcite and fired to high temperatures. Similarly, variations in post-firing behaviour between briquettes tempered with sea and fresh water also tend to confirm the results of Rye (1976) - briquettes tempered with sea water generally suffer less post-firing damage than those tempered with fresh water. However, it must be noted that the post-firing behaviour of the two groups did not replicate the distinct difference recorded by Rye (*op. cit.* 128, fig. 114).

The results of experiment two are probably of greater significance. In general, they confirm that increasing percentages of calcite and increasing firing temperatures lead ultimately to the highest degrees of post-firing damage. Most interesting, however, is the effect of quenching and water soaking upon the four sample pastes. Damage to the Iž samples began at 800°C, and to the Kaluderovac and Potravlje samples at 850°C, probably reflecting the higher percentage of calcite in the Iž paste. However, those treated with flour and water survived until 950°C, with the notable exception of the Iž sea water sample which capitulated from 850°C. Similarly, the water soaked samples survived to 900°C, and in the case of the Kaluderovac sample to 950°C, but the Iž sea water sample showed damage from as low as 800°C.

3.5 Conclusions and Implications

Experiment two in part replicates the results of experiment one. In part it reveals that the post-firing decomposition of calcite-tempered ceramics proceeds more rapidly when the ceramics in question contain higher proportions of calcite and are fired to higher temperatures. Second, it also seems to confirm Rye's (1981) finding that the addition of salt water as an agent of rehydration allows the sample to be fired higher before post-firing damage ensues. Most significantly, however, it also suggests that a fired calcite-tempered fabric rehydrated with salt water is highly susceptible to damage when exposed to water, certainly much more so than calcite-tempered ceramics rehydrated with fresh water.

It is also apparent that the strategy most likely to protect ceramics containing a high percentage of calcite fired to temperatures within the critical range (for open firings) of 750°C to 900°C is quenching, though for less heavily tempered fabrics water soaking is perhaps preferable. The implications of this are very important. It is commonly noted that the maximum temperature achievable through open firing is normally in the region of 850°C (the upper limit can be controlled to some extent by using particular kinds and sizes of wood, and by modifying firing practices). It is also recognised that calcium carbonate ($CaCO3$) in a fabric will dissociate (to CaO and $CO2$) from around 750°C, depending on the period of exposure, although the sizes and concentration of particles used may effect the extent of subsequent damage. Therefore, it may be suggested that the quenching process is a means of protecting pottery within the danger zone between 750°C and the maximum temperature likely to be achieved in controlled open firing, around 850°C, thereby offering insurance against over-firing caused by accident, negligence or ignorance, or indeed by climatic factors.

In summary, amongst several different reasons perceived or given for the process of dipping pots in solution immediately after firing, its role in ameliorating the potentially detrimental effects of over-firing is paramount. Undoubtedly, the resultant colouration of vessels is also important, as also may be its symbolic value as a transformational event, but these are secondary to its functional role in allowing potters to use higher firing temperatures than would otherwise have been possible. It is clear that the joint strategy of open firing and quenching is ideally suited to the production of calcareous fabrics. Therefore, far from being the 'primitive' firing technology noted by generations of fieldworkers (c.f. Popović 1959 and Randić-Barlek 1990), it is a subtle and highly adapted response to particular physical constraints imposed by chosen materials. Rather than the available firing technology dictating the materials to be used in the paste, as so often presumed, particularly with regard to the use of calcium carbonate (c.f. Shepard 1965; Rye 1981; Rice 1987), the converse appears more likely. Above all, however, it is increasingly apparent that firing and post-firing strategies are part of a delicate balance of complementary factors which make sense only when perceived holistically. There is evidence that the way pots are used is also of significance in this regard - buyers of pots are variously urged to treat their aquisitions before use by boiling fresh water inside or cooking fatty or starchy dishes, which, as well as aiding the removal of any hydrating CaO, serves to clog pores in the fabric with food residues, thereby improving the liquid retentive qualities of the vessel.[21]

3.6 Wider Implications

Further implications of this reflect upon the perceived technological conservatism of firing practices and their spatial organisation. Charges of primitivism against West Balkan pottery-making, often couched in terms of 'conservatism' or 'tradition', levelled from an essentially evolutionary perspective which sees progress in terms of increasing technological complexity, are flawed. Technologically advanced firing strategies are not universally employed, not because they are unknown or beyond the technological competence of some potters - after all, potters at Višnjica and Liješevo in Central Bosnia, and Kompolje in west Croatia, have practised with double-chambered kilns within the 'hand-wheel zone' for centuries, as have potters on its western and northern fringes (Carlton 1999b), but because they are considered inappropriate. Whereas open firing is perfectly suited to the use of the preferred, calcareous pastes - at least in part, as shown above, because over-firing can be guarded against by quenching - more advanced firing technologies would potentially take firing temperatures beyond the range of the potters' control. Furthermore - and perhaps just as significant in the overall balance of factors - open firing requires no investment or maintenance of

[21] A subsidiary effect of this is to imbue the taste of the food residues to foods cooked (or stored) in the vessel as flavours previously absorbed and retained within the porous walls leach out (Arnold 1985: 138; Deal 1998: 90). The ability of pottery to influence the taste of food is a phenomenon referred to repeatedly by informants in the West Balkans, but in ceramic studies remains, as Barley notes, a neglected consideration (Barley 1995: 76).

structures, nor does it compromise other domestic or economic activities, since it occupies space - a highly prized and often restricted commodity in peasant communities - only temporarily. Arnold (1992) has argued just the converse, that:

> "The process of activity restriction (results in) the transition from open-air firing to kiln firing" (Arnold 1991: 120).

This is not born out by examples from the West Balkans, where the location of open firing may be invisible, or can easily be made so, within an hour of the event, thereby imposing no restrictions on the use of that space for other activities. Other, broader social factors, notably the inherent insecurity of land tenure and related transience of settlement may also play a part in promoting the use of portable pottery-making technologies, including bonfire firing. In the borderlands of Bosnia-Hercegovina and Croatia, long-term permanency of occupation of particular houses or places is a concept alien to many communities, most of which have at least an inherited family tradition of migration caused by military or economic invasion.

Ultimately, it can be argued that it is the quality of thermal shock resistence in terms of vessel use that is most significant here, since potters can and do get away with firing coarse pastes comprising other non-plastic materials. The intended use of vessels for cooking, or otherwise in association with heat, is the main factor influencing the choice of calcite as the non-plastic component of pottery fabrics.

4 Field Survey

With regard to the potential applications of information derived from ethnoarchaeological study at Veli Iž, two classes of evidence can be noted. First, the remains of pottery-making activities may be revealed by the structures, tools and waste materials associated with them. Second, the pottery products themselves, whether in complete or fragmentary form, may be interpreted by technological analysis to shed light upon the organisational context of their production.[22]

4.1 Ethnoarchaeological Assessment

Hitherto, ethnoarchaeological studies, and archaeological applications of ethnoarchaeological data, have tended to concentrate on the first of the above classes of evidence. This reflects the tendency in archaeological pottery studies to concentrate on, and work towards defining, fixed-point production units materially and stylistically associated with time and place. In the present case, the investigation of pottery-making activities and their remains by observation and exploratory excavation of workshop areas and the firing site, concluded that Veli Iž probably would not in the normal course of analysis be identified as a production site on the basis of structural and depositional remains alone (Carlton 1988a).[23] In particular, firing remains, the most likely source of positive identification, are too ephemeral and too far from centres of occupation to be detected. Furthermore, even if remains of pottery-making could be detected, or if production was known to have occurred from documentary or oral sources, little could be deduced about its organisational state or economic importance. Similar studies of non-industrial pottery-making elsewhere reveal similar, slight or ephemeral remains, but tend to be over-optimistic or non- committal about the (very poor) likelihood of detecting such remains archaeologically and characterising the nature of the industry responsible for them (cf. Deal 1998). Technological analysis of the pottery itself, therefore, may be the only useful source of information in this regard.

4.2 Field Survey

Archaeological field survey in the neighbourhood of Veli Iž was carried out in order collect a sample of pottery from earlier phases of production for technological comparison with the sample collected from active potters.
Fieldwalking was carried out over a pre-defined area using controlled fieldwalking methods and finds collection strategies devised for similar terrain on the mainland (see Batovic and Chapman 1985; Chapman & Shiel 1988; and Chapman, Shiel and Batovic 1996: 47-50). It resulted in two assemblages of finds, one from the hinterland of the modern village, the other from the Iron Age settlement of Košljin, both of which were dominated by coarse pottery. The two assemblages, Veli Iž and Košljin, were analysed from the perspectives of fabric and technology in order to compare them with the modal pattern established by the ethoarchaeological study of modern pottery-making and subsequent fabric analysis of its products.

Amongst the questions investigated in this study were the following:

Can individual potters/pottery-making concerns (micro-traditions) be detected by technological analysis?
Is the pottery technologically distinct from the products of living potters?

Additional, specific questions with respect to the Košljin assemblage included the following:

Can the hill-fort be identified from its remains as the site of pottery-making?

[22] The preceding experimental investigations into the behaviour of calcareous pastes can also be regarded as broadly ethnoarchaeological (following Gould and Watson 1982), since they ultimately have the same impact upon archaeological reasoning, essentially using the same mechanism of observation and analogy.

[23] This is corroborated by the results of ethnoarchaeological enquiry elsewhere in the region, which have concluded that not only pottery making places within a settlement, but also the settlements themselves may be fluid and unfixed.

On the basis of similarities/differences in the technology of pottery-making, can the modern and Iron Age pottery-making traditions be regarded as part of a continuous tradition?

4.3 Analysis of the Veli Iž Assemblage

The majority of sherds collected during field survey, and the largest scatters, were found in association with sheep huts and on field clearance walls, particularly those used as refuse dumps close to houses (Fig. 16). Many individual sherds, however, were found in isolated locations in the countryside, usually on field walls, rarely on the earth. Older village informants suggested the likely origin of such sherds was not through manuring or dispersal of 'wasters' by potters, but as vessels taken into the countryside as cooking pots or food containers which were discarded when they broke.

The overwhelming majority of sherds collected in the Veli Iž survey, though varying in colour from orange or buff to red, brown, grey or even black, were of the calcite-rich fabric probably produced by modern potters. Traces of manufacturing techniques found on many of the sherds in the assemblage suggested a similar range of forming techniques to those of modern potters, but the majority of sherds were too worn for any detailed interpretations of forming techniques to be made. Macroscopic analysis of the material collected during the survey confirmed that all of the coarse wares, with the exception of the three sherds found at find-spot 10, were calcite-rich and very coarse, indivisible in nature from the material currently manufactured at Veli Iž.

A total of 17 sherds and prepared clay samples from the Veli Iž survey were included in a programme of thin-section analysis. The results suggested that, while textural and some mineralogical variations are apparent within the assemblage, these probably reflect different clay sources (probably the change from Soline to Privlaka sources) rather than contemporaneous micro-traditions (different groups of potters).

4.4 Analysis of the Košljin Assemblage

The Košljin hill enclosure field survey recovered some 107 sherds, with distribution being fairly uniform around the perimeter ramparts, except for a pronounced concentration towards the south-east corner of the enclosure (fig. 17).

Fig. 16. Iž discard. Discarded pottery associated with a sheep hut and refuse dump close to settlements at Veli Iž

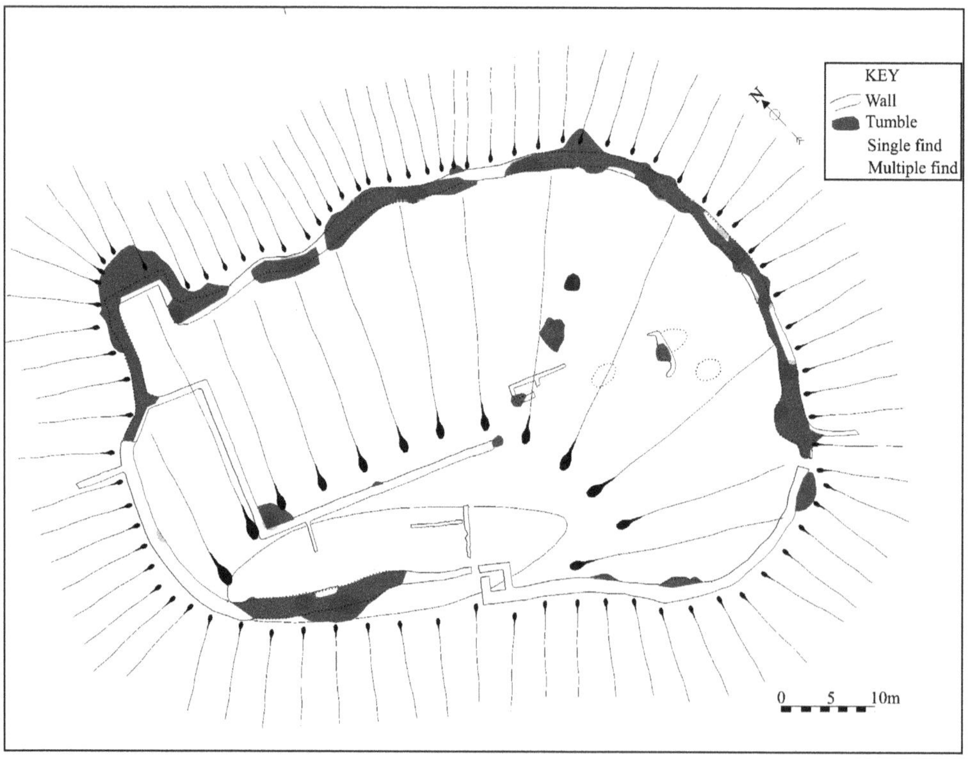

Fig. 17. Plan of Košljin enclosure, Veli Iž, showing pottery find spots

Within the enclosure, pottery was mainly recovered from walls and exposed patches of bedrock, though several sherds were recovered from one small platform of bare earth, moss and sparse grass, suggesting that more survive underground elsewhere. Although the great majority of pottery consisted of friable coarse-wares, a few brown or transparent-glazed sherds were also found. The appearance of lichen on two of these indicated considerable, but otherwise indeterminate, age.[24]

Initial macroscopic sorting of the Košljin hill field collection led to the identification of nine fabric groups, but it was clear during the sorting process that considerable grey areas existed between them. Although perhaps in this respect typical of most archaeological pottery assemblages, it is worth making the observation explicit that in terms of fabric this is a very mixed assemblage, with considerable, perhaps continuous variation apparent through macroscopic analysis alone. In order to clarify, and assess the validity of the macroscopic grouping, a total of 43 sherds from Košljin enclosure were selected for thin-section analysis.

The principal features of the macro- and microscopically-derived groupings within the Košljin hill assemblage may be summarised as follows:

	Microscopic grouping		*Macroscopic grouping*
1a	Calcite-rich; quartz absent or v. sparse	1	Coarse, calcite-rich
1b	Calcite-rich; moderate quartz up to 1%	2	Less calcite (c.20%) some possible grog
1c	Calcite-rich; abundant quartz up to 5%	3	Calcite c.2-10%
2	5-10% calcite; c.1% quartz	4	Abundant sandy; non-calcite incls. c.20-30%
6a	Incl. coagulates; c.15% calcite; c.1% quartz	5	Sparse calcite and ?sand incls. >2%
6b	Incl. coagulates; c.3-7% calcite	6	Abundant crushed shell
7	Coagulates and c.10-12% quartz	7	Abundant light brown incls.
8a	Fine	8	Fine fabric
8b	Fine with apparent grog/clay inclusions	9	Vitrified appearance

What do these categories mean? Whallon's summary of the nature and meaning of the type concept, in which he supports the definition of a type as "*a group of objects exhibiting interrelated similar features which have temporal and spatial significance*" (Whallon 1972: 13-14), retains currency, although it is also broadly accepted that any classification, constructed by the producer or the analyst, "*captures only a part of the order embedded in material categories and is always from a particular perspective*" (Miller 1985: 11).

In general, whilst the degree of variation recorded macroscopically is supported by microscopic thin-section analysis, the definition of groupings is not. With regard to the total assemblage of pottery from Košljin enclosure it is suggested that the most appropriate modification of the original macro. groupings to account for the findings of microscopic analysis, would allow for four main groupings based on macro. groups 1, 2, 3/4/5/7 and 6. These groupings best represent the combined findings of macro- and microscopic analyses.

Very few of the Košljin sherds revealed traces of forming technology during analysis: four exhibited traces consistent with wheel forming, while 11 sherds (representing a maximum of 10 vessels) displayed signs of hand-building. Probably the most distinctive amongst the latter were rim and body sherds from the same vessel which indicated that the vessel was constructed in at least two stages, the main body being allowed to harden before the addition of a final roll, or ring, of clay to form the rim. That the vessel was allowed to harden considerably before this final stage is clear from the nature of the fracture between the two sherds, which reveals a well formed, virtually unaltered rim in the upper cross-section of the lower sherd. Such a clean fracture would not occur - certainly none have been noted during the examination of thousands of sherds from various locations - using the techniques of ring-building and diagonal smoothing practised by modern hand-wheel potters, especially when using the very soft pastes preferred at Veli Iž. Partial drying prior to completing vessels is also suggested by other sherds in the assemblage.

With regard to firing technology, it may be noted that all of the sampled sherds are soft and highly porous, therefore may be interpreted as low-fired. The prevalence of calcite and other calcareous materials as inclusions places firing temperatures below around 800C (see above). The range of colours exhibited by sherds - mainly dull combinations of buff, brown, grey and orange - is also consistent with firing temperatures between 600-800C. The grey or black cores which prevail in all fabric groups except 1 and 5, probably result from the incomplete burning of carbonaceous material within the fabric, although it is also likely that they were fired in conditions of partial reduction. The nature of most fabrics, being less coarse than modern fabrics from Veli Iž, makes them unsuited to the relatively uncontrolled, highly oxidising bonfire firing of the sort currently practised. We may postulate, on the basis of fabric and colour characteristics, that most pottery recovered from Košljin was fired for short periods in small, low-temperature scooped-hearth fires, being denied complete oxidation and/or combustion of organic inclusions by a combination of low temperature and partial enclosure by fuel, ash or a pit structure. The friable nature and black, carbonaceous appearance of some sherds is also consistent with pit firing, although the oxidised appearance of most sherds from large vessels suggests that scooped hearths were probably used, with smaller pots becoming buried in ash at ground level. The possibility that the black appearance of some sherds may be due to carbonisation or soot impregnation during use should not, however, be discounted.

[24] None of the glazed sherds found on Iž could be provenanced to the nearest modern glazed-ware producers in the Lika region to the north-west of Zadar.

Recent ethnographic study of West Balkan potters has shown that individual potters may produce a range of fabrics according to consumer demand and other factors. It may be suggested, therefore, that most or all of the fabrics recovered from Košljin **could** be products of a single group of local potters and there may be no need to look beyond very local sources of production to account for the variety of fabrics displayed.

Two factors may be presented in support of this. First, Košljin and its immediate vicinity are rich in supplies of pottery-making raw materials, including clay sources on the slopes of Korinjak, at Soline and at Mali Iž, all of which are suitable for the production of low-fired, porous vessels of the kinds retrieved during fieldwork at Košljin. Furthermore, materials suitable for use as inclusions in clay pastes are available locally in the form of calcite, limestone, shell, coral, sand and grog, all of which, with the exception of sand, were located during fieldwork upon or near Košljin hill. Indeed, a large lump of friable calcite - the preferred non-plastic of both ancient and modern potters - was found upon the site in direct association with two pieces of worn grindstone. Second, fabric variability is so great, and clay preparation so lax - certainly in macro groups 2-7 - it seems likely that whoever was making the pottery found on Košljin was making use of materials lying to hand, without always precisely selecting particular materials and quantities.

Very few sherds give much indication of overall vessel form, which would be of assistance in assessing potential function. Consideration of fabric and form characteristics suggest, however, that most of the pottery recovered from Košljin - specifically macro. fabric groups 1, 2, 4, 6 and 7 - may be regarded as fairly-highly suitable for use in culinary activities, principally cooking, but unsuitable for long-term liquid storage, due to its high porosity, or functions involving high mechanical stress - drinking and eating vessels, for example - due to its structural weakness (although this weakness can be mitigated by increasing wall thickness, as in the case of sherds 102c/e).

4.5 Interpretation of the Košljin Assemblage

Broad conclusions drawn from the analysis of pottery from Košljin are that it was a location of fairly intensive activity, at least some of which was domestic, as evidenced by large amounts of pottery and two grindstone fragments. There is nothing to suggest that the pottery found recovered from Košljin was not produced there, since all necessary materials are available locally. That said, however, it should be noted that some of the ceramic material found at Košljin could be modern, being indistinguishable from calcite-tempered cooking-wares produced locally over the last 250 years or more.

The use of calcite amongst a range of other, locally available tempering materials in Iron Age contexts is significant, since it implies that calcite, though considerably more difficult to prepare than other locally available materials, was deliberately preferred over materials such as sand, dung, chaff and grog. As discussed above, recent ethnographic information suggests that the adoption and continued use of calcite as a non-plastic component of ceramic fabrics has more to do with the subsequent survival value of vessels in use than their successful firing.

Although a number of important similarities can be noted between ancient and modern production, notably with regard to the use of calcite inclusions and low-temperature (probably open hearth) firing, it is clear that modern production is much more standardised. The range of fabric variation exhibited by the ancient material defies simple categorisation and stands in contrast with assemblages of modern pottery collected during field survey and directly from producers. Furthermore, the chance discovery of two conjoining sherds from Košljin, exhibiting distinctive traces of a stationary ring-building forming process, indicates that forming techniques were markedly different from those of modern potters. It is notable that stationary ring-building is a much more laborious technique than the rotational variant currently used, thus supporting the view, suggested by the diverse range of poorly mixed fabrics, that Iron Age pottery-making was an occasional, non-specialist, domestic activity, wherein the manufacture of a single vessel took considerably longer than a modern equivalent, but the total time any given individual spent pottery-making annually was much less. In addition to the use of wheel technology (signs of which are visible on a high proportion of sherds collected), the relative uniformity, in petrological and textural terms, of the pottery collected in the Veli Iž survey is indicative of a higher level of specialisation than can be sustained for the Iron Age material.

Thus, although we can not refer with certainty to the continuity of a tradition of pottery-making, in terms of organisation or technique, between the Iron Age and modern periods at Veli Iž, it is possible to suggest the continuity of certain technological traits, notably preference for calcite-rich fabrics, probably in association with certain traditions of use. And as argued above, the use of calcite imposes technological constraints which have a strong bearing in determining other aspects of the manufacturing process, such as methods of firing and post-firing treatments. Calcite may be viewed as the pivotal part of a pottery-making package attested since the sixth millennium B.C. (Chapman *et. al.* 1996: 189), which reappears at Košljin and in bronze age and Iron Age contexts on the mainland (Griffiths 1988: 77), in which form it is a direct precursor of modified, later and continuing traditions in the same region (Carlton 1996: 89-90). While, some aspects of the Iron Age tradition, which was probably a non-commercial, household activity, may be regarded as survivals from preceding periods, they may also be regarded as having survived into the present era. Changing contextual factors and economic imperatives, together with differences in the accumulated reservoir of technical knowledge (of wheel technology, for example) from period to period, lead to different developments in pottery-making practice, organisation and products over time.

Bibliography

ADAMS, W. Y.

1979 On the argument from ceramics to history: a challenge based on evidence from Medieval Nubia. *Current Anthropology* 20: 727-734.

ARNOLD, D. E.

1985 *Ceramic Theory and Cultural Process.* Cambridge: Cambridge University Press.

ARNOLD, D. E.

1992 *Ecology and ceramic production in an Andean Community.* Cambridge: Cambridge University Press.

ARNOLD, P. J. III

1991 *Domestic Ceramic Production and Spatial Organization. A Mexican Case Study in Ethnoarchaeology.* Cambridge: Cambridge University Press.

BALFET. H.

1965 Ethnographic Observations in North Africa and Archeological Interpretation: The Pottery of the Mahgreb, pp. 161-177. In Matson, F R. (ed.) *Ceramics and Man.* Chicago: Aldine.

BATOVIĆ, S.

1974 Prapovijesni ostaci na zadarskom otočju; *Zbornik*, Zadar.

1977 Caracteristiques des agglomerations fortifiees dans la region des Liburniens, *Godisnjak; Centar za Balkanoska Ispitivanja*, 15: 201-225.

BATOVIĆ, S. and CHAPMAN, J.

1985 The Neothermal Dalmatia Project, pp. 158-195. In Macready S. and F. H. Thompson (eds.), *Archaeological Field Survey in Britain and Abroad.* London: The Society of Antiquaries of London. Occasional Papers, 6.

BENAC, A.

1985 *Utvrdjena Ilirska Naselja* (I), Sarajevo.

BRACEWELL, C. W.

1992 The *Uskoks of Senj: Piracy, Banditry, and Holy War in* the *Sixteenth Century Adriatic.* Cornell University Press.

BRAKESPEAR, R.

1986 Technological Analysis of Prehistoric Pottery from Dalmatia: towards an understanding of ceramic technology in Dalmatian prehistory. Unpublished B.A. dissertation, University of Newcastle upon Tyne.

CARLTON, R. J.

1988a The ethnoarchaeological study of pottery-making on the Dalmatian island of Iž. pp. 101-123. In Chapman, J. C., J. Bintliff, V. Gaffney and B. Slapsak (eds.), *Recent Developments in Yugoslav Archaeology.* Oxford: BAR International. Series 431.

1988b Ethnoarchaeological study of pottery production in Dalmatia, *University of Durham and University of Newcastle upon Tyne Archaeological Reports Series: Archaeological Reports for 1987*, Durham, 50-55.

1989 An ethnoarchaeological study of hand-wheel pottery industries in Bosnia and Croatia, *University of Durham and University of Newcastle upon Tyne Archaeological Reports Series: Archaeological Reports for 1988*, Durham, 56-61.

1990 A continued ethnoarchaeological study of hand-wheel pottery industries in Bosnia and Croatia, *University of Durham and University of Newcastle upon Tyne Archaeological Reports Series: Archaeological Reports for 1989,* Durham, 65-67.

1996 Post-Medieval and Modern Pottery (AD1750-1990); In *The Changing Face of Dalmatia - Archaeological and Ecological Studies in a Mediterranean Landscape,* Leicester University Press/The Society of Antiquaries of London, 89-90.

1999a The Past and Future of Traditional Pottery-Making in Bosnia, *The Studio Potter* 27 (1).

1999b The Pirocanci pottery-making tradition in former Yugoslavia, *The Studio Potter* 27 (2): 98-99.

2000 A response to the term 'Interpreting Ceramics', *Interpreting Ceramics* (e-Journal). UCWS (original paper presented at the *Interpreting Ceramics* Symposium, V & A Museum, September 2000).

CHAPMAN, J. C. and S. BATOVIC

1987 Neothermal Dalmatia Project - Fifth (1986) Season, *Archaeological Reports (Universities of Durham and Newcastle) for 1986*, 3-8.

CHAPMAN, J. C., J. BINTLIFF, V. GAFFNEY and B. SLAPŠAK (eds.)

1988 *Recent developments in Yugoslav archaeology*, Oxford: BAR International Series 431: 85-99.

CHAPMAN, J. C. and R. SHIEL

1988 The Neothermal Dalmatia Project - archaeological survey results. pp. 1-30. In Chapman J. C., J. Bintliff, V. Gaffney and B. Slapšak (eds.), *Recent developments in Yugoslav archaeology*, Oxford: BAR International. Series 431.

CHAPMAN, J., R. SHIEL and S. BATOVIĆ

1996 *The Changing Face of Dalmatia: Archaeological and Ecological Studies in a Mediterranean Landscape.* Leicester University Press/The Society of Antiquaries of London.

COURTNEY, P.

1997 Ceramics and the history of consumption: pitfalls and prospects, *Medieval Ceramics* 21: 95-108.

ČOVIĆ, B.

1983 Eneolitski substrat. *Praistorija Jugoslavanskih Zemalja* IV, Bronzano doba: 103-112, Sarajevo.

CVITANOVIĆ, V.

1964 Otoci Iž i Premuda. *Radovi Instituta Jugoslavenske Akademije Znanosti I Umjetnosti u Zadru*, 1: 69-108.

DEAL, M.

1998 Pottery Ethnoarchaeology in the Central Maya Highlands. Salt Lake City: University of Utah Press.

EHRLICH, N.
1985 Lončarstvo ručnog kola na otoku Veli Iž; Diplomski rad, Etnološki seminar, Filozofski Fakultet u Zagrebu. *Unpub.*

FINE, J.
1991 The Early medieval Balkans: a critical survey from the sixth to late twelfth century. The University of Michigan Press.

FORENBAHER, S.
1996 Velebit: 9,000 years of people, goats and sheep on an Adriatic coastal range; *Das Altertum*, Vol. 41: 211-226.

GAVAZZI, M.
1978 Kaluderovački lončari; *Lički Zbornik*, 1: 25-34.

GOULD, R. A. and WATSON, P. J.
1982 A dialogue on the meaning and use of analogy in ethnoarchaeological reasoning; *Journal of Anthropological Archaeology* 1: 355-381.

KALMETA, A.
1954 O seljačkom lončarstvu u srednjoj i zapadnoj Bosni; *Glasnik Zemaljskog Muzeja u Sarajevu* (Etnologija), nova serija, 9: 127-168.

KRAMER, C.
1979 *Ethnoarchaeology: Implications of Ethnography for Archaeology*. New York: Columbia University Press.

KRAMER, C.
1985 Ceramic Ethnoarchaeology. *Annual Review of Anthropology* 14: 77-102.

LOCKWOOD, W. G.
1975 *European Muslims: economy and ethnicity in western Bosnia*. Academic Press.

LULIĆ, J.
1988a Tradicijski život na zadarskim otocima. pp. 32-41. In Uranija, V. *Čovjek, More, Otoci - prisustvo čovjeka na zadarskim otocima, katalog izložbe*, Narodni Muzej Zadar.
1988b Man, Sea, Islands - the presence of man in the Zadar islands. pp. 41-43. In Uranija, V. *Čovjek, More, Otoci - prisustvo čovjeka na zadarskim otocima* (katalog izložbe). Narodni Muzej Zadar.

MARIJAN, L.
1990 *Tradicijska kultura u Velom Ižu*. Etnografski zbirka, Veli Iž.

MASON, P.
1988 Coming down from the hills: A preliminary report on the intensive surface collection of a lowland early Iron Age activity area in Slovenia. pp. 240-46. In Chapman, J. C., J. Bintliff, V. Gaffney and B. Slapsak (eds.), *Recent developments in Yugoslav archaeology*, Oxford: BAR International. Series 431.

MATSON, F. R.
1965 Ceramic ecology : An approach to the study of the early cultures of the Near East, pp. 202-217. In Matson F. R. (ed.) *Ceramics and Man*. Viking Publications in Anthropology 41. Chucago: Aldine.

MIĆEVIĆ, L.
1933 Negdašnje lončarstvo u selu Baljivac; *Glasnik Etnografskog Muzeja u Beogradu*, XX: 108-110.

MILLER, D.
1985 *Artefacts as Categories - a study of ceramic variability in Central India*. Cambridge: Cambridge University Press.

NANDRIS, J.
1988 Ethnoarchaeology and Latinity in the mountains of the southern Velebit. pp. 125-43. In Chapman, J. C., J. Bintliff, V. Gaffney and B. Slapsak (eds.), *Recent developments in Yugoslav archaeology*, Oxford: BAR International Series 431.

ORTON, C. R., P. A. TYERS and A. G. VINCE
1993 *Pottery in Archaeology*. Cambridge: Cambridge University Press.

OŠTRIĆ, O.
1979 *Narodno lončarstvo* (katalog izložbe), Narodni Muzej Zadar, Etnografski odjel.

PEACOCK, D. P. S.
1981 Archaeology, Ethnology and Ceramic Production. pp. 187-194. In Howard, H. and E. Morris (eds.) *Production and Distribution: A Ceramic Viewpoint*. Oxford: BAR International Series 120.
1982 *Pottery in the Roman World: an ethnoarchaeological approach*. London: Longman.

PETRICIOLI, S.
1988 Zadarski otoci od početka naseljenja do pada Napoleona. pp. 9-19. In Uranija, V., *Čovjek, More, Otoci - prisustvo čovjeka na zadarskim otocima, katalog izložbe*. Narodni Muzej Zadar.

POPOVIĆ, C.
1956 Lončarstvo u Bosni i Hercegovini I; *Glasnik Zemaljskog Muzeja u Sarajevu* (Etnologija), nova serija, 11: 95-122.
1957 Lončarstvo u Bosni i Hercegovini II; Glasnik *Zemaljskog Muzeja u Sarajevu* (Etnologija), nova serija, 12: 17-46.
1959 Thenika primitivnog lončarstva u Jugoslaviji; *Glasnik Zemaljskog Muzeja u Sarajevu* (Etnologija), nova serija, 14: 25-59.

RANDIĆ-BARLEK, M.
1990 *Lončarstvo ručnoga kola* (Katalog izložbe). Etnografski Muzej Zagreb.

RICE, P. M.
1987 Pottery *Analysis: a sourcebook*. Chicago: University of Chicago Press.

RYE, O. S.
1976 Keeping your temper under control: materials and the manufacture of Papuan pottery; *Archaeology and Physical Anthropology in Oceania* 11: 106-137.
1981 *Pottery Technology: Principles and Reconstruction*. Manuals on Archaeology 4. Washington D.C.: Taraxcum.

SARASWATI, B. and N. BEHURA
1966 *Pottery Techniques in Peasant India*. Calcutta: Anthropological Survey of India.

SHEPARD, A. O.
1965 Ceramics *for the Archaeologist*, 5th printing; Washington D.C.: Carnegie Publication 609.
SINGLETON, F.
1985 *A Short History of the Yugoslav Peoples*. Cambridge: Cambridge University Press.
STARK, M. T.
1991 Ceramic production and community specialisation; a Kalinga ethnoarchaeological study. *World Archaeology* 23(1): 63-77.
STIPČEVIĆ, A.
1980 *The Illyrians: history and culture*. New Jersey: Noyes Press.
TOMIĆ, P.
1983 *Grnčarstvo u Srbiji*. Etnografski Muzej u Beograd.
VAN DER LEEUW, S. E.
1976 *Studies in the Technology of Ancient Pottery* (two vols.). Amsterdam.
VAN DER LEEUW, S. E. and A. C. PRITCHARD (eds.)
1984 *The many dimensions of pottery: Ceramics in archaeology and anthropology*. CINGULA 7. Amsterdam: Institute for pre- and proto-history, University of Amsterdam.
VAN DER LEEUW, S. E., A. J. SPRUIJT and V. A. SHELTON-BUNN
1987 Ceramic Production – Assendelver Polders, pp. 225-264. In Brandt, R. W., W. Groenman-van Waateringe and S. E. Van der Leeuw (eds.), *Assendelver Polder Papers I*. Amsterdam.
VELDE, B. and I. C. DRUC
1998 *Archaeological Ceramic materials: origin and utilisation*. Springer.
WHALLON, R. Jr.
1972 A new approach to pottery typology; *American Antiquity* 37 (1) : 13-33.
WILKES, J. J.
1969 *Dalmatia*. London: Routledge and Kegan Paul.
1992 *The Illyrians*. Oxford: Blackwell.
WOODS, A. J.
1986 Form, fabric and function: Some observations on the cooking pot in antiquity. In Kingery (ed.), *Technology and Style - Ceramics and Civilisation* 2. Columbus, Ohio: American Ceramic Society.

FIRE AND AIR DRAUGHT: EXPERIMENTING THE CHALCOLITHIC PYROINSTRUMENTS

Dragos Gheorghiu

1. Introduction

The study of fire in prehistoric societies is a very important issue for archaeological studies, since for the populations of traditional societies fire represented a vital instrument and a daily ritual experience. At the same time phenomenon and material culture, fire imposes a special archaeological approach involving the analysis of physical processes combined with the study of material culture.

1.1. Experiment as a Compensating Approach

Since in archaeological record fire remains are not to be sufficient to understand plenary the phenomenon, a complementary mode of investigation could be experimentation that can reproduce processes as the control of air or the burning up of fuel. This is the reason why, in explaining prehistoric pyrotechnologies, I believe that experimentation should be employed as a basic method to materialize invisible processes for the archaeologist (see Andrieux 1997: 693) and therefore should be studied through the different relationships they had with the material structures that generated and supported combustion. In the present text these material supports of fire were labeled pyroinstruments, i .e., objects designed for the management, or in other words "domestication" of fire.

The present paper will discuss some of these pyroinstruments used by Chalcolithic societies of Eastern Europe for the management of fire by means of reconstructions after archaeological remains, with the intention to analyse the relationship existing between fire, materials, gas dynamics and the shape of these tools.

A direct relationship with (the replica of the) pyroinstrument could help archaeologist to offer a more sensible interpretation of archaeological record and of the physical phenomenon of combustion determined by the relationship between temperature and shape of objects, fuel, and colour of flames. All the experiments described in the present paper took place in the village Vadastra, in the Olt County, south Romania, between July 2000 and November 2001.

2. Fire and Domestication

Control of animal or of caloric energy seems to have been perceived as being analogous in the past, since historical data from ancient cultures equate fire with a wild animal (see Herodotus, III, 16), possessing a voracious character (Vernant 1979: 66), hence the need to domesticate (cf. Vernant 1979: 64), or to control it.

One of the (many) techniques of controlling materials would be sieving, a process that allows the separation, division (or disintegration) of a whole into parts that can recompose (or not) a new whole. Sieving can be applied to liquids and solids, as well as to fire, and a minute analysis of the material culture of Chalcolithic societies of Eastern Europe reveals a large number of ceramic sieves, both for food and fire management (for the use of the ceramic sieves as "Bunsen lamps", see Wood 1999: 185), in favour of this hypothesis.

3. Identification of Fire Relationships in Archaeological Record

The identification of the complex relationships of fire with cultural environment that generated it is not an easy matter, because only some of them are visible in the archaeological record, as those between fire and built environment.

In the prehistoric buildings of the Lower Danube Chalcolithic there are substantial data in favour of the existence of a symbolic relationship between the orientation of inner space and pyroinstruments as ovens or hearths (see Berciu 1937: 5, fig. 3; Comsa 1996: 78), as well as between whole buildings and fire (see Tringham 1992, Stevanovic, this volume).

Other relationships still visible in the archaeological record but not studied in depth, are those connecting fire with materials.

3.1. Relationships with Materials

Part of the complex pyroinstruments, to mention only kilns, was made of composite materials. Generally, composite materials as wattle and daub have special thermal qualities, as a high thermal inertia, due to the empty spaces into the fabric of clay left after the combustion of organic materials.

Better performances for rising temperatures in pyroinstruments resulted when composite materials were plastered with a slip that forms rapidly an extra insulation ceramic layer.

3.2. Relationship with Air

There is a relationship that was practically neglected in current studies on prehistoric fires, due to its "virtual" character, that is the relationship with air.

This study will focus only on this immaterial relationship with air in pyroinstruments, air being at the same time support and generator for fire, all pyroinstruments, or air draught objects, functioning with the help of an airflow circulating through them. One cannot study in archaeological record the virtual element "fire" without studying the virtual element 'air", therefore a classification of pyroinstruments can be made depending on their degree of control of the air. The most advanced pyroobjects were those with air absorption areas (characterized by perforated surfaces) on their exterior or interior surfaces. A history of pyroinstruments seems to have been the one of the development of automata functioning on the principle of air draught. The effect of convection of hot air was exploited in ceramic design, ceramic pyroinstruments with a chamber (braziers or kilns) utilizing the phenomenon of up-draught airflow to produce a vacuum effect that aspired fresh air through the perforations positioned at their base or on their walls, thus controlling and feeding fire.

3.2.1. Burned Houses or Firing Chambers with Draught Airflow

One of the most suggestive archaeological examples of the relationship between fire, airflow, and materials seems to be that of burned houses.

East European Chalcolithic is characterized by the frequent presence of fired houses or settlements, a process that was attributed by some archaeologists to a ritual purpose (Tringham 1992, Stevanovic, this volume).

Probably some of the arsons were accidental or the result of warfare, and only a small part were the consequence of an intentional, ritualistic, action, as the presence of fired fragments of buildings as structured deposition in pits, along other fragments of objects (see also Comsa 1990), prove it. Part of ceramic objects and fragments of architecture display traces of vitrification, as the result of a firing over 1150°C.

The firing of a 6 x 3 x 2 m room with the walls made of wattle and daub, even filled with textiles and wooden objects, could not reach temperatures over 1150°C able to vitrify ceramic material. In the course of the elaboration of the present text, a fire devastated part of the Village Museum in Bucharest, consuming a large number of peasant houses before the fire brigade could intervene; but in spite of the intensity of the fire, the result of the disaster was only the consuming of wooden and vegetal structures without crumbling or vitrifying clay walls. Then how vitrification was possible in Chalcolithic burned houses?

In the conditions of a narrow space resulted by the demolishing of the houses' walls, to control the spreading of fire over the settlement, and in the conditions of a powerful airflow through the resulted channel, a dramatic rise of temperature could be reached and could be maintained for a long period of time, because of the insulating qualities of the composite material made of clay mixed with straws that did not allow convected and radiated heat to dissipate.

An experiment to rise temperature similar to that in fired houses was made by accident, by the author, together with Dr. Alex Gibson, in the village of Vadastra, southern Romania, in August 2000, using a subterranean up-draught kiln of 1.5 m deep and a vent aperture large of 40 cm covered with a sieve-like layer of ceramic shards. A sudden strong air stream led to a dramatic rise in temperature in the kiln, from 800 to 1150°C, in less than 15 minutes, when the thermocouples were removed for safety. The base of the ceramic loading in contact with the platform of the kiln was melted and displayed traces of vitrification.

3.3. Pyrotechnologies

A taxonomy of pyroinstruments from simple bonfires to subterranean kilns (cf. also Sinopoli 1991: 31 ff) would therefore be made according to the relationship of these objects to airflow, soil surface (cf. Woods 1982: 16, Table 1), and material, following the degree of complexity (see Rhodes 1981: 3 ff) of this relationship.

The study of a technological growing complexity in firing strategies (cf. Rice 1997: 252), starting with bonfires and ending with up-draught kilns did not discuss the performances in temperature between pyroinstruments, since "firing strategies could not be differentiated on the basis of temperatures, because all reached overlapping temperature ranges of 600 – 900°C" (Rice 1997: 253), but according to their degree of efficiency in the mode of ventilation, and duration of firing. The present study tries in discussing various study cases of Chalcolithic pyroobjects to discover similar designs in the control of air.

4. Study Cases

4.1. Surface Bonfires

(Experiments by Drs. Alex Gibson and Dragos Gheorghiu, May 2001, by Dr. Dragos Gheorghiu and Cristian Doros, July-August 2001 and Dr. Dragos Gheorghiu and Ion Cococi, October 2001)

Ceramic vases were placed on a grid of thin wood splinters,

which allowed air to be draught through this sieve-like structure, the same time insulating the ceramic objects from the ground. In the wet season this method protected the vases from soil moisture.

The advantages of this mode of firing objects is the low cost (in energy and time consuming), compared to the construction of a kiln (see also Rice 1987), the fuel positioned over the objects (see Gosselain and Smith 1995) producing also a hot protection against cold uncontrolled airflows (see also Rice 1987: 156). The fuel used was cattle dung, compressed under the shape of thin bricks, that burned longer and more constant that wood.

A main disadvantage of bonfire is that it could not be controlled by the experimentalist, because there are many variables that cannot be monitored. Other disadvantages are the greater loss of heat, the greater weather dependency (Rice 1987: 156) since the uncontrolled air currents are its most dangerous enemies, the diminished airflow through the piled objects, and subsequently the unevenness of heat (Rice 1987: 156), that produces easy cracks, dent, or over- or underfired objects (Rice 1987: 156), as well as fire clouding (Gibson and Wood 1997: 156; Rice 1987: 155).

Generally the temperatures reached are around 650–900°C (Gibson and Woods 1997: 27), and sometimes higher (see Rice 1987: 156; Gosselain and Smith 1995), this performances depending also on the nature of fuel and the dimension of objects fired (Rice 1987: 157).

In the experiments conducted in May 2001 by Dr. Alex Gibson (see Gibson 2001, Report No. 71, Department of Archaeological Sciences, University of Bradford), at a 20°C air temperature, the vases positioned on a three-layered lattice of split timber (to allow hot air to circulate from base to top) were covered with straw and split wood in a pyramid-like structure, to direct the air flow. The temperature reached 800°C in 25 minutes (see Diagram 1). A slight breeze produced uneven firing and enforced experimentalists to put additional fuel on the uncovered vases, to protect them.

In July 2001, at an air temperature of 30.5°C, and after changing the fuel (wood was replaced with dung), a temperature of maximum 865°C was reached after sixty minutes (Fig. 1)

The disadvantages of uncontrolled air currents, especially in the cold season, were compensated by quenching the combustion area with earth after the wooden structure covering the vases transformed into amber, and leaving the top of the pile uncovered. In November 2001, using as fuel willow wood, at an air temperature of 15°C, within 40 minutes 791°C were reached in its middle and 120°C on the cold soil surface.

A preliminary conclusion could be that air temperature depending on season has little influence on the combustion process, compared to the nature and position of fuel on ceramic objects that allows a good draught. The greatest risk for objects fired in bonfires are air currents, the loss of vases being due to their uncovering by air currents during the process of cooling.

Fig. 1. Vases positioned on a lattice of split timber and covered with dung.

Fig. 2. A large vase (made by student Madalina Mihaila), fired in a shallow–pit.

4.2. Shallow Pits

(Experiments by Dr. Dragos Gheorghiu, Professor Ralph Rowlett and Liana Svintiu, August 2000)

Unlike the bonfire case, the practice of firing ceramic objects in shallow pits was archaeologically identified in many places in the Muntenia Plain and the Balkans (Comsa 1976: 29). For example, in the Vadastra Chalcolithic site (5th millennium BC), near which experiments took place, a pit of 0.75 m in diameter and 0.80 m deep used to fire ceramic objects was discovered (Mateescu 1959: 68-69).

Even if fuel, temperatures reached, and positioning of objects and fuel are similar, an advantage of pits upon bonfires is an improved protection of the combustion process from uncontrolled air currents (see Diagram 2).

In the case of objects positioned in shallow pits and covered completely with dung, combustion generated a reduced atmosphere, as shown by the brown and black colours of the objects (Munsell 5YR 3/2 – 5YR 2.5/2) (Fig. 2).

4.3. Corbelled-shaped Deep Pit

(Experiments by Dr. Dragos Gheorghiu, with the help of Ion Cococi and Florin Alexa, August 2001)

Experiments to observe the combustion in closed spaces, supposed to produce incomplete draught ventilation, were carried on in deep corbelled pits, similar to those used for food storage or to some up-draught kilns. The most efficient shape of the container, both for grain storage (for which there are many historical examples) or pit firing seems to be the corbelled pit, with a small aperture to avoid heat loses, and a large base.

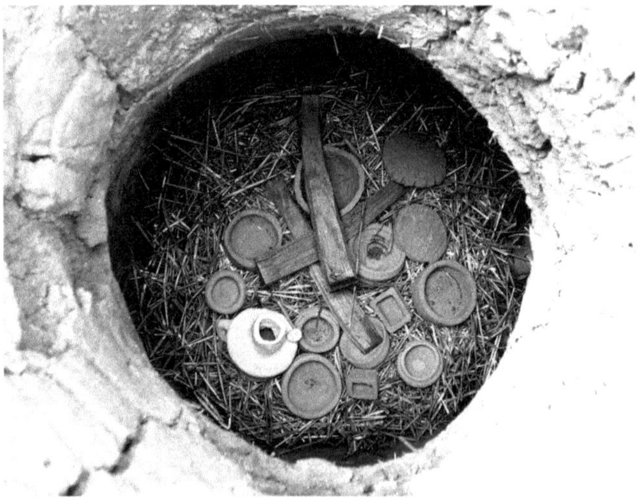

Fig. 3. The positioning of vases on a layer of straw.

A corbelled pit of 2 m depth and 1.5 m width at the base, with a top aperture of 0.4 m, was filled successively with layers of straw (Fig. 3), vases, dung and split timber of 0.8 m high (Fig. 4). After combustion started, the aperture was covered with tiles and earth, leaving small orifices to allow air suction and smoke elimination.

Fig. 4. The covering of vases with split timber.

This ceramic covering functioned as a sieve for air draught, part of the orifices acting as aspiring pumps for the down draught air, since the other acted as propel pumps, as proved by a strong emission of white smoke.

Fig. 5. The process of combustion.

After 2 hours of combustion (Fig. 5), when temperature reached the maximum of 735°C, these orifices were covered with earth too, consuming all oxygen inside. The ceramic production of such combined pyroinstrument, half bonfire, half kiln, was very diverse, from black (Munsell 5YR 2.5/2) to spotted, with fire clouds to yellow white (Munsell 7.5Y/8/2). As one can conclude, the results of this experiment warns us about the dangers of identifying pyrotechnologies in archaeological record, since the objects of a single load were fired in reduced as well as in oxidized atmosphere.

4.4. Fixed Ovens

(Experiments by Dr. Dragos Gheorghiu with the help of Ion Cococi, November 2001)

In traditional societies an indoor, as well as an outdoor pyroinstrument designed for cooking food and ceramics and to protect fire from air currents, as well as to protect the building from the danger of ignition, was the oven. Eastern Europe Chalcolithic buildings used to have these pyroinstruments modeled as small houses (see Civilizatia Boian, cat., 1999: XX, Deshayes 1974: 72, fig. 5) with one to many openings (see the Gumelnita A2 model from Izvoare in Serbanescu 1997: fig. 3/6) or as semi-spherical domes (see cat. Boian 1999) (Fig. 6).

Fig. 6. Model of an oven from Vadastra, Vadastra culture.

If the semi-domed ovens could stand without an internal wooden structure (as contemporary ethnologic examples show), those in the form of houses with roofs with large openings needed at least a wood structure to support the clay roof.

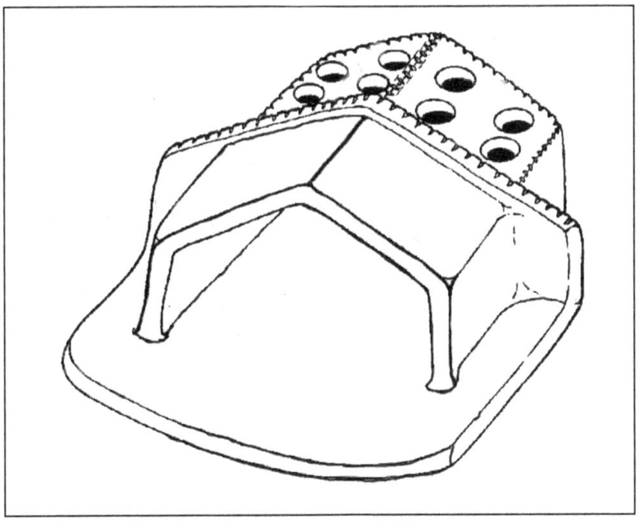

Fig. 7. Model of an oven from Drama, Gumelnita culture (after Lichardus *et al.* 1996).

When designed to be an indoor object, the oven should protect from fire the interior of the household, through the control of absorbed air. A ceramic model from Gumelnita culture found in Bulgaria at Drama (see Lichardus *et al.* 1996: 30, fig. 12) representing a two-roofed construction with large perforations on posterior of the roof, could be a model for the type of oven that used air draught to entertain fire and to protect the interior of the house (Fig.7). The elimination of hot gases through the perforated roof allows a good ventilation of the inner space of the oven, and subsequently, a good combustion. It is possible that air control could be improved through the filling up of the roof's perforations with ceramic blockers, or small vases that could at the same time heat liquids or solid food. Fire could be lit on the platform in front of the architectural structure, that would have the role of a hearth, and be absorbed through the ceramic container and exit, filtered, through the perforated roof.

A number of small ceramic vases could be fired in this type of oven, parallel to food preparation and house heating, at an average temperature of 550°C.

The large opening allows beside a fair air circulation an easy removal of ashes and incandescent vases that can be thus wrapped with straws, organic waste used as fuel or with the ashes pulled out the oven, creating a black lustre colour on their surface.

4.5. Vases-kilns, Portable Ovens, Braziers and Food Heating Instruments

(Experiments by Dr. Dragos Gheorghiu, Dr. Alex Gibson, May 2001; Dr. Dragos Gheorghiu, and Florin Alexa, July 2001)

There are a lot of Chalcolithic ceramic objects whose function was not established yet, having as common trait the principle of air suction through various perforations.
These categories include a lot of ceramic pyroinstruments of different shapes, whose role is to use the caloric energy of embers to transfer it to environment or to objects positioned on them. The most efficient, acting as a sort of pump for air absorption are those with perforated walls (the so believed to be "sieves" for cheese-making) which produce a controlled airflow, aspiring cold air through the perforations at the base and expelling hot air through those on their upper part. I don't reject the categorization of "sieving" assigned until today to such objects, since domestic fire was sieved, and therefore controlled by the use of such objects, as experiments seem to prove (Fig. 8).

Fig. 8. Replicas of Boian sieves (made by students Alina Tudor and Madalina Mihaila).

Fig. 9. Sieve – like stand supporting a small vase.

Wood (1995: 185) labelled the corbelled sieves "Bunsen lamps", since their shape could produce a relatively good flame on the top perforation when positioned on ambers.

Generally the form of the objects does not follow function, but was influenced by symbolism; therefore there are Gumelnita culture heaters in the form of houses (that could be also models of ovens; see Serbanescu 1997; 251; fig. 2), settlements as the architectural model from Cascioarele, Gumelnita culture, for example (see Miclea and Florescu 1980: fig. 220), stands, or rectangular bottomless boxes with perforations that evoke furniture. In Cucuteni culture (5th–4th millennia BC), there are many examples of perforated stands, shaped as cylinders, with perforations, or as iconic images of group of people positioned in the round, the human silhouettes being separated by openings with the role of absorbing air (see Cucuteni – the Last Great Chalcolithic Civilization [cat.] 1999: 127, fig. 51).

Part of these objects put in this category could have also the function of heaters (Fig. 10), part could also have been used to as portable ovens, since experiments showed they could bake small ceramic objects or dry salt solutions, at an average temperature of 550°C.

Fig. 10. Replica of a vase from Draguseni, Cucuteni culture (made by student Vera Toma).

Perforations could also have the role of an easy manipulation and transport of the pyrointruments when hot.

4.7. Surface Single-chambered Kiln

(Experiments by Dr. Dragos Gheorghiu and Dora Morcov, August 2001)
Archaeological record in Chalcolithic Transylvania reveald traces of single compartment pear-shaped kilns (Comsa 1976: 30), whose syncretic role was at the same time of fire and combustion chamber.

As every clay construction on soil surface it could have used twigs as internal structure to reinforce the walls, and a corbelled profile, to reduce heat losses (Fig. 11).

Fig. 11. Single – chambered kiln (made by student Dora Morcov).

The ceramic objects positioned inside could be partly covered with shards, to impel the contact between vases and fire, as well as the opening of the vent, in the case of producing a reduced atmosphere. A kiln of 1m high, with a vent aperture diameter of 0.20 m, could create a very good absorption of air and reach temperature of cc. 800°C with a minimal rate of losses, compared to bonfires.

To start the air absorption of a single chambered kiln, the interior shall be filled out with an easily ignited material, whose combustion would produce a rise in temperature inside de corbelled chamber, starting the process of air draught.

Fig. 12. The inner twig structure remained unburned.

Experimental kilns built on soil surface practically disappeared after less than half year due to weathering and animal action, the composite material made of clay mixed with straws having a poor resistance to weathering, due to many empty spaces resulted after the combustion of vegetal components. In spite of the light temperature reached, the internal twig structure of the kiln remained intact (Fig.12).

4.8 Subterranean Up-draught Kilns

(Experiments by Drs. Dragos Gheorghiu and Alex Gibson, July 2000, May 2001, and by Dr. Dragos Gheorghiu and Florin Alexa, July-August 2001)

One of the obvious examples of pyroinstruments with the role of absorption and at the same time of filtering air, and subsequently fire, is the up-draught kiln with platform, a tool able to produce a new ceramic material, better fired, i.e. ceramics.

Early attested examples of subterranean up- draught kilns come from Mesopotamia (Simpson 1997a: 39), to cite the Yarim Tepe I kilns (Hassuna period) that "were circular, measuring 1,2–2 m (3,6–6 ft) across, consisting of a sunken firebox capped by a permanent clay grate – up to 10 cm (4 in) thick and perforated by up to fifty holes 15 cm (6 in) across – with sloping stoke-hole at one end." (Oates and Oates 1976: 42; Merpert and Munchaev 1993: 76-7, 84-5; both cited in Simpson 1997a: 39). The temperatures attained were "estimated at between 850 and 1050°C" (Simpson 1997a: 39). In the Samarra period (c. 6300–6000 BC) circular kilns could reach 850–1050°C, in Ubaid (c. 6000 BC – 4000 BC), temperatures between 1050–1200°C (Simpson 1997a: 39-43) and in Uruk (4th millennium BC) kilns were "fired within a temperature range of 850–1050°C" (Postgate and Moon 1982: 105; Killick *et al.* 1988: 18-21; both cited in Simpson 1997b: 50, note 4).

In Eastern Europe well preserved subterranean up-draught were attested in Cucuteni culture (5th–4th millennia BC) to cite only the site Glavanestii Vechi (see Comsa 1976: 25), or Costesti (Markevic 1981).

4.8.1 Functioning of the Up-draught Kiln

There seems to be a sound relationship between the general shape of the up-draught kiln and the temperatures reached in a determined interval of time.

As demonstrated by experiments, a corbelled chamber produces a superior absorption of the hot gases and of the flames compared to the cylindrical shapes from recent ethnographic examples, the latter passing through the load of stacked objects, after the interior temperature reaches a temperature over 800 C, due to the small aperture of the exhaust vent which has the complex function of preserving and adjusting temperature (Kingery 1997: 14).

A difference between kilns and other pyroinstruments is the strong directional control of air, the absorption of flames of the fire lit in front of the opening of the fire tunnel being very strong. The perforated platform with the role of screen between the chamber of combustion and the chamber of vases, of partition the air flow and flames in a number of current tubes (or absorption tubes), whose length depends of the platform's thickness, and whose addition forms a balanced surface of absorption of fluids (gases and flames).

Other difference is the possibility of adding fuel during the firing process, and to put ambers over the obstructed vent, to compensate the differences in temperature between the bottom and the top of the ceramic load at the end of the firing process. Specific for kiln design, are heat generation, heat transfer to ware, heat containment, ware support (Kingery 1997: 17) and the movement back and forth of fire in the combustion chamber, to regulate temperature. In this perspective, compared to other pyroinstruments, kilns are interactive instruments, since they could keep temperature at a constant level, to allow molecular water to evaporate without damaging clay objects, and to allow the sintering of internal grains in clay (Rye 1981: 106), this two temperature levels being very visible in firing diagrams.

A disadvantage of the up-draught kiln is the difference in temperature between bottom and top mentioned above.

The kiln's efficiency improves with every number of firings by forming an insulating ceramic crust that conserves radiated and convection heat (compare the two diagrams the first and the second fire in the same kiln), which sometimes is perceived by archaeologists as being the "walls" of the kiln. Due to the high temperatures attained, this ceramic interior surface of the chambers attains a thickness of up to 0.20 m. One can infer that the first ceramic object fired by a kiln is the kiln itself, but in spite of the thermal protection of the fired surfaces, a large part of energy, between 30% and 40% is lost through radiation, convection, and conduction (Rice 1987: 162).

Although red heat is visible in daylight at 550–625°C (Rice 1987: 157), kilns are night instruments because the different nuances of the flames inform the potter about the stages of the process inside the firing chamber. It is also an oriented instrument, the opening of the combustion chamber being positioned according to the dominant winds.

Shards positioned as a sieve-like structure on the exhaust vent, to control air-draught, allow the potter to see the colouration of the interior load, this informing about the level of temperature attained. Control of openings implies a control of air, therefore up-draught kiln can create oxidized as well as reduced atmosphere. In spite of its complexity, the up-draught kiln needs only one person to control the firing process, through direct action and indirect signals, as flame colour, smoke, smell. Experiments demonstrated that the overall dimensions of the kiln, the technology of making the platform inside the corbelled space, the filling up and emptying the firing chamber with vases, as well as the repairing of the firing tunnel and the lower part of the platform were designed at a scale that hardly allowed an adult person to perform the basic operations mentioned.

In consequence we asked children to perform these operations, and the results were excellent.

Generally temperatures attained in up-draught kilns are under 1100°C, the rarely cases of vitrification of the ceramic structure could be due to the local clay properties (as a high illite content of clays; see Miller 1997: 66), or to violent atmospheric movements, an air turbulence being able to rise the temperature in a 1.5m deep up-draught kiln from 800°C to 1150°C in less than 15 minutes, vitrifying the lower part of the ceramic load (experiments by Drs. Alex Gibson and Dragos Gheorghiu, August 2000).

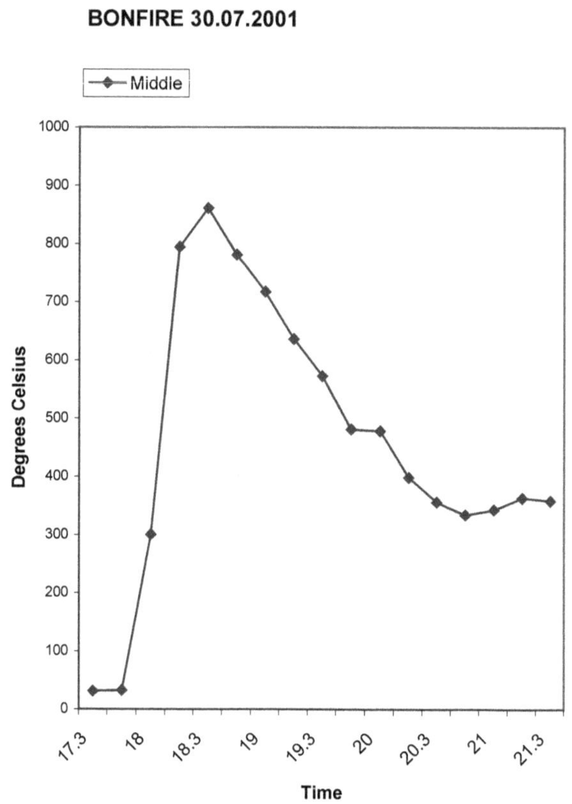

Diagram 1

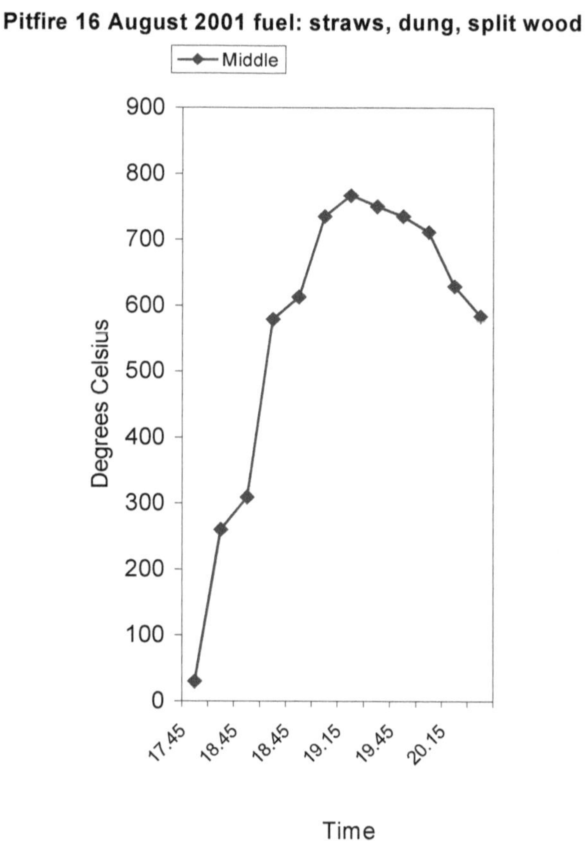

Diagram 2

4.8.2 Description of the Replica of a Chalcolithic Kiln

Experiments were carried in a 2 m deep subterranean kiln with a diameter of the firing chamber of 1.5 m, a combustion chamber of 0.50 m high, and a wattle and daub perforated platform of 0,20 m thickness positioned on a clay tongue-support, the design being copied after the Chalcholithic kilns of Cucuteni culture phase B (4th millennium BC)(Fig. 13).

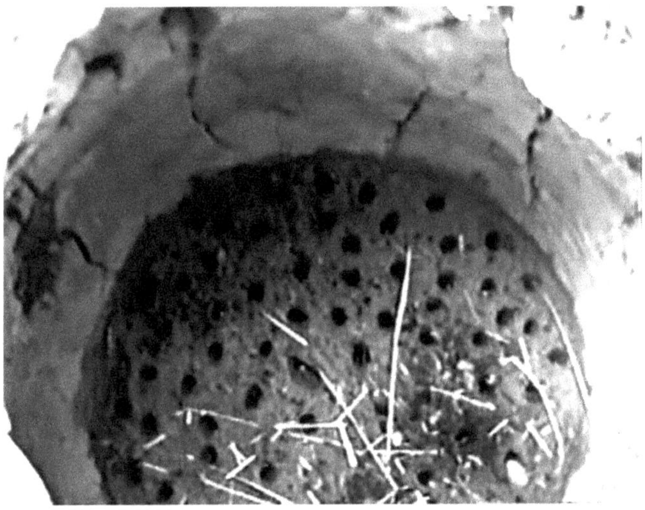

Fig.13. View of the perforated platform.

The kiln was dug having the mouth of the fire tunnel oriented to be protected against dominant local winds, with the fire box at the lower level (see Rice 1987: 159), and the large corbelled chamber plastered with a fine coating made of clay mixed with dung and straws. A rougher material was used to model the platform, positioned on a robust radial twig structure, and perforated from outside the kiln with a long stick.

After the completion of the work, the kiln was filled up with straws and dried dung and fired for three hours to force out the water from clay, but the process was insufficient, and the first firing was not effective.

4.8.3 Experiments

E/21.05.01 (Drs. Dragos Gheorghiu, Alex Gibson, with the assistance of the potter Adrian Tambrea)
A first experiment was to fire the kiln with a new built in platform, made of clay mixed with straws. The dried platform was fired for one hour, but because of the excess of humidity in the porous material, the performance of this first fire were very low: after a 6.30 hours firing the maximum temperature attained on the platform's surface was 517°C with 219°C at the middle of the ceramic load.

E/24.05.01 (Drs. Dragos Gheorghiu, Alex Gibson, with the assistance of the potter Adrian Tambrea)
The first firing helped the kiln to dry completely, consequently improving its functioning, by forming a 0.4 m thickness ceramic crust on the internal surface of the chambers. A load of 200 objects was fired after three days, reaching 1070°C on the platform's surface with 714°C at the middle of the load in 3 hours (Fig. 14).

Fig.14. Vases positioned in the kiln.

E/30.07.01 (Dr. Dragos Gheorghiu, Florin Alexa, with the assistance of the potter Adrian Tambrea)
The kiln with a new made (and therefore damp) platform was fired for 7.45 hours and reached only 811°C on the platform and 508°C at the middle.
The diagram of firing shows that the platform acted here as a pump that drained, the internal water under the form of vapours, through an alternation of high and low temperatures, as the evaporation cooled rhythmically the material.
Could this pattern be a macro model for the process of elimination of water from small ceramic objects? Only detailed further research on micro-phenomenon level could answer this question (see Diagram 3).

E/04.08.01 (Dr. Dragos Gheorghiu, Florin Alexa, with the assistance of the potter Adrian Tambrea)
A second fire with now a dried up kiln components, reached 1050°C on the platform and 700°C at the middle of the ceramic load in 7 hours.
Comparing the diagrams of the two firings, one can observe that the performance of the kiln (i.e., the relationship between temperature and time) increases with the degree of desiccation (see Diagram 4).

E/11/08.01 (Dr. Dragos Gheorghiu, Dr. Vasilica Lungu, Florin Alexa, with the assistance of the potter Adrian Tambrea)
The experiment took place on a windy weather, with the wind blowing on the mouth of the fire tunnel, and in 2.15 hours the temperature reached 911°C on the platform and 660°C in the middle of the ceramic load. Due to the weather activity the kiln was sealed at this level of temperature.

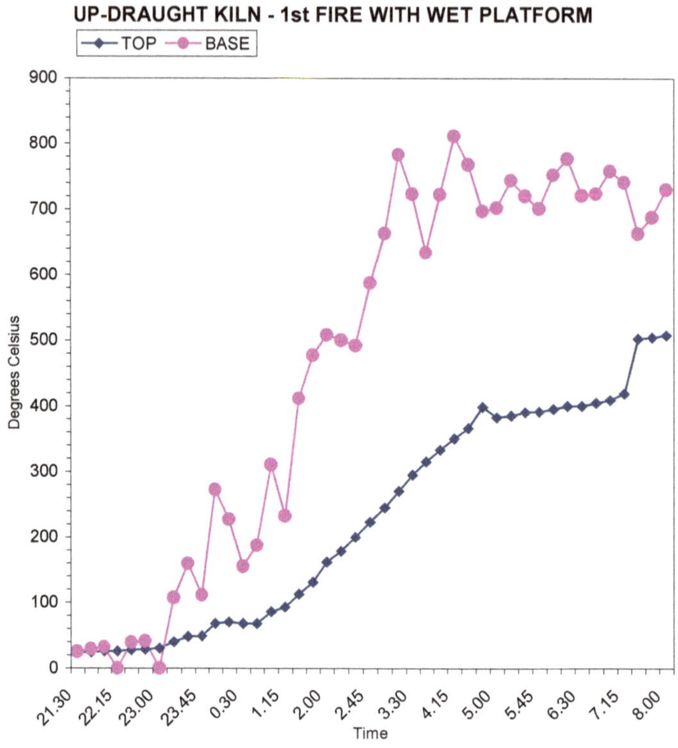

Diagram 3

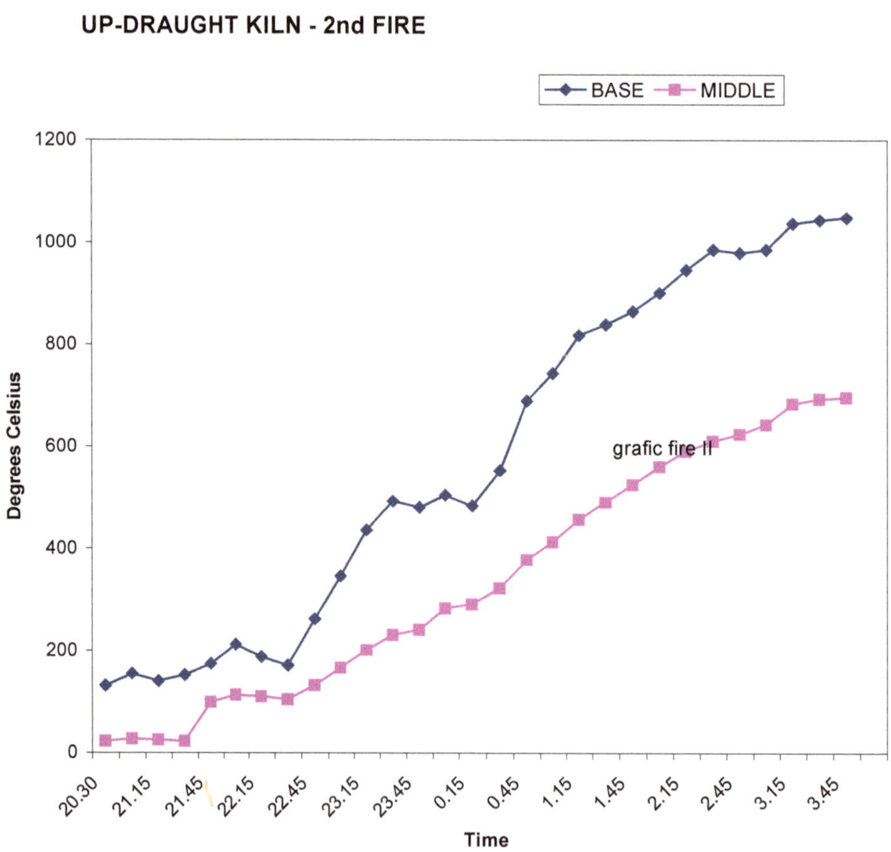

Diagram 4

4.8.4. Discussion

From the above mentioned pyroinstruments, the up-draught kiln demonstrates to have been the most effective tool to control fire.

Fuel nature has also an importance in temperature rising, in all experiments we used mostly willow and poplar wood, since these species did not produce a high temperature, and were easily available and processed in the marshy environment of the majority of Chalcolithic settlements.

As one could see from the firing diagrams, except the temperature floors at 120°C and 550–573°C, representing human interference when removing ambers to keep a constant temperature for some time to allow internal processes in clay (see Hamer 1975: 270 and 314), the heating and cooling the kiln seems to follow an ideal bell-shaped curve, depending only on the specific features of each kiln (proportions, number of firings) as well as on the degree of humidity and atmospheric activity. The specific features of each kiln are not all available to the archaeologist, hence the need for experimentation.

In a good designed kiln, the ascending airflow that transports the flames is so powerful that it prevents the convection heat to dissipate, therefore the core of the walls of a (surface) kiln will be heated later after putting out the fire from the combustion chamber, fact that can be experimented on surface up-draught kilns.

To prevent weather action from damaging the process of firing, it is possible, as shown by ethnologic examples, that kilns could be partly covered by light shards, to produce a sieving absorbing surface that protected thermally the vases inside and at the same time letting the potter to see inside the kiln. Being a ceramic object, its archaeological remains are quite well preserved in time.

5. Concluding Remarks

A reconstruction of the relationship between man and fire, through the testing of the different kinds of pyroinstruments and of different fuels, demonstrated the importance of experimentation to understand air control in Chalolithic pyrotechnology, and led to the conclusion that fire domestication was possible through different airflow "sievings". A common feature of all pyroinstruments, as bonfires, braziers or up-draught kilns, seems to be the air suction grid-structured (resulted from the positioning of the fuel or of shards), or perforated internal or exterior surfaces. The materials of the firing chamber or of filters also influence the temperature of the process of combustion, the special relationship existing between fire, materials, and the proportions of air draught containers.

Although fire is perceived as a unique physical phenomenon, differences in temperature, filtering of air, type of fuel used and time of combustion, distinguish different varieties of fire, from "uncontrolled", to "tamed", as those produced by kilns. Since, at the end of Chalcolithic, prestige ceramic vases were fired in up-draught kilns, compared to the kitchenware fired in bonfires, shallow or deep pits, one can conclude that these *different* fires, creating different materials (ceramics is qualitatively different from terracotta), produced by the different kinds of pyroinstruments described in the paper, use to have different roles and different social uses, and their technological *chaines operatoires* used to be perceived as rituals.

In concluding, an archaeology of fire as that brought to discussion by the present book, would insist on the importance of the role of experiment in analyzing the phenomenon; therefore the study of the complex control of fire in prehistoric societies could be inferred through replication of the design (i.e., proportions, scale, functioning, materials) of the pyroobjects found in archaeological remains. Experimentation is a process that involves, beside the ethical approach of the observer-archaeologist, an emic approach of the participant-archaeologist too, whose accounts should be united in one single narrative.

As experimentalists, our activity was not only to identify the real shapes and the functioning of the objects, but also to *live* the past by embodying the ritual experience of domesticating fire.

Acknowledgements

The experiments carried on in the village Vadastra, Olt County, South Romania, were possible due to the kind help of many experts and of the villagers and authorities. The financial support was due to a Minstry of Culture grant for 2000 and of a CNCSIS/World Bank grant for 2001 and 2002 (No. 112). I would like to thank Dr. Alex Gibson for his kind guidance and remarkable help during two campaigns, Professor Ralph Rowlett for the technical assistance with instruments and analysis, Liana Svintiu for support in starting experiments, the students Robert Neacsu, Madalina Mihaila, Alina Tudor, Florin Alexa, Tudor Mirescu, Catalin Oprea, Dora Morcov and Chris Doros for the diagrams, photos, splendid ceramic vases and efficient kilns made during the two years and for the sustained help in carrying out experiments.

A special thanks to professor Dorin Theodorescu, Director of the Department of Culture Slatina and to Marin Batranca the Major of Vadastra, for their continuous material support of the project. My gratitude goes to the villagers Ion Cococi, Adrian Penescu, the potter Adrian Tambrea, and to the many and very talented children-potters we were able to identify in the village of Vadastra, whose names, due to the lack of space, are not mentioned here.

As always, thanks to Ruth Tomlinson who improved the English of the text.

Photographs by Florin Alexa and Dragos Gheorghiu.

Bibliography

ANDRIEUX, P.
1996 Experimenter la metallurgie: le lien entre la dynamique thermique et le vestige archeologique, pp. 81-107. In Ofer Bar-Yosef, L. L. Cavalli-Sforza, R. J. March, M. Piperno (eds.), *The Lower and Middle Palaeolithic, Colloquium IX* (5), Forli: ABACO Edizioni.

BERCIU, D.
1937 Sapaturile de la Petru Rares (1933 si 1935), *Buletinul Muzeului Judetului Vlasca*, II.

COMSA, E.
1976 Caracteristicile si insemnatatea cuptoarelor de ars din aria culturii Cucuteni-Ariusd, *SCIVA*, 27, 1, pp. 23-33.
1996 Vetrele si cuptoarele din locuintele neolitice din Muntenia, *Istorie si traditie in spatiul romanesc*, III, Bucharest.
1990 Complexul neolitic de la Radovanu. *Cultura si civilizatie la Dunarea de Jos*, VIII, Calarasi.

DESHAYES, J.
1974 Fours Neolithiques de Dikili Tash, *Melanges Helleniques offerts a Georges Daux*, Paris.

GIBSON, A. and A. WOODS
1997 *Prehistoric Pottery for the Archaeologist*, 2nd ed., London and Washington D.C.: Leicester University Press.

GOSSELAIN, O. P. and A. L. SMITH
1995 The Ceramic and Society Project: An Ethnographic and Experimental Approach to technological Choices, *KVHAA Konferenser 34*: 147-160.

HAMER, F.
1975 *The Potter's Dictionary of Materials and Techniques*, London: Pitman Publishing, New York: Watson-Guptil Publications.

KILLICK, R. G. *et al.*
1988 *Tell Rubeidheh; an Uruk Village in the Jebel Hamrin*, Warminster.

KINGERY, W. D.
1997 Operational Principles of Ceramic Kilns, pp. 11-19. In Rice, P. M. (ed.) *The Prehistory and History of Ceramic Kilns*. The American Ceramic Society.

LICHARDUS, J., A. FOL, L. GETOV, F. BERTHEMES, R. ECHT, R. KATINCAROV and I. KRASTEV ILIEV
1996 *Bericht ueber die bulgarish-deutchen Ausgrabungen in Drama (1989 – 1995)*, Mainz am Rhein: Phillip von Zabern.

MANTU, C., G. DUMITROAIA and A. TSARAVOPOULOS (eds.)
1997 *Cucuteni. The Last Civilization of Europe* [cat.] Bucharest: Athena Publishing House.

MARKEVIC, V. I.
1981 *Pozdne-Tripolskie plemena severnog moldavii*, Kishinev.

MATEESCU, C.
1959 Sapaturi arheologice la Vadastra. *Materiale si cercetari arheologice*, V: 61-74

MERPERT, N. I. and R. M. MUNCHAEV
1993 Yarim Tepe I, pp. 73-114. In Yofee, N. and J. J. Clark (eds), *Early Stages in the Evolution of Mesopotamian Civilisation: Soviet Excavations in Northern Iraq*, Tucson and London.

MICLEA, I.and FLORESCU, R.
1980 *Preistoria Daciei*, Bucharest: Meridiane.

MILLER, H., M.-L.
1997 Pottery Firing structures (kilns) of the Indus civilization during the third millennium BC, pp. 41-66. In Rice, P. M. (ed.), *The Prehistory and History of Ceramic Kilns*. The American Ceramic Society.

OATES D. and J. OATES
1976 *The Rise of Civilisation*, Oxford.

POSTGATE, J. N. and J. MOON
1982 Excavations at Abu Salabikh 1981, *Iraq* 44: 103–136.

RHODES, D.
1981 *Kilns. Design, Construction, and Operation,* (second ed.), Randor, PA: Chilton Book Company.

RICE, P. M.
1987 *Pottery Analysis: A Sourcebook*, Chicago: University of Chicago Press.

RYE, O. S.
1981 *Pottery Technology: Principles and Reconstruction*, Washington D.C.: Taraxacum Manuals on Archaeology 4.

SERBANESCU, D.
1997 Models of Eneolithic dwellings and sanctuaries, *Cultura si civilizatie la Dunarea de Jos*, XV: 240 – 251 Calarasi: Caro.

SIMPSON, J. S.
1997a Prehistoric Ceramics in Mesopotamia, pp. 38-43. In Freestone, I. and D. Gaimster (eds.), *Pottery in the making. World Ceramic Traditions*, British Museum Press.
1997b Early Urban Ceramic Industries in Mesopotamia, pp.50-55. In Freestone, I. and D. Gaimster (eds.), *Pottery in the making. World Ceramic Traditions*, British Museum Press.

SINOPOLI, C. M.
1991 *Approaches to Archaeological Ceramics*, New York: Plenum Press.

TRINGHAM, R.
1992 Households with faces: The Challenge of gender in prehistoric architectural remains, pp. 93-131. In Gero J. and M. Conkey (eds.) *Engendering Archaeology. Women in Prehistory*. Blackwell.

VERNANT, J.-P.
1979 A la table des hommes, In Detienne, M. and J.-P. Vernant, *La cuisine du sacrifice en pays grec*. Paris: Gallimard.

WOOD, J.
1999 Bunsen burner or cheese mould? A new re-interpretation of a Bronze Age ceramic. *Abstracts*. 5th EAA Meeting, Bournemouth.

- - -
1999 *Civilizatia Boian pe teritoriul Romaniei* [cat.], Calarasi: Daim.

LES NÉCROPOLES À INCINÉRATION DE L'ÂGE DU BRONZE RÉCENT (XIV-XIII s. av. J.-C.) EN ITALIE NORD-OCCIDENTALE

Laura Domanico

Abstract

The Necropolis of the Recent Bronze Age (14th – 13th century BC) in North-West Italy

Cremation rituals in NW Italy appear as early as the end of the Early Bronze Age and the start of the Middle Bronze Age, as demonstrated by the Vallare tomb recently discovered in Casale Monferrato in Piedmont. As from the end of the Middle Bronze Age and the beginning of the Recent Bronze Age, cremation begins to be more widespread, to the point where it becomes prevalent. The tombs are small pits dug in the earth with the human remains either laid in this directly or in an urn, often together with grave treasure and accompanying materials. The analysis taking place on all the necropoleis known so far in this territory has shown a great variety of solutions in terms of the internal organisation of the burial sites, the spatial relations between the tombs, the grave structures and the funeral ceremonies. The identification in some necropoleis – Canegrate and Premeno, for example – of areas in which the dead were burned has enabled us to partially reconstruct the ceremony that took place, from the funeral to the laying of the remains in the grave, often together with the earth burnt in the fire. A number of the personal belongings of the deceased were also entrusted to the fire, while bronze offerings on the funeral pyre indicate that rituals not directly linked with the burial of the dead also took place in the area of the necropolis.

Introduction

Pendant le Bronze Moyen, l'Italie nord-occidentale fait partie d'un vaste ensemble culturel, où les rapports avec les groupes occidentaux de la Civilisation des Tumulus sont prédominants. À partir de cette époque, et pour tout l'âge du Bronze Moyen et Récent, un réseau de communications et d'échanges commerciaux a été mis en place, grâce notamment à la disponibilité de ressources minières en cuivre, l'accroissement de la demande de matières premières et d'objets travaillés et à la morphologie du territoire, avec des cols alpins faciles à franchir et un réseau de voies fluviales en direction soit nord-sud, soit est-ouest.

Un horizon culturel homogène peut être maintenant envisagé dans toute l'Italie nord-occidentale - Piémont, Ligurie, Lombardie et Émilie occidentales - caractérisé par la diffusion de la céramique à cannelures et à cupules au centre relevé et l'introduction de typologies métalliques d'origine centre-européenne.

Les nombreux sites découverts font preuve d'un développement soit démographique soit de l'organisation du territoire finalisé aux échanges et au contrôle des voies de communication et la continuité dans le temps de la plupart des habitats du Piémont et de la Ligurie est un témoin de la stabilité de l'occupation du sol à partir de cette époque.

La découverte d'une tombe à incinération à Casale Monferrato (Piémont), datant de la transition Bronze Ancien-Bronze Moyen, fait preuve de la précocité de ce rituel et du fort dynamisme culturel caractérisant l'Italie nord-occidentale à cette époque.

Le rite de l'incinération entraîne une série de pratiques autour du feu qui n'intéressent pas seulement la crémation du corps des défunts, mais aussi les objets de parure et d'habillements les accompagnant. La présence d'épées et de poignards ayant subi l'action du feu est un autre indice de la valeur symbolique attribuée au feu et du rôle représenté dans la communauté par les porteurs d'armes. Les traces des bûchers trouvées dans les nécropoles, les structures rituelles et les zones de passage entre les tombes, l'offrande de bronzes brûlés et la présence de la terre du bûcher à l'intérieur des fosses nous permettent enfin d'avoir une idée des cérémonies funéraires accompagnant l'enterrement des défunts.

Le cadre chrono-culturel

La connaissance archéologique des cultures de l'âge du Bronze Moyen et Récent en Italie nord-occidentale remonte au milieu du XIXe siècle. Après la découverte de la nécropole de Rovio dans le Tessin (1846) et des tombes de Rogorea di Rogoredo (1861), Cattabrega di Crescenzago (1868) et Coarezza (1873) en Lombardie, devient de plus en plus évidente l'existence d'une phase chronologique typique de l'âge du Bronze et spécifique de ce territoire. Cette phase ne fut cependant définie que vers la fin du siècle, grâce notamment aux études de Pompeo Castelfranco (Castelfranco 1875; 1891).

La découverte de la nécropole de Canegrate (1926), fouillée à la moitié des années 1950, mit au jour un répertoire typologique bien caractérisé et qui trouva des comparaisons avec le Tessin et la civilisation des Champs d'Urnes européenne. L'importance de cette nécropole étant évidente, une analyse des matériels datés de cette période et conservés près des musées de la Lombardie et de la Suisse fut entreprit par F. Rittatore Vonwiller. Avec une série d'ouvrages visant à offrir un cadre historique et chronologique de Canegrate, F. Rittatore aboutit à la définition de cette manifestation culturelle, en la appellant "culture de Canegrate" (Rittatore 1953-1954; 1956-57; 1975), Selon F. Rittatore, à une èpoque datée du 1400-1200 av. J.-C., des gens appartenant à des groupes celtiques et provenant des terroirs au nord des Balkans et au sud de l'Allemagne, auraient traversé les cols des Alpes, en particulier ceux du Saint Bernardin et Splügen, et en suivant la Vallée de la Moesa et le cours du fleuve Tessin, se seraient installés en Italie nord-occidentale sur un territoire bien délimité entre les fleuves Tessin et Olona. Ce groupe aurait introduit auprès des populations indigènes le rite funéraire de l'incinération et des nouvelles technologies métallurgiques et céramiques.

Au sud de cette zone, les nécropoles de Scamozzina, Monza, Crescenzago, Rogorea et Garlasco représentent le substrat culturel local qui absorbe le rite de l'incinération transalpin diffusé par le groupe de Canegrate. Il s'agit donc de gens indigènes, probablement d'origine ligurienne, supposés contemporains. Le répertoire céramique de ces derniers paraissant différent de celui de Canegrate, Rittatore pensa de pouvoir différencier ces groupes du point de vue soit typologique soit culturel. La fusion entre ces deux cultures (Canegrate et Scamozzina), ayant perdu leurs caractères spécifiques, déterminera ensuite la formation de la culture de l'âge du Fer de Golasecca.

À partir des années 1970, les études de R. De Marinis et de M. Primas ont modifié le cadre chrono-culturel de ces faciès, en proposant une succession dans le temps de Canegrate par rapport à la culture indigène, dite "Scamozzina-Monza ", datée de la fin du Bronze Moyen. D'après De Marinis, la culture de Canegrate ne se serait développée qu'à partir d'une phase avancée du Bronze Récent (De Marinis 1981; 1994; 1998), tandis que M. Primas a proposée une phase du début du Bronze Récent (ersten Phase der spätebronzezeitlichen Urnengräber; Primas 1975), en y attribuant récemment la plupart des tombes de la nécropole de Canegrate (Primas 1998).

Pendant les années plus récentes, des nouvelles découvertes effectuées en Lombardie occidentale, Ligurie et notamment dans le Piémont, ont contribué à mieux définir les phénomènes historiques intéressant l'Italie nord-occidentale pendant l'âge du Bronze. Au même temps, il apparaît évident que le groupe de Canegrate n'était pas intrusif, mais une manifestation culturelle caractérisant ce territoire en continuité avec les cultures présentes ici dès le début du Bronze Moyen (Del Lucchese-Odetti-Maggi 1989; Del Lucchese-De Marinis-Gambari 1991-92; Gambari 1998; De Marins 1998; Bernabò Brea 1997 et 1998).

Les affinités culturelles avec le domaine circum-alpin, notamment le Valais suisse et la France sud-orientale et alpine, ont permis de sortir la culture de Canegrate de l'étroit territoire auquel on la croyait confinée et de la replacer dans un contexte "occidental" plus étendu (David Elbiali 1997 et 1998; Vital 1989 et 1999; Bocquet 1997). D'autre part, les comparaisons réalisées dans un cadre de compréhension plus large, comprenant l'Italie nord-orientale et centrale et les régions au-delà des Alpes, nous ont permis de proposer une nouvelle lecture chrono-culturelle et d'établir un parallèle entre l'âge du Bronze Moyen nord-italien et les phases Bz B, C1 et C2 de la chronologie centre-européenne. Une séquence en deux phases pour l'âge du Bronze Récent (BR1 et BR2), correspondant au Bz D1 et D2, et la datation de la culture de Canegrate du BR1 et BR2 ont été également proposées (Domanico 1998a; 2000a; 2000b).

Les sites funéraires

Les sites funéraires datant de l'âge du Bronze Moyen et Récent découverts jusqu'à présent en Italie nord-occidentale sont 43, dont 24 nécropoles et 16 tombes isolées. Dans 3 cas il s'agit de matériels sporadiques trouvés dans des nécropoles plus récentes 1. Les sites funéraires sont répartis de façon homogène dans la partie centrale du territoire ici pris en compte, les vides dans les zones occidentales et au sud du Pô étant dûs notamment au manque de recherches systématiques (Fig.1).

La plupart des nécropoles semble située en position élevée, probablement près des habitats correspondants, bien que l'état insuffisant de la documentation rend difficile toute conclusion à ce propos. Cependant, ne manquent pas des nécropoles sur pente ou sur des terrasse fluviales (Fig.2).

La faible profondeur des tombes (40-50 cm)[2], la présence de marques symboliques signalant les sépultures et les structures consacrées aux rites funéraires indiquent que les nécropoles étaient visibles sur le territoire. Les zones réservées aux morts étaient donc organisées comme des unités topographiques, à savoir comme un espace différencié de celui des vivants, aucune sépulture n'étant trouvée à l'intérieur des habitats.

[1] Pour la discussion critique de ces attributions et de la position topographique des sites funéraires, voir Domanico 2000b.

[2] La profondeur des tombes indiquée en bibliographie en général 40-50 cm environ. Il y a aussi des tombes plus profondes (100-150 cm: Canegrate, Zeri, Rovio), ou de 25-30 cm (Appiano Gentile, Premeno), ou 15 cm à peine (Scamozzina; Canegrate t.157). Ces ciffres sont en relation avec le niveau du terrain au moment de la découverte, à l'exception d'Alba San Cassiano (paléosol entre 173,15 et 173 m, Venturino Gambari 1995, Fig.79).

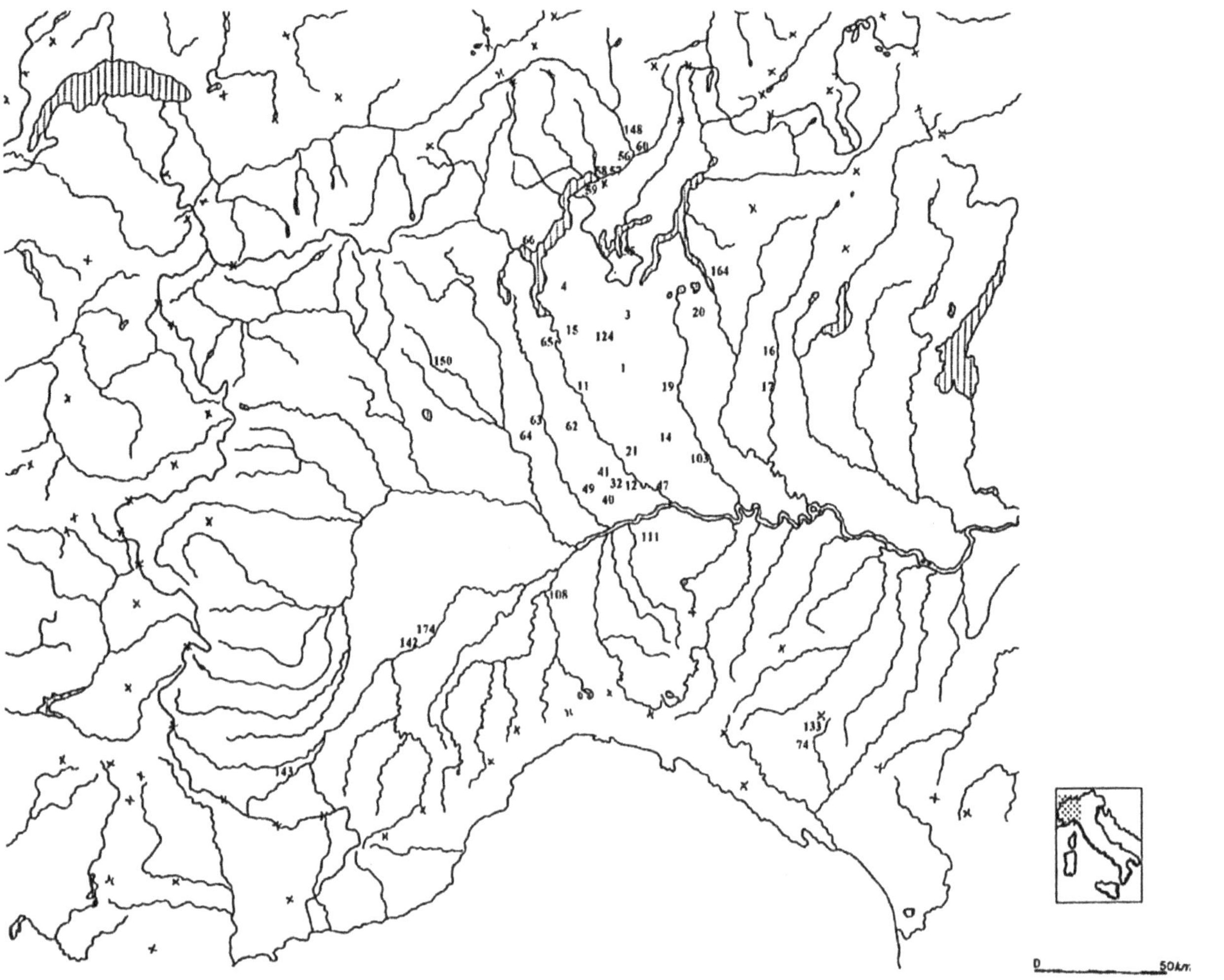

Fig.1 – Carte de répartition des sites funéraires de l'âge du Bronze Moyen et Récent en Italie nord-occidentale.

Dans la nécropole d'Alba San Cassiano, la découverte du paléosol originel entre 173,15 et 173 mètres d'altitude a permis d'observer qu'a ce niveau étaient visibles en surface la marque symbolique de la t.22 (une pierre de granite) et les dalles de couverture des tombes 20, 21, 24 et 25. Une structure de poteaux disposés en forme de demi-cercle autour de la t.17 de la même nécropole, peut être interprétée comme un petit enclos ou une palissade construite en matériel périssable signalant la sépulture (Venturino 1995, Fig.84). Même à Canegrate les tombes étaient visibles en surface. Des cumulus de pierres étaient au-dessus des tombes 9, 98 et 125. Des pierres étaient également entre les tombes 13 et 15, et en général les dalles et les pierres de couverture des sépultures devaient se trouver au niveau du terrain (Rittatore 1953-1954).

Mode de groupement des tombes

Dans les nécropoles ici analysées, les tombes semblent réparties par groupements. Le manque de planimétries ne favorise pas une étude de détail de la répartition des tombes, très peu de données étant publié à ce sujet. Néanmoins l'analyse de la documentation disponible nous révèle des modalités assez variables de l'organisation de l'espace dans les nécropoles. Quatre types de groupement des tombes peuvent être envisagés:

1 - tombes réparties sur lignes

D'après les descriptions publiées, il semble que les tombes de Zeri, et une partie des tombes d'Appiano Gentile étaient alignées régulièrement, mais on ignore la distance inter-

Site funéraire	*Lieu-dit et province*	*Région*	*Type*	*Position topographique*
ALBA	San Cassiano (Cuneo)	Piémont	nécropole	petits reliefs?
ALBA	Mokafé (Cuneo)	Piémont	nécropole	terrasse fluviale
APPIANO GENTILE	Monte di Mezzo (Como)	Lombardie	nécropole	élevée?
ARBEDO	Cerinasca (Bellinzone)	Tessin	nécropole	?
BISOGNINA	Bisognina (Novare)	Piémont	nécropole	?
BORGO SAN SIRO	inconnu (Pavie)	Lombardie	nécropole?	?
CANEGRATE	Canegrate (Milan)	Lombardie	nécropole	petit relief ou pente
CATTABREGA DI CRESCENZAGO	Crescenzago, via Trasimeno (Milan)	Lombardie	nécropole	petit relief?
CILAVEGNA	Case Nuove (Pavie)	Lombardie	nécropole	petit relief
COAREZZA	Vigna della Crocetta in Pelizzolo (Varèse)	Lombardie	nécropole	pente
COLOGNO AL SERIO	Fondo "Palazzo" (Bergamo)	Lombardie	nécropole	?
FORNOVO SAN GIOVANNI	Castellotto, propriété Sansoni (Bergamo)	Lombardie	nécropole	élevée?
GAMBOLO'	Garbana - Dosso della Guardia (Pavie)	Lombardie	nécropole	petit relief
GLISENTE	Glisente (Novare)	Piémont	nécropole	pente
GRAVELLONA LOMELLINA	Monte Giano (Pavie)	Lombardie	nécropole	petit relief
LOCARNO	Via S. Jorio (Locarno)	Tessin	nécropole	élevée
MONZA	Monza (Milan)	Lombardie	nécropole	?
PRADELLINARA DI ROSSANO	inconnu	Ligurie	nécropole	?
PREMENO	Premeno (Novare)	Piémont	nécropole	pente
ROVIO	Selvone, Val Boasca (Lugano)	Tessin	nécropole	élevée
SCAMOZZINA	Cascina Scamozzina, Pisani-Dossi (Milan)	Lombardie	nécropole	petit relief?
TENERO	Tenero (Locarno)	Tessin	nécropole	pente?
VICOLUNGO	Regione Bostangone (Novare)	Piémont	nécropole?	terrasse fluviale
ZERI	Rossano (La Spezia)	Ligurie	nécropole	élevée?
ALESSANDRIA	Cascina Chiappona (Alessandria)	Piémont	tombe	terrasse fluviale
BALLABIO	Ballabio (Côme)	Lombardie	tombe	?
CASALE MONFERRATO	Vallare (Alessandria)	Piémont	tombe	?
CESTO	Cesto (Novare)	Piémont	tombe	terrasse fluviale
CLARO	Alla Monda (Bellinzone)	Tessin	tombe	?
COSSILA	Serre (Verceil)	Piémont	tombe?	?
GARLASCO	Cavo Bareggia (Pavie)	Lombardie	tombe	?
GUADO DI GUGNANO	Guado di Gugnano (Milan)	Lombardie	tombe?	petit relief?
LIGURNO	Ligurno (Varèse)	Lombardie	tombe	?
MARCIGNAGO	Cascina Remondò (Pavie)	Lombardie	tombe?	?
ROBECCHETTO CON INDUNO	Cascina Paradiso (Milan)	Lombardie	tombe	terrasse fluviale
ROGOREA DI ROGOREDO	Castellazzo (Côme)	Lombardie	tombe	élevée?
SANTO STEFANO ARNO	Santo Stefano Arno (Varèse)	Lombardie	tombe	?
"territoire de Casteggio"	inconnu (Pavie)	Lombardie	tombe?	non déterminable
"territoire de Crema"	inconnu (Cremone)	Lombardie	tombe?	non déterminable
"territoire de Pavia"	inconnu (Pavie)	Lombardie	tombe?	non déterminable
VALDIERI	Via alle Ripe (Cuneo)	Piémont	matériel ssporadiques	terrasse fluviale
GUDO	Campicc (Bellinzone)	Tessin	matériels sporadiques	terrasse fluviale
ASCONA	Ascona (Ascona)	Tessin	matériels sporadiques	pente?

Fig.2 – Liste des sites funéraires avec l'indication du type et l'indication de la position topographique proposée sur la base des données publiées (d'après Domanico 2000b).

tombes (Formentini 1941; Ghislanzoni 1932). En ce qui concerne Scamozzina, le seul dessin publié, bien que schématique, nous indique un alignement des tombes sur deux rangs en direction nord-sud. La distance inter-tombes semble être régulière (Castelfranco 1909). À Monza, les tombes étaient alignées avec une distance de 1 m (Castelfranco 1891), tandis qu'à Locarno-S. Jorio les sépultures étaient sur deux lignes en direction est-ouest, avec 2 m de distance inter-tombes (Crivelli 1943).

2 - tombes sur une ligne à demi-cercle
Un seul cas documenté à Premeno (Canetta 1970).

3 - tombes à groupes avec des alignements internes
Un type de répartition des tombes sur lignes, mais par groupements, peut être observé dans les nécropoles de Cilavegna et Canegrate.

À Cilavegna les tombes étaient presque complètement détruites au moment de la découverte, mais d'après la planimétrie publiée il semble qu'elles étaient organisées sur rangs selon la pente du terrain et divisées en deux groupes, sauf la tombe C.CA/1 qui apparaît isolée (Vannacci 1988).
À Canegrate les tombes semblent réparties de façon irregulière, mais une analyse de détail indique l'existence d'au moins 11 groupements de sépultures. Dans chaque groupe, les tombes étaient souvent réparties sur lignes (Fig.4A). À l'intérieur des groupes les sépultures sont très serrées, mais entre les groupes il y a des zones vides, souvent avec des tessons.

4 - tombes à groupes avec des zones vides
On peut constater ce type de répartition des tombes dans les nécropoles de Gambolò, Alba San Cassiano, Appiano Gentile (fouilles 1932) et, peut-être, Rovio. À Alba on peut

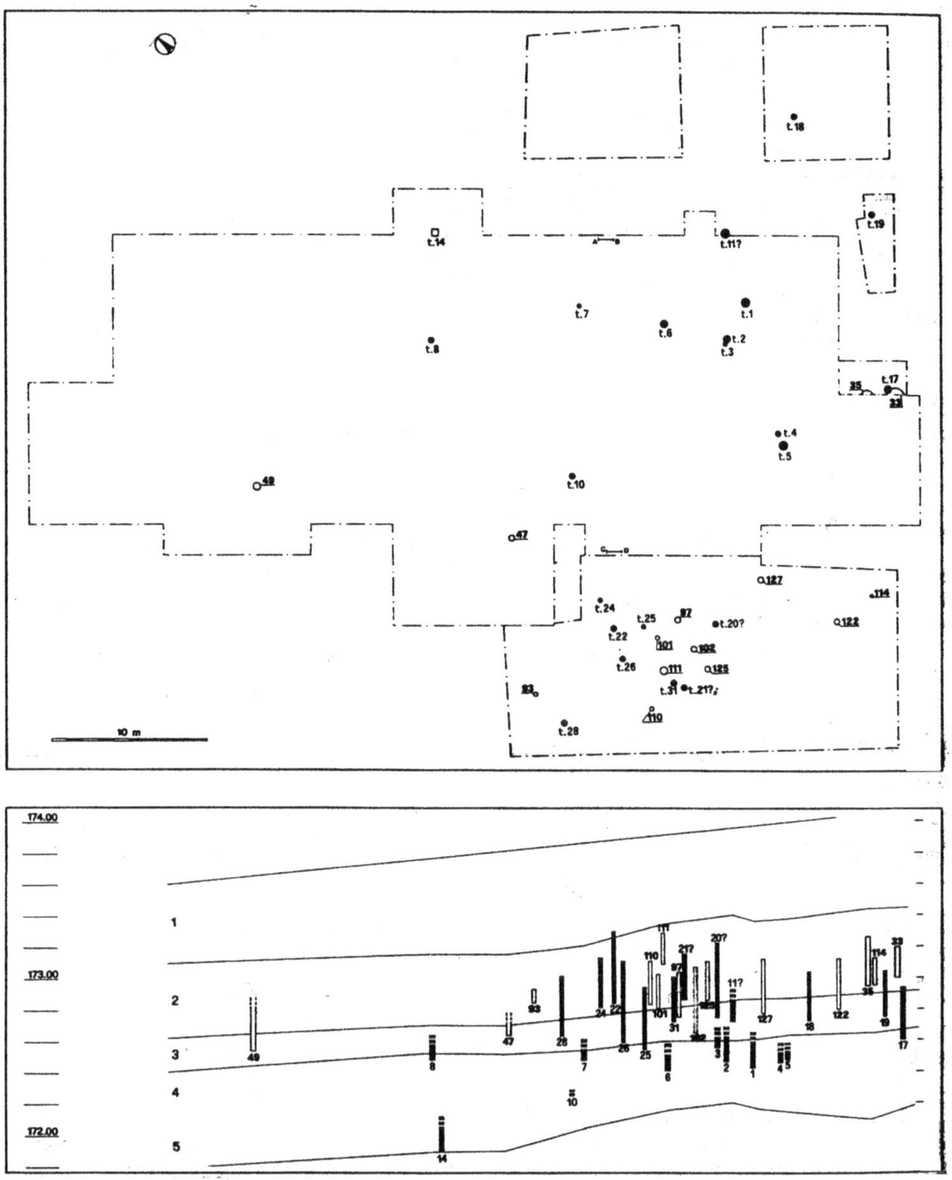

Fig.3 – Planimétrie et coupe de la nécropole de Alba San Cassiano (d'après Venturino 1995).

observer un groupement serré de tombes dans la zone sud-ouest de la nécropole et un ou deux groupes plus dispersés dans la partie est, avec des zones vides entre les groupes. À proximité des tombes il y avait des structures probablement rituelles (Fig.3).

Au moins cinq groupements peuvent être observés dans la nécropole de Gambolò (Fig.4B). L. Simone a constaté dans chaque groupe la présence de tombes datant des différentes périodes de la nécropole (fin BM, BR1 et BR2), ce qui lui a permis de proposer une organisation sur la base des clans familiaux (Simone 1990-91 : 93). La même hypothèse a été proposée pour la nécropole d'Alba San Cassiano (Venturino 1995 : 98).

En ce qui concerne Appiano Gentile, E. Ghislanzoni nous informe que les 12 tombes fouillées étaient réparties en petits groupements avec une distance de 5-6 mètres inter-groupes (Ghislanzoni 1932). Il paraît par contre que d'autres tombes détruites (70 environ) étaient alignées, mais sur une aire de 3,50x20 mètres, ce qui nous autorise à supposer une majeure concentration des sépultures.

En ce qui concerne les autres nécropoles, toute indication de l'organisation interne et de la répartition des tombes fait défaut. Dans le cas de Vicolungo, les auteurs nous informent que les sépultures étaient "les unes près des autres" (Rittatore-Schramli 1970 : 8).

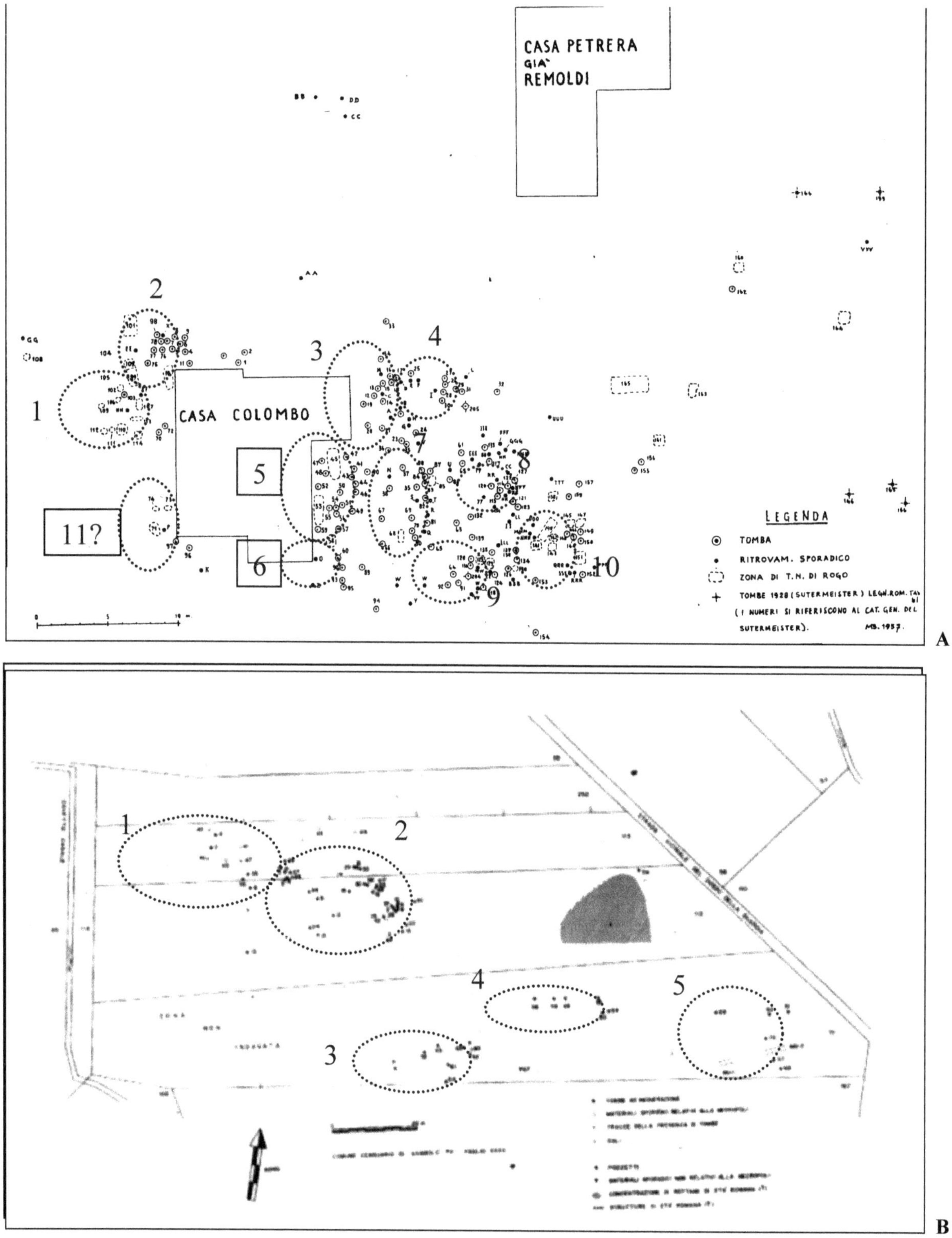

Fig.4 – Planimétries des nécropoles de Canegrate (**A**) et de Gamboló (**B**) avec l'indication des groupements des tombes (réélaborées d'après Rittatore 1956-57; Simone 1992-93).

Les structures et les dépôts rituels dans les nécropoles

Des rares descriptions publiées à ce sujet, il ressort qu'il y avait, au moins dans quelque cas, des structures ou des traces sur l'aire des nécropoles qui n'étaient pas directement reliées aux tombes, leur sens étant à rechercher dans les rituels d'accompagnement des défunts.

À Alba San Cassiano, on a constaté la présence de nombreux tessons de céramique grossière, restes d'animaux, fragments en terre cuite, débris en grès et cailloux. Ces restes étaient concentrés sur les aires vides entre les groupements des tombes, qui peuvent être interprétées comme des zones de passage. Près des tombes il y avait des fosses dépourvues de restes humains, mais remplies avec des tessons appartenant à un seul récipient, probablement intentionnellement cassé, des restes brûlés d'animaux (suidé dans la str.47) et des petits charbons. Ces fosses avaient la même forme des sépultures, leur profondeur étant légèrement inférieure (Venturino 1995).

Le dépôt de vases comme offrande est documenté également dans la nécropole de Gambolò, où des récipients étaient enterrés isolés (str.49), ou près des tombes (str.65) (Simone 1990-91 et 1992-93). À Canegrate, la str.70 était une aire de terre brûlée avec les fragments d'un vase en céramique grossière entourés de pierres (Rittatore 1953-54). Des vases fragmentés étaient également dans les structures 19, 71, 154 et en général la pratique d'enterrer des tessons semble assez diffusée dans cette nécropole. Il en va de même pour des bronzes, parfois déposés dans la terre sans aucune trace d'ossements humains (str.26, 33, 42, 44, 102 et 106), et ayant subi l'action du feu (str.102). S'agit-il de dépôts rituels, à savoir d'offrandes accompagnant l'enterrement du défunt, ou faut-il penser à un manque accidentel de restes humains, dû aux modalitées de la découverte et aux méthodes de fouille employées? Sauf dans le cas de la str.33, isolée dans la partie nord de la nécropole, les autres structures, de même qu'à Alba, étaient toutes près des tombes (Fig.4A). Leur valeur rituelle peut être supposée, bien que difficilement confirmée. La découverte de la structure 49 de Alba San Cassiano peut cependant nous offrir quelques éléments de plus à ce sujet. Ici, une épée intentionnellement cassée en deux parties était placée verticalement dans la fosse avec un pendentif triangulaire ajouré. La structure fut ensuite remplie avec de la terre pressée et couverte avec deux dalles (Fig.5). Aucun reste d'ossements humains n'a été trouvé, la valeur d'offrande des bronzes étant donc évidente dans ce cas.

Dans la nécropole de Premeno a été découverte une structure en pierres soigneusement rangées. Sur cette structure, de 230x150 cm, il y avait des fragments de bois carbonisés et des restes d'ossements humains (Canetta 1970). L'interprétation comme d'un bûcher semble la plus probable. Il s'agit alors du seul exemple connu jusqu'à présent d'une structure construite avec la fonction de bûcher, les autres cas documentés en Italie septentionale n'étant que des simples aires de terre brûlée (par ex. dans les nécropoles de Canegrate et de Montata à Reggio d'Émilie, Degani 1962).

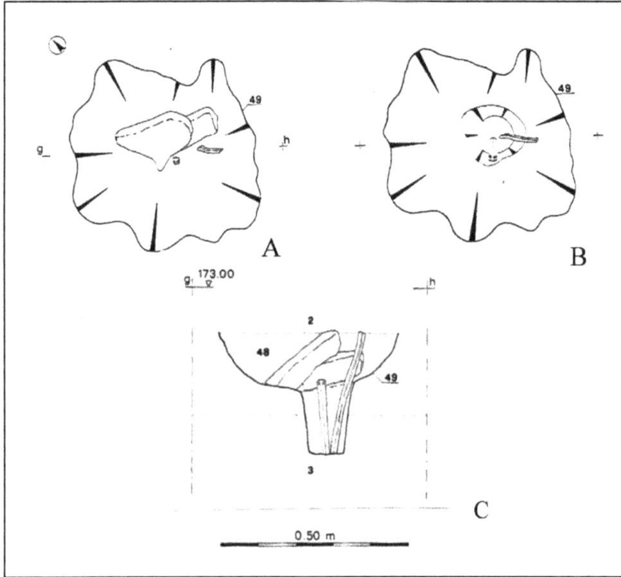

Fig.5 – Plans (A et B) et coupe (C) de la structure rituelle 49 de la nécropole de Alba San Cassiano (d'après Venturino 1995).

Les structures tombales et les rituels funéraires

Dans toutes les zones ici prises en compte, les tombes à incinération sont sans exception des structures à fosse creusée dans le terrain, de forme d'un petit puits circulaire, quadrangulaire ou elliptique à l'embouchure. Comme nous l'avons déjà mis en évidence dans une étude spécifique (Domanico 1998b), on peut distinguer des tombes à fosse simple et des tombes à fosse structurée avec des pierres ou des dalles de couverture, de revêtement ou à la base. Il ne manquent pas des coffres (Appiano Gentile, Locarno-S.Jorio, Tenero) et des cumulus de pierres et de terre au-dessus des sépultures (Canegrate). L'analyse de chaque élément de construction des tombes révèle un nombre très haut de variantes et d'associations, nous donnant l'impression qu'il y avait à cette époque une véritable architecture tombale, avec des sépultures offrant des caractéristiques techniques particulières, par exemple à Canegrate, où il y a des tombes avec deux, trois ou quatre éléments différents associés (Fig.6). Pourtant, bien que documentées, les tombes structurées ne sont pas très diffusées. Leur répartition dans les nécropoles peut être mise en relation avec l'âge ou le sexe des défunts ou leur position sociale et/ou familiale au sein de la communauté, le matériel d'accompagnement étant dans tous ces cas très riche. Cependant, toute conclusion à ce propos demeure difficile, faute de données suffisantes.

Il semble que les fosses structurées soient présentes notamment dans les zones de la Lombardie occidentale et du Piémont, aucune mention n'étant faite de ce type de structure en Lomellina[3], tandis que dans le Tessin sont signalés le

[3] Il faut cependant remarquer que la plupart des tombes découvertes en Lomellina (province de Pavie) était détruite dans la partie supérieure.

revêtement en dalles et les coffres (Locarno et Tenero; Baserga 1936).

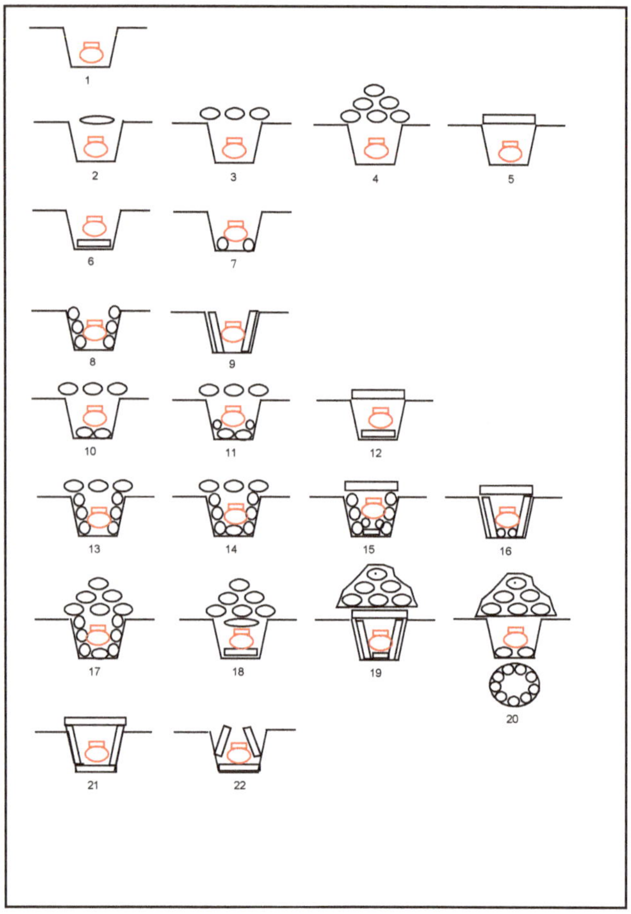

Fig.6 - Les caractéristiques techniques des tombes structurées et avec contenant. **1.** urne directement dans la fosse simple sans éléments structurels; **2.** une pierre de couverture; **3.** plusieurs pierres de couverture; **4.** cumulus de pierres de couverture; **5.** une dalle de couverture; **6.** une dalle à la base; **7.** pierres de renforcement à la base; **8.** pierres de revêtement; **9.** dalles de revêtement; **10.** pierres de couverture et à la base; **11.** pierres de couverture et à la base et pierres de renforcement à la base; **12.** dalle de couverture et à la base; **13.** pierres de couverture et de revêtement; **14.** pierres de couverture, de revêtement et à la base; **15.** dalles de couverture et à la base, pierres de revêtement et pierres de renforcement à la base; **16.** dalles de couverture et de revêtement et pierres à la base; **17.** tumulus en pierres de couverture et pierres de revêtement et à la base; **18.** tumulus de pierres et une pierre de couverture, une dalle à la base; **19.** tumulus de pierres et de terre de couverture, revêtement de dalles et dalle à la base; **20.** tumulus de pierres de couverture et cercle de pierres à la base; **21.** coffre lithique quadrangulaire; **22.** coffre lithique en forme de "cabane".

Les tombes sont en majorité individuelles, les cas de tombes doubles et multiples étant très rares[4]. Dans la plupart des cas les restes humains incinérés étaient déposés, avec les objets les accompagnant, à l'intérieur d'une urne en céramique fine, souvent décorée, placée dans la fosse en position droite ou renversée. Il ne manquent pas des cas où les restes étaient directement dans la fosse, hors contenant (Appiano Gentile, Gambolò, Canegrate), bien que la présence d'un contenant en matériel périssable ne peut pas être exclue à priori. Parfois les restes humains étaient au-dessous d'une pierre ou de fragments en céramique (Fig.7). S'agit-il de ce qui reste du contenant, ou faut-il penser à la déposition rituelle d'un seul fragment à la place du vase? Le cas de déposition symbolique d'un seul tesson sont en effet largement documentés dans les nécropoles ici prises en compte, notamment Canegrate.

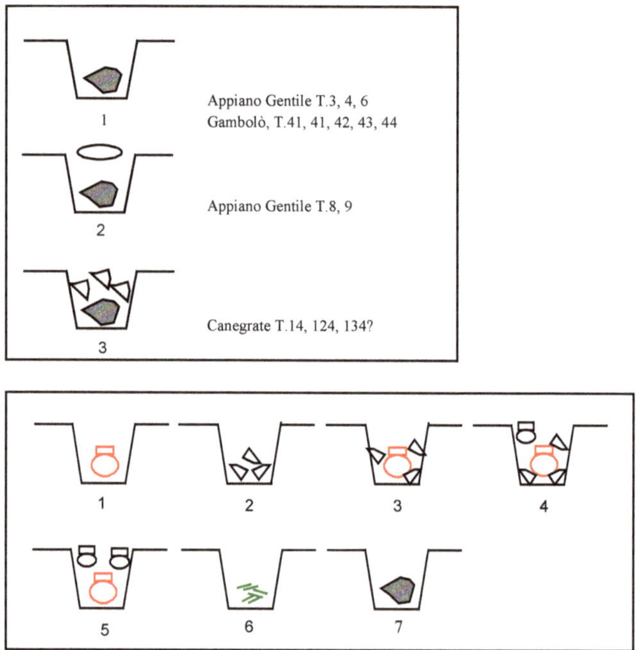

Fig.7 – À gauche: les structures tombales avec les restes humains directement dans la fosse, hors contenant. 1. fosse simple sans éléments structurels 2. avec une pierre de couverture; 3. avec des fragments céramiques au-dessus des restes humains. À droite: types de dépôt du matériel d'accompagnement du défunt: 1. dépôt du seul cinéraire; 2. dépôt de fragments de céramique grossière, sans cinéraire; 3. dépôt du cinéraire et de fragments de céramique grossière; 4. dépôt du cinéraire, de fragments de céramique grossière et de vases accessoires; 5. dépôt du cinéraire et de vases accessoires, sans fragments de céramique grossière; 6. dépôt de bronzes; 7. dépôt sans rien (restes humains directement dans la fosse).

Dans nombre de sépultures on a constaté la présence de la terre du bûcher à l'intérieur de la fosse5. On peut donc

[4] Les tombes doubles sont 6: Gambolò t.4 (un adulte masculin et une adolescente féminine), Alba t.3 (adulte masculin et enfant de 6-9 ans), Canegrate t.25 (enfant de 10 ans et adulte de plus de 25 ans) et probablement Zeri. Dépôts multiples: Canegrate t.9 (2 petits-enfants et 3 enfants de 1 et 9 ans) et Gambolò t.6 (probablement 3 individus dont deux adultes et un adolescent) (Rittatore 1953-54 et 1956-57; Simone 1990-91 et 1992-93; Formentini 1941).

[5] Cette pratique est signalée dans les nécropoles d'Alba (tombes 1, 4, 5, 6, 7, 8, 17 et str.47), Appiano Gentile (tombes 1, 1SP, 3, 4, 6, 8, 9, 11), Cilavegna (tombes CCA2, CCA3A, CCA5A, CCA5B, CCA5C, CCA6, CCA7A-B), Gambolò (tombes 6, 11, 13/16, 20, 22, 23, 27, 31, 33, 39, 40/44, 48, 59, 63 et structures 45, 46, 47), Monza, Guado di Gugnano, Rogorea di Rogoredo, Garlasco?, Scamozzina (tombes A, B, C, D, E, F, G, H, I, J), et notamment Canegrate (tombes 1, 4, 7, 9, 10, 19, 22, 23, 24, 27, 28, 29, 30/45, 48/55, 58, 59, 63/68, 70/74, 76-8, 77/80, 82, 83, 85/91, 93/97, 99, 100, 101, 102,

imaginer que les restes incinérés des défunts étaient recueillis sans distinction sur la pire et placés dans la tombe, au moins de supposer un rituel comprenant le dépôt intentionnel de la terre brûlée.

D'après les descriptions publiées par F. Rittatore, à Canegrate est documentée la présence, à côté des fosses, de nombreuses "aires de terre noire" avec des restes osseux incinérés (96 cas sur 200 tombes environ). Dans la plupart des cas il est fort probable qu'il s'agit de tombes bouleversées dont les restes de l'incinération et les matériels d'accompagnement ont été dispersés (par exemple tombes 85, 87 et 88). Il n'est cependant pas à exclure la possibilité qu'il y avait dans cette nécropole un type différent de traitement du corps et d'enterrement des défunts, à la fois rituel et funéraire. La t.73, par exemple, était une grande aire de terre noire de 1,60 mètres de longueur, avec d'abondants restes humains appartenant à un individu masculin adulte. Dans la terre on a trouvé des fragments de céramique grossière et des parures de bronze (épingle, torques, bracelets, dont un avait des bavures de fusion et donc n'avait été jamais utilisé) (Rittatore 1953-54a, p.24). Il s'agit cependant d'un cas isolé, la plupart des "aires de terre noire" n'ayant que très peu d'ossements humains.

La t.89 était elle aussi une aire de terre noire de 60 cm de diamètre avec de nombreux tessons de céramique grossière, deux fragments d'urne cinéraire avec décor de cannelures, des bronzes (un bracelet, un torque et un pendentif circulaire ajouré) mais très peu de restes humains, non analysés (Rittatore 1953-54a, p.28).

Très peu d'ossements humains étaient également dans les structures 24, 34, 35, 36, 67, 82, 100 (60x60 cm), 112, 128, 130 et dans les autres structures mentionnées sur la Fig.8. Il a été possible de déterminer l'âge du défunt seulement dans les cas des structures 132 (80 cm de largeur; individu indéterminé, mais pas un enfant), et 151 (70x40 cm; probable grand-enfant ou adolescent).

La faible quantité de restes humains pose la question du type de traitement du corps du défunt et des rituels accompagnants son enterrement. Est-il possible d'interpréter au moins quelques-unes de ces structures comme des bûchers, à savoir des lieux où les morts étaient incinérés? Sauf pour la t.73, où les nombreux restes appartenant à un seul individu nous font supposer qu'il s'agissait effectivement d'une tombe, cette hypothèse nous paraît fort probable.

La t.10, par exemple, de 180x90 cm à nord, 40 cm à sud et de 30-40 cm de profondeur, avait une grande quantité de fragments de céramique grossière et fine et des bronzes (épingle, torques, bracelets). Aucune mention n'est faite des restes humains incinérés et selon F. Rittatore est probable qu'il s'agissait d'un bûcher (*ustrinum*) (Rittatore 1953-54a : 17). La structure 101 était une grande aire de terre noire de 160x90 cm, avec quelques tessons de céramique grossière et un seul fragment d'ossements humain. Dans la même nécropole, la structure 140, de 170x40 cm d'ampleur, avait des tessons de céramique grossière et fine, des bronzes (bracelet, anneau) et 3 petits fragments d'os humains (Rittatore 1956-57 : 23 et 27).

À Canegrate les restes humains ont été analysés seulement dans 65 tombes. Le classement des individus en fonction de l'âge indique une haute proportion d'enfants entre 10/13 et 1 ans (27 individus, dont 5 dans la t.9). Dans la t.126 il y avait les restes d'un nouveau-né âgé de quelques mois. Les adultes sont 14, tous interrés individuellement, sauf dans la t.25 où il y avait aussi un enfant. Les adolescents de 18/19 ans sont 11. Les individus âgés sont bien représentés (4 individus entre 45 et 70 ans) par rapport à la faiblesse de cette classe d'âge dans les autres nécropoles. D'après cet échantillon, on peut supposer une répartition démographique de 38% d'adultes et adolescentes, 6% d'âgés et 41,5% d'enfants (grands-enfants, petits-enfants et nouveau-nés). En ce qui concerne le sexe des défunts, il est trop rarement déterminé pour autoriser toute observation, seulement 10 étant les individus masculins et 2 les féminins. Aucune différence entre les catégories des défunts en fonction de l'âge n'a été envisagée en ce qui concerne le matériel d'accompagnement. Les enfants n'avaient pas des traitements particuliers à eux, ni étaient isolés dans la communauté. Comme les adultes et les adolescents, ils étaient enterrés à l'intérieur de la nécropole, dans des tombes structurées et parfois avec des riches parures. Au contraire, un traitement particulier semble réservé aux défunts âgés, bien que les 4 tombes qui ont livré ces individus ne sont pas représentatives. Néanmoins, il faut observer que trois d'elles étaient dépourvues de tout matériel d'accompagnement et seulement la t.161 avait une épingle et un bracelet.

À Gambolò, de même qu'à Canegrate, aucune distinction n'est faite entre catégories de défunts, un traitement analogue étant réservé aux adultes et aux enfants.

À Alba San Cassiano la t.17, avec un riche mobilier d'accompagnement, appartenait à un adolescent de 15/17 ans. Comme nous avons déjà eu l'occasion de souligner, il s'agit d'une tombe à coffre avec un probable enclos en bois. La disposition du matériel révèle un rituel complexe, avec les objets de parure déposés en partie à l'intérieur de l'urne, en partie sur la dalle de couverture et en partie dans la terre de la fosse. Une épingle et des perles en pâte de verre avaient subi l'action du feu et il a été observé aussi le bris intentionnel de la céramique. Bien que ces éléments du rituel sont difficiles à interpréter correctement, il semble évident qu'il s'agit dans ce cas d'un individu ayant une rôle particulier dans la communauté. Étant un adolescent, il est possible de supposer son appartenance à un groupe éminent du point de vue sociale.

D'autres éléments rituels accompagnant l'enterrement des défunts peuvent être envisagés. Dans cinq cas (Canegrate tombes 69, 75, 83, 84; Rovio) le fond du cinéraire était percé d'un petit trou d'un centimètre environ. En Italie du nord la présence de céramique funéraire percée est rare. Les seuls cas

104/115, 116-119, 117, 118, 120, 121, 123/126, 128, 129, 131-133, 132, 134/149, 151, 152, 155/161, 163, 164, 164S, 165, 165S, 199, Asilo).

Type de dépôt des restes humains	**Nécropoles et tombes**
dépôt d'os incinérés dans contenant	Alba 1, 6, 7, 8, 10 Appiano Gentile 1, 5, 7, 10, 11, 12 Canegrate 1, 2, 3, 4, 13, 14, 15, 16, 17, 27, 39, 40, 47, 52, 62, 69, 77, 78, 80, 81, 83, 84, 87, 97, 103, 117, 118, 121, 122, 123, 127, 135, 136, 138, 146, 148, 150, 157, 164, 165, 199, 204, 205, Asilo, 116/119 Cilavegna CCA1, CCA2, CCA3A, CCA3B, CCA4, CCA5A, CCA5B, CCA5C, CCA6, CCA7A-B, CCA8 Gambolò 1/39 , 48, 49, 50/59, 62, 63, 64, 100, 49, 65 Gravellona Lomellina 1/7, 11 Scamozzina A, B, C, D, E, G, H, J, F Monza B, Robecchetto con Induno; Rogorea di Rogoredo; Zeri; Coarezza; Cologno al Serio; Garlasco; Cesto; Guado di Gugnano; Alessandria
dépôt d'os incinérés hors contenant	Gambolò 40, 41, 42, 43, 44 Appiano Gentile 3, 4, 6, 8*, 9* Canegrate 14, 124, 134
"aire de terre noire"	Canegrate 19, 20, 23, 24, 26, 28, 29, 30, 31, 32, 33, 34, 35, 36, 37, 41, 42, 44, 45, 48, 49, 50, 51, 53, 54, 56, 58, 60, 61, 63, 65, 66, 67, 68, 70, 71, 72,74, 79, 82, 85, 86, 88, 89, 90, 91, 93, 94, 96, 99, 100, 102, 104, 105, 106, 107, 108, 109, 112, 113, 115, 129, 130, 134, 139, 141, 142, 143, 144, 145, 146, 147, 151, 153, 154, 155, 158, 163, 165
incinération sur place (bûcher)?	Canegrate 10? 101?, 140?

Fig.8 – Types de dépôt des restes humains incinérés dans les nécropoles du Bronze Moyen et Récent en Italie nord-occidentale. L'astérisque indique la présence d'une fosse structurée avec une pierre de couverture (d'après Domanico 2000b).

signalés sont dans la nécropole *terramaricola* de Copezzato (Parme), où le fond d'un cinéraire et la parois d'une jatte de couverture avaient des petits trous (*Le Terramare* 1997 : Fig.387, 4 et 5). Le sens pratique de ces trous étant difficile à comprendre[6], une fonction reliée aux rituels funéraires peut être supposée (Bietti Sestieri 1992 : 500). Il en va de même pour la pierre brûlée trouvée dans l'urne de la t.84 de Canegrate, dépourvue de bronzes et appartenant à un enfant masculin de 10 ans; les petites pierres ou dalles placées comme "fermeture" sur l'embouchure de l'urne, documentées à Canegrate, Monza, Alba, Coarezza, Zeri et Cologno al Serio, et pour la position du contenant qui ne semble avoir aucune relation avec la structure tombale, le sens étant dans les rituels accompagnant l'enterrement des morts (Domanico 2000b).

Le matériel d'accompagnement du défunt

Six grandes catégories de matériel d'accompagnement sont attestées dans les tombes ici prises en compte:

- céramique (cinéraire, vases accessoires, fragments de céramique grossière)
- parures et objets d'habillement (bronze, os, pierre, pâte de verre, bois de cerf)
- outils et ustensiles (bronze)
- armes
- pains de bronze/débris de fusion
- restes de faune

Tous ces matériels se trouvent également dans les habitats contemporains du même territoire et ne semblent donc pas réservés au seul usage funéraire, bien que le vase de forme biconique semble être préféré comme cinéraire. Il semble évident que le type de dépôt, de même que la disposition des objets dans la fosse, sont à rapporter aux rituels accompagnant le défunt, mais le rôle précis de ces objets dans le rituel funéraire demeure encore en grande partie à comprendre. Le type de dépôt du matériel d'accompagnement dans les tombes est assez varié dans toutes les nécropoles ici analysées, mais les exemples les mieux documentés sont dans la nécropole de Canegrate. Les cas observés sont (Fig.9):

1 - dépôt du seul cinéraire
2 - dépôt de fragments de céramique grossière, sans cinéraire
3 - dépôt du cinéraire et de fragments de céramique grossière
4 - dépôt du cinéraire, de fragments de céramique grossière et de vases accessoires
5 - dépôt du cinéraire et de vases accessoires, sans fragments de céramique grossière
6 - dépôt de bronzes
7 - dépôt sans rien (restes humains directement dans la fosse)

Les tombes dépourvues de tout matériel ne sont pas nombreuses, mais il est intéressant d'observer qu'il s'agit toujours de cas où les restes humains étaient directement dans la fosse simple, sans éléments structurels (Appiano Gentile tombes 3, 6, 8, 9; Gambolò t.44)[7], ou bien dans des "aires de terre noire" (Canegrate str.20, 32, 34, 35, 36, 37, 45, 74, 109).

[6] Les trous sur le bord du vase biconique de la t.9 de Canegrate pouvaient avoir eu une fonction pratique (suspension du récipient?). Le vase a été utilisé comme contenant funéraire pour les restes de 3 enfants et de 2 petits-enfants (Rittatore 19554 : 16).

[7] A Gambolò les tombes 45, 46 et 47 avaient de la terre noire à l'intérieur mais elles étaient presque détruites au moment de la découverte (Simone 1990-91 : 124).

Type de dépôt du matériel d'accompagnement	*avec objets en bronze*	*sans objets en bronze*
dépôt du seul cinéraire	Canegrate 2, 9, 13, 15, 17, 22, 64, 69, 75, 92, 111, 118, 126, 127, 159 Gambolò 3A-B+, 4+, 5+, 6A-B+, 7+, 8+, 9, 10, 13, 14B+, 15+, 16+, 21, 22, 24, 33, 34, 51, 54, 57+, 58+, 59+, 100+ Scamozzina A+, D, I Alba 1, 2, 4-5, 10, Appiano Gentile 1SP, 2, 5, 7 Gravellona Lomellina 2, 3, 5+ Arbedo 7, 8, 9 Claro 11 Coarezza 1 Monza, Premeno, Rogorea di Rogoredo, Tenero, Cologno al Serio, Garlasco, Guado di Gugnano, Lgurno	Canegrate 1, 3, 4, 8/76, 11, 12, 16, 18, 77, 78, 80, 81, 87, 91, 95, 122, 147, 164S, 165S, 205 Gambolò 1, 2, 11+, 12, 13, 17+, 18, 19, 23+, 25+, 26, 27, 28, 29, 30+, 31, 32, 35, 36, 37, 38, 39, 48, 50+, 52, 53+, 55+, 56, 60, 61, 62, 63, 64+ Scamozzina C, E, F+, G, H, I Alba 3, 6, 11 Appiano Gentile 1, 10 Gravellona Lomellina 6+, 7, 11 Cilavegna CCA5B Robecchetto con Induno Cesto
dépôt de fragments de céramique grossière, sans cinéraire	Canegrate 29, 51, 54, 73, 79, 82, 86, 88, 105, 128, 129, 151, 165 Gambolò 40, 41, 42, 43 Cilavegna CCA2	Canegrate 19, 23, 24, 48, 49, 60, 61, 66, 67, 70, 71, 72, 85, 90, 94, 96, 100, 101, 108, 113, 124*, 130, 154, 158, 163 Gambolò 49, 61, 65 Alba 33, 35, 47
dépôt du cinéraire et de fragments de céramique grossière	Canegrate 10, 28, 39, 40, 50, 52, 53, 55, 56, 57, 65, 89, 99, 103, 104, 107, 112, 115, 123, 125, 134, 135, 140, 143, 149+, 153, 161, 164, 116/119, 131/133 Alba 17 Cilavegna CCA1, CCA4, CCA7A-B	Canegrate 14, 21, 27, 30, 31, 41, 58, 59, 63, 68, 84, 93, 97, 98, 116/119, 117, 136, 137, 138, 139, 141, 142, 145, 148, 150, 152+, 160, 199, 204 Appiano Gentile 11, 12 Cilavegna CCA3A, CCA3B, CCA5C, CCA6, CCA8
dépôt du cinéraire, de fragments de céramique grossière et de vases accessoires	Canegrate 5, 6, 25, 38, 62, 83, 121, 132, 145, 157, Asilo+ Alba 8	Canegrate 7, 47, 157
dépôt du cinéraire et des vases accessoires, sans fragments de céramique grossière	Canegrate 43, 46, 120, 146, 156 Gambolò 14A+, 20+ Scamozzina B, J	
dépôt de bronzes	Canegrate 26, 33, 42, 44, 102, 106, 144, 155 Appiano Gentile 4 Alba 7, 49	
dépôt sans rien		Canegrate 20, 32, 34, 35, 36, 37, 45, 74, 109 Gambolò 44, 45, 46, 47 Appiano Gentile 3, 6, 8, 9

* fragment en silex; + avec une coupe de couverture du cinéraire

A

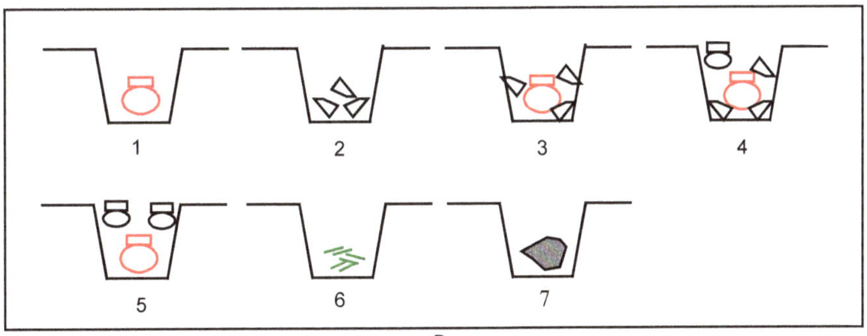

B

Fig.9 – Types de dépôt du matériel d'accompagnement des défunts dans les nécropoles du Bronze Moyen et Récent en Italie nord-occidentale (d'après Domanico 2000b).

Catégories	Type d'objet	fragmentation	action du feu	pliure	bavures de fusion
parures/ habillement	épingle	Canegrate 25, 159; Cilavegna CCA1	Alba 17; Cilavegna CCA1; Canegrate 15, 69?, 103; Gambolò 3A-B, 51, 59; Alessandria	Cilavegna CCA1; Gambolò 59; Alba 10; Appiano Gentile 4, 7; Canegrate 2, 6, 28, 62,73, 146; Gambolò 5, 6, 10, 51, 58, 59, GDG	
	bracelet		Appiano Gentile 2; Monza+; Gambolò 21, 43	Appiano 2; Gambolò 5, 6	Canegrate 73, 83
	torque	Ligurno	Canegrate 75	Canegrate 64; Gambolò 5, 6, 54	
	pendentif		Gambolò 5	Gambolò 5, 21, GDG	
	anneau		Canegrate 83	Canegrate 75	
	ressort à boudin			Gambolò GDG	
	élém. ceinture				
	jambière				
	boucle d'oreilles				
	tube à spiral		Canegrate 69, 75, 134		
	lamelle		Gambolò 15	Canegrate 110; Gambolò 34, 57	
	bouton/bossette			Gambolò 9	
	disque percé				
	perle pâte de verre		Alba17; Canegrate 75; Gambolò 15, 20		
	perle/pendentif os		Canegrate 6		
	perle stéatite				
	bracelet bois de cerf				
	bracelet os		Canegrate 83		
outils/ ustensiles	couteau				
	hache				
	fil				
	pincette				
	aiguille				
	barrette				Canegrate 121
	lame				
armes	épée	Alba 49; Canegrate 6, 25, 92; Cilavegna CCA1; Cologno al Serio; Ligurno	Canegrate 6, 92; Cilavegna CCA1; Ligurno	Canegrate 6, 92; Cilavegna CCA1; Cologno al Serio; Ligurno	
	poignard	Canegrate 129, 132; Coarezza; Monza	Canegrate 129, 146?	Gambolò 9	Canegrate 75
	pointe de lance				
divers	fragment		Alba2; Arbedo 7; Cilavegna CCA2; Canegrate 54, 151	Appiano Gentile 7	Canegrate 40
	écuelle				
	pain/débris de fusion		Alba7; Gambolò 14B		

+ hors contexte

Fig.10 - L'état de dépôt du mobilier d'accompagnement dans les tombes du Bronze Moyen et Récent en Italie nord-occidentale (d'après Domanico 2000b).

Les céramiques utilisées dans les sépultures sont notamment jarres, pots, gobelets, coupes/coupelles et jattes. Les gobelets et les jattes en céramique fine, souvent décorés, sont normalement utilisés comme cinéraires, mais il y a aussi des cas où les restes humains étaient dans des pots ou des vases tronconiques en céramique grossière[8]. Les vases accessoires semblent répéter, en petites dimensions, la même forme et souvent le même décor des cinéraires, ce qui est évident notamment à Canegrate (tombes 25, 43, 46, 83, 157) et Scamozzina (t.B). Ils accompagnent toujours le cinéraire,

[8] Par exemple à Canegrate (tombes 2, 57, 126, 164, 165), Gambolò (tombes 11, 12, 14, 15, 16, 17, 25, 26, 35, 38, 39, 53), Alba (tombes 2, 5, 8, 17) et Appiano Gentile (tombes 1, 7).

aucun cas de restes humains directement dans la fosse avec vase accessoire n'étant documenté. La présence de vases d'accompagnement a été réputée l'un des traits caractéristiques de la culture de Scamozzina (De Marinis 1981). Au contraire, ils sont signalés à Canegrate, mais dans les autres nécropoles sont rares, à l'exception d'Alba t.8, leur absence étant due probablement à la nature des découvertes. Des petits vases sporadiques, hors contexte (nécropoles de Scamozzina, Premeno, Cilavegna, Fornovo S. Giovanni, Glisente?, Rovio, S. Jorio di Locarno, Gudo) pouvaient avoir eu en effet la fonction de vases accessoires, indiquant que ce type de vases d'accompagnement était plus diffusé de ce qu'on avait supposé. Il est aussi possible que les coupes de couverture du cinéraire, parfois documentées, aient eu un usage accessoire, pour le voyage du défunt dans l'au-delà, et non seulement fonctionnel comme fermeture de l'urne.

Les récipients en céramique grossière sont presque toujours brisés, parfois représentés par un unique tesson, la forme originelle étant donc difficile à reconstruire. Là où il a été possible, on a pu documenter des jarres et des pots, souvent avec décor plastique de cordon, ou avec la superficie "rustiquée", caractéristique notamment du répertoire de Canegrate (Fig.9A).

Les objets de parure et/ou d'habillement trouvés jusqu'à présent dans les tombes documentées comprennent des séries différentes du point de vue soit du matériel soit de la typologie. Les objets de bronze sont notamment épingles, bracelets, torques, souvent associés en groupes de deux ou plus, pendentifs et anneaux, mais on a trouvé aussi des éléments de ceinture, des boucles d'oreilles, des boutons et des bossettes, bien que dans très peu de tombes. D'autres objets (ressort à boudin, tubes à spiral, lamelles, disques percés) peuvent également être interprétés comme des éléments de parure et/ou d'habillement[9]. La tombe d'Alessandria a livré la seule jambière rubanée connue jusqu'à présent dans notre territoire (Fig.11). Dans quelques cas on a trouvé des fils ou des fragments de bronze qui sont difficiles à interprétér correctement. Il est probable qu'ils avaient fait partie d'objets de parure/habillement, ou bien d'outils/ustensiles, leur typologie n'étant plus possible de reconstruire.

Les objets en os sont également à interpréter comme des pièces d'ornement des défunts. Il s'agit de pendeloques (Gambolò t.6) et notamment de petites perles circulaires ou rectangulaires, faisant partie probablement de colliers ou de bracelets mais aussi des vêtements des défunts (Scamozzina t.B, Gambolò t.5, Canegrate t.6).

Un fragment rectangulaire aplati en os avec décor de poinçon de la t.83 de Canegrate peut être interprété comme un bracelet. Un probable pendeloque en os poinçonné était dans la tombe de Zeri. Un bracelet en bois de cerf était enfin dans la t.9 de Canegrate.

La même fonction d'objets d'ornement peut être attribuée aux perles de stéatite, trouvées jusqu'à présent seulement dans la nécropole de Gambolò (tombes 4 et 13), mais provenantes aussi des habitats (Rocche di Drusco et Castellaro di Zignago, couche G, pendeloque), et aux perles de pâte de verre, trouvées à Canegrate t.75, Alba t.17, Gambolò t.15 en forme de petits anneaux, et Appiano Gentile t.1SP.

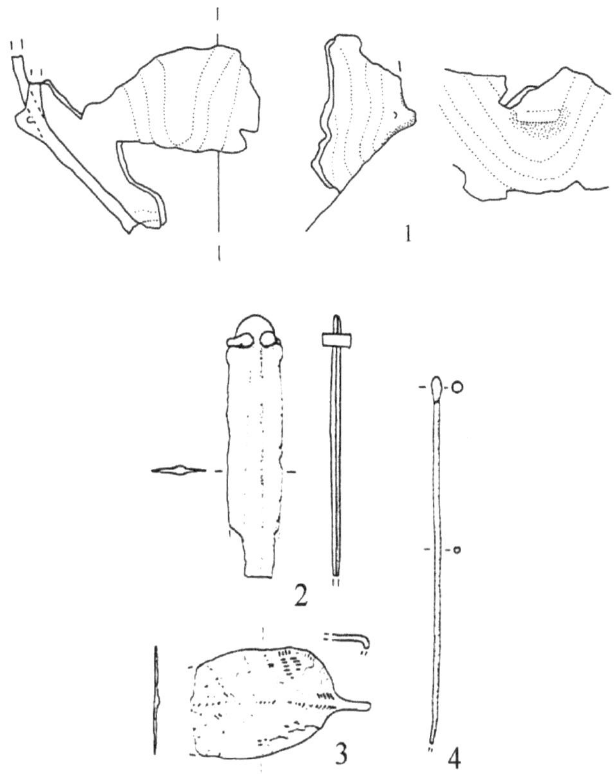

Fig.11 –Matériels d'accompagnement de la tombe d'Alessandria Cascina Chiappona. 1: cinéraire; 2: poignard à deux rivets; 3: jambière rubanée; 4: épingle à tête en massue(échelle 1:3 céramique, 1:2 bronzes) (d'après Venturino-Giaretti-Oberti 1995).

21 nécropoles ont livré des armes. Dans la plupart des cas il s'agit d'objets hors contexte ou isolés, attribuables à des tombes, notamment épées (Monza, Cattabrega di Crescenzago, Cossila, "territoire de Pavie"; "territoire de Crema") et poignards (Scamozzina, Monza, Coarezza, Rovio, S. Jorio di Locarno, Glisente, Appiano Gentile, Fornovo San Giovanni), mais il y a aussi des tombes à armes bien documentées. Les tombes 6, 25, 92 et 153 de Canegrate, la t.CCA1 de Cilavegna et les tombes de Cologno al Serio et Ligurno avaient des épées parmi les objets d'accompagnement. Des poignards étaient à Canegrate (tombes 75, 110, 129, 132, 144, 146); Garlasco; Monza (t.1889); Scamozzina t.J; Mercurago; Coarezza; Alba (t.8); Gambolò (tombes 4, 9, 14A, 51); Fornovo San Giovanni; Rogorea di Rogoredo; S. Stefano Arno; Alessandria. Les seules pointes de lance trouvées jusqu'à présent dans des sépultures étaient à Canegrate (t.110) et à Cologno al Serio. Dans les nécropoles les tombes à arme, caractérisant le rôle du guerrier, sont bien réparties et jamais plus d'une tombe à arme a été trouvée dans le même groupement.

[9] Pour la typologie et la chronologie des objets en bronze, voir Domanico 1998a; 2000a et 2000b.

Les outils et les ustensiles sont très rares dans les tombes. Les seuls objets de ce type jusqu'ici documentés sont trois couteaux, une hache, deux pincettes, une aiguille et des fils dont la fonction comme outils/ustensiles n'est pas évidente. La plupart d'eux était hors contexte. Les seules nécropoles qui ont livré ces objets sont Canegrate (t.111, couteau), Rovio et Locarno-S.Jorio (couteaux hors contexte), Cattabrega di Crescenzago (hache hors contexte), Gravellona Lomellina t.3 et Marcignago hors contexte (pincettes), Scamozzina t.J (aiguille) et Canegrate tombes 5, 6, 13, 39, 50 (fils).

Dans deux cas des débris de bronze ayant subi l'action du feu étaient dans des tombes. Il s'agit de la t.14B de Gambolò, où il y avait trois petits fragments - probables pains de bronze - et de la t.7 d'Alba, avec une probable goutte de fusion. Les débris de Gambolò étaient dans l'urne cinéraire avec les restes incinérés d'un petit enfant, ce qui pose beaucoup de problèmes d'interprétation (Simone 1990-91, p.111). Dans le cas d'Alba, la tombe avait été bouleversée anciennement, ne livrant que très peu de fragments. Il est néanmoins intéressant d'observer qu'il y avait aussi des restes en terre cuite.

Les restes d'animaux sont presque complètement ignorés dans la bibliographie archéologique concernant les sites funéraires nord-occidentaux. Il ne sont jamais pris en compte, ni analysés ou étudiés. Par conséquence, aucune information n'est disponible en ce qui concerne les offrandes et/ou les banquets rituels qui auraient dû se dérouler pendant les funérailles ou les rites accompagnant les défunts. Le seul cas où des restes d'animaux ont été enregistrés est celui de la probable structure rituelle 47 d'Alba San Cassiano, où une mandibule de suidé était dans une fosse circulaire avec des fragments de céramique grossière (Venturino 1995 : 96)

L'état du dépôt du matériel d'accompagnement

L'observation du matériel d'accompagnement indique quatre types d'état:

- fragmentation
- pliure
- bavures de fusion
- action du feu

Il convient d'observer que l'état de conservation du matériel dépend largement des modalités de la découverte, la plupart des sépultures étant bouleversée et les dépôts incomplets. Pourtant, la présence d'objets fragmentés porte à nous interroger sur la nature réelle des dépôts funéraires et sur l'état originel des matériels placés dans les tombes (objets entiers? bris intentionnel? fragment symbolique d'un vase?). On a vu que les objets intentionnellement brisés, notamment céramique grossière et armes, sont bien documentés dans les nécropoles ici prises en compte. Il ne fait pas de doute que ces cas sont à rapporter aux rituels funéraires dont le sens est difficile à interpréter correctement. La fragmentation des épées en deux, trois jusqu'à sept parties (Cologno al Serio) est de même largement documentée.

Dans les tombes ici analysées, toutes les épées paraîent intentionnellement brisées, sauf l'épée de la t.153 de Canegrate dont le mauvais état de conservation empêche toute observation à ce propos. Sur cette base, nous avons pu attribuer à des tombes les épées isolées de Cossila et du “territoire de Crema”, brisées elles aussi en trois et quatre parties respectivement. Le dépôt rituel de l'épée dans la str.49 d'Alba, cassée en deux parties et placée verticalement dans la fosse, donne un autre exemple indicatif. Le bris intentionnel étant documenté aussi pour les poignards (Canegrate tombes 129, 132; Coarezza, Monza), il en ressort que probablement, parmi les bronzes, cette pratique était réservée aux armes et le bris des épingles et des torques était donc casuel (Fig.10).

Souvent les objets étaient aussi repliés. De même que la fragmentation, il est difficile d'apprécier l'intentionnalité des pliures. Il convient toutefois d'observer que les épées repliées étaient toujours aussi brûlés et/ou brisées (Canegrate t.92, Cologno al Serio, Cilavegna t.CCA1, Ligurno). Il en va de même pour le poignard de la t.129 de Canegrate (brisé et brûlé). Les bavures de fusion observées à Canegrate sur le poignard de la t.75, les bracelets de la t.73 et de la t.83 et sur des fragments des tombes 40 et 121 nous indiquent que ces objets étaient neufs au moment de leur dépôt dans la tombe.

Pourtant, c'est le feu l'élément le plus important du rituel funéraire. Il est frappant d'observer la trace de l'action du feu sur les objets accompagnant les défunts. Souvent l'arme était brûlée, comme indiquent les épées de Canegrate (tombes 6 et 92), Cilavegna (t.CCA1) et Ligurno et les poignards de la t.129 et probablement de la t.146 de Canegrate. Plus souvent étaient les objets de parure et d'habillement à subir l'action du feu. Nombre d'épingles, bracelets, un pendentif à Gambolò (t.5) et un torque à Canegrate (t.75), étaient brûlés avec le mort, aussi que des ornements en os et pâte de verre. Le défunt était donc vêtu et équipé avec ses armes et ses ornements, lors de sa crémation. On peut observer par contre que aucun des outils/ustensiles n'a été plié ou fragmenté intentionellement, ni avait été brûlé avec le défunt.

Conclusions

Les aspects concernant les nécropoles à incinération de l'âge du Bronze Moyen et Récent en Italie nord-occidentale - organisation des nécropoles et mode de groupement des tombes, présence de structures dans les nécropoles, types d'architecture tombale, mode de dépôt des restes humains, matériel d'accompagnement - posent des problèmes d'interprétation en grande partie à rapporter aux rituels funéraires et à l'attitude des communautés vis-à-vis à la mort. Les données dont nous disposons ne sont pas abondantes, à cause de la destruction de la plupart des sites funéraires, mais il est aussi vrai que nombre de ces rituels ne laissent aucune trace matérielle. Le danger d'en proposer une interprétation est donc énorme. Néanmoins, il convient en conclusion de poser quelques questions qui peuvent alimenter, peut-être, la problématique de la recherche.

Catégories	Type d'objet	fragmentation	action du feu	pliure	bavures de fusion
parures/ habillement	épingle	Canegrate 25, 159; Cilavegna CCA1	Alba 17; Cilavegna CCA1; Canegrate 15, 69?, 103; Gambolò 3A-B, 51, 59; Alessandria	Cilavegna CCA1; Gambolò 59; Alba 10; Appiano Gentile 4, 7; Canegrate 2, 6, 28, 62,73, 146; Gambolò 5, 6, 10, 51, 58, 59, GDG	
	bracelet		Appiano Gentile 2; Monza+; Gambolò 21, 43	Appiano 2; Gambolò 5, 6	Canegrate 73, 83
	torque	Ligurno	Canegrate 75	Canegrate 64; Gambolò 5, 6, 54	
	pendentif		Gambolò 5	Gambolò 5, 21, GDG	
	anneau		Canegrate 83	Canegrate 75	
	ressort à boudin			Gambolò GDG	
	élém. ceinture				
	jambière				
	boucle d'oreilles				
	tube à spiral		Canegrate 69, 75, 134		
	lamelle		Gambolò 15	Canegrate 110; Gambolò 34, 57	
	bouton/bossette			Gambolò 9	
	disque percé				
	perle pâte de verre		Alba17; Canegrate 75; Gambolò 15, 20		
	perle/pendentif os		Canegrate 6		
	perle stéatite				
	bracelet bois de cerf				
	bracelet os		Canegrate 83		
outils/ ustensiles	couteau				
	hache				
	fil				
	pincette				
	aiguille				
	barrette				Canegrate 121
	lame				
armes	épée	Alba 49; Canegrate 6, 25, 92; Cilavegna CCA1; Cologno al Serio; Ligurno	Canegrate 6, 92; Cilavegna CCA1; Ligurno	Canegrate 6, 92; Cilavegna CCA1; Cologno al Serio; Ligurno	
	poignard	Canegrate 129, 132; Coarezza; Monza	Canegrate 129, 146?	Gambolò 9	Canegrate 75
	pointe de lance				
divers	fragment		Alba2; Arbedo 7; Cilavegna CCA2; Canegrate 54, 151	Appiano Gentile 7	Canegrate 40
	écuelle				
	pain/débris de fusion		Alba7; Gambolò 14B		

La visibilité des nécropoles sur le territoire fait preuve que la zone réservée aux morts était organisée comme une unité topographique, à savoir comme un "espace funéraire" différencié de celui des vivants, avec à l'intérieur des structures (zones de bûcher, aires de passage, fosses rituelles) consacrées aux cérémonies d'accompagnement des défunts. Dans cet espace, les tombes sont souvent organisées par groupements ou alignements et dans quelques cas (Cilavegna, Canegrate) avec des alignements à l'intérieur des groupes. L'interprétation de la répartition des tombes dans les nécropoles nous a permit d'individualiser quatre types de groupements, le sens de cette organisation étant difficile à apprécier. Est-elle à rapporter aux groupes familiaux, aux différentes catégories de défunts, ou est en fonction de la chronologie?

Dans les nécropoles de Gambolò, Alba et Canegrate la présence dans chaque groupe individualisé de tombes datant de différentes phases (BM3, BR1, BR2) semble confirmer l'hypothèse d'une organisation par groupes familiaux, bien que les tombes datées ne sont pas nombreuses par rapport à la totalité des sépultures découvertes. Une interprétation en fonction de l'âge et du sexe des défunts trouve un obstacle dans le très peu d'analyses des ossements humains. On peut cependant observer qu'à Canegrate les catégories des défunts sont présentes dans presque tous les groupes de tombes individualisés, avec une possible tendance à grouper ensemble les sépultures d'enfants. L'hypothèse que la répartition des tombes est à rapporter aux groupes familiaux semble donc confirmée, au moins dans cette nécropole.

En ce qui concerne la position sociale du défunt, l'étude des structures tombales et du matériel d'accompagnement fait suposer qu'il existait, au moins dans les nécropoles les mieux connues, une différenciation entre les morts. De cette analyse se dégage en effet une impression de très grande variété, voire de diversité, entre les types de sépultures, avec des cas uniques et exceptionnelles à la fois du point de vue de la structure tombale, des éléments d'accompagnement et des rituels. Est cette inégalité à rapporter à la position des défunts pendant leur vie? Bien que d'autres raisons peuvent être apportées (personnalité des défunts, modalités de leur déces, idée des vivants des circonstances de la mort; Dedet 1992) cette possibilité mérite d'être mise en évidence.

Les différences entre les tombes ne semblent pas à corréler avec les catégories des défunts en fonction de l'âge, des traitements particuliers étant réserves soit aux adultes, soit aux adolescents, soit aux enfants. Dans les nécropoles analysées, il semble que les enfants bénéficient d'une sépulture du même type de celle des adultes, le matériel d'accompagnement étant autant riche. Ils sont admis dans les nécropoles dès la naissance, comme témoignent les restes du nouveau-né de la t.126 de Canegrate et des petits-enfants, aucun cas de squelettes d'enfants dans les habitats n'étant documenté en Italie nord-occidentale pendant le Bronze Récent[10].

La présence importante, dans toutes les nécropoles ici analysées, de tombes structurées avec des pierres ou des dalles de couverture ou à la base, ou avec des coffres ou des revêtements, nous suggère que la sépulture était conçue comme la maison définitive du mort, à savoir une structure solide et destinée à durer dans le temps. Certains éléments (couverture de la tombe, fermeture de l'urne, enclos, urne renversée, cinéraire percé, pierre brûlée à l'intérieur de l'urne) peuvent être interprétés en fonction du rituel, d'après le modèle proposé par R. Hertz. L'espace bien délimité, domaine du défunt, serait alors une "protection" pour les vivants de la condition "dangereuse" du mort: la tombe doit permettre à son âme le repos définitif et l'accès au pays des morts et, au même temps, assurer qu'il ne revient pas sur la terre. Le feu, les cérémonies d'enterrement, la fermeture de l'urne et de la tombe, aident les morts et les vivants à surmonter le moment difficile de la séparation (Hertz 1907, Chapman *et alii* 1981, Dedet 1992, Bietti Sestieri 1992).

Les tombes sont en majorité individuelles et les cas de tombes doubles et multiples sont très rares. On ignore si les dépôts doubles et multiples dans la même sépulture étaient simultanés ou successifs[11]. En tous cas ils appartiennent toujours à la même phase chronologique. Quel peut être le sens des tombes doubles et multiples? Le bas nombre de tombes de ce type (6 doubles et 2 multiples) révèle leur nature exceptionnelle. La connaissance de la répartition par âge indique que, sauf dans les tombes 4 et 6 de Gambolò où il y a respectivement un et deux adultes et un adolescent, dans les autres tombes il y a toujours des enfants/petits-enfants, seuls (Gambolò t.14; Canegrate t.9) ou avec un adulte (Alba t.3, Canegrate t.25, Zeri). L'hypothèse qu'il s'agissait de la sépulture des parents avec leurs enfants, et donc de consanguins, n'est pas à exclure, bien que le manque d'analyses et souvent de la détermination du sexe des adultes ne favorise pas une conclusion à ce propos[12].

Le placement des restes incinérés dans la fosse, soit du type simple soit structuré, implique que jamais le dépôt était primaire. Il faut donc imaginer un rituel d'incinération hors de la tombe elle même, mais peut-être près d'elle, de toute façon sur le sol de la nécropole. Le cas des "aires de terre noire" de Canegrate nous offre une idée à ce propos[13]. D'autres exemples nous donnent des indices des rituels funéraires, comme par exemple la plate-forme de pierre de Premeno, dont la fonction de bûcher semble la plus probable.

[10] Le nombre de sépultures d'enfants dans le groupe 2 de Canegrate peut suggérer l'existence d'un coin spécial qui leur était réservé dans cette nécropole (Fig.4A). Cependant, la présence de tombes d'enfants aussi dans les autres groupes, bien que moins concentrée, rend cette interprétation incertaine.

[11] Pour les dépôts doubles de la nécropole de Gambolò, L. Simone a suggéré l'existence du sacrifice de l'épouse, ce qui peut être supposé uniquement pour la t.4 (1 adulte masculin et une adolescente féminine) et peut-être la t.6 multiple (un adulte masculin, un adulte féminin et un adolescent non déterminable); au contraire R. De Marinis a supposé la pratique de conserver les restes du défunt hors de la tombe, en attendant la mort d'un autre membre de la famille à ensevelir dans la même sépulture, les deux hypothèses n'étant pas en tous cas vérifiables (Simone 1990-91, p.91).

[12] Dans ce cas il faut imaginer la reouverture de la tombe ou la conservation des restes hors de la tombe.

[13] À Canegrate des "aires de terre noire" sont à l'intérieur de chaque groupe de tombes envisagé (Fig.4A).

Le manque de données sur la quantité et le volume des lots d'ossements trouvés dans les tombes nous empêche d'apprécier si la totalité ou seulement une partie du défunt incinéré était placé dans l'urne[14]. Cependant, le très peu d'ossements trouvé sur les "aires de terre noire" ou sur la structure de Premeno nous indique que les restes du défunt étaient recueillis, bien que pas soigneusement.

Dans presque toutes les sépultures le défunt était enseveli avec son matériel d'accompagnement, les cas de tombes dépourvues d'objets étant rares. Dans la plupart des cas il s'agit d'éléments de parure et d'habillement, notamment épingles, torques et bracelets, souvent en groupes de deux, trois ou plus. Le défunt était donc déposé dans la tombe avec ses objets personnels. La présence d'épingles ayant subi l'action du feu peut indirectement témoigner qu'il était brûlé avec ses vêtements. Dans quelques cas le défunt était brûlé avec son arme, comme indiquent les traces de l'action du feu notamment sur les épées. À Canegrate, sauf dans les goupes 2 et 9, on ne retrouve jamais plus d'une tombe à épée par groupe, et pas dans tous les groupes, de même que les tombes à poignard[15]. Comment interpréter ces données? La possibilité qu'il s'agissait de personnages de rang est difficile à prouver, néanmoins mérite d'être mise en valeur. Une interprétation en fonction de la position sociale de ces défunts peut être donc supposée, bien qu'elle se heurte à divers problèmes, notamment le manque de données suffisantes concernant le sexe et l'âge.

Souvent les armes étaient intentionnellement brisées lors de leur dépôt dans la tombe, la valeur symbolique et rituelle de cette action étant évidente, bien que difficile à interpréter correctement. Peut-être que les armes, de même que les objets de parure et d'habillement ayant appartenu au défunt pendant sa vie, lui étaient nécessaires dans son existence dans l'au-delà, mais le bris intentionnel a été aussi interprété comme une action des vivant pour réduire les "risques" pour la communauté dérivants du *status* pas encore définitif du défunt en tant que porteur d'armes de son vivant (Bietti Sestieri 1992). Le bris des objets dans les sépultures peut d'ailleurs être interprété de manières diverses[16].

Il est intéressant de constater que les outils et les ustensiles, présents bien que rarement dans les tombes, ne révèlent aucune trace de fragmentation et/ou de pliure intentionnelles, ni ont jamais subi l'action du feu. Les peu d'exemplaires trouvés étaient donc dans leur état "fonctionnel". Est-ce que cet état est à rapporter à la nécessité du défunt "d'utiliser" les outils tels quels dans l'au-delà? Ou faut-il plutôt penser qu'ils étaient les symboles du métier du défunt? Les très peu de cas d'outils dans les tombes empêchent toute conclusion à ce propos, bien que ces possibilités méritent d'être prises en considération[17]. La fragmentation intentionnelle est documentée également pour la céramique, notamment les récipients en céramique grossière. À Canegrate on retrouve des tessons dans presque toutes les tombes, ce qui nous permet de supposer qu'il s'agissait ici d'un élément en commun pour toutes les catégories de défunt, ou de quelque sorte "obligatoire" pour le rituel funéraire. La disposition de ces tessons dans les tombes est très variée et fait preuve de leur valeur rituelle. Des descriptions publiées il ressort que dans cette nécropole la céramique intentionnellement brisée était déposée à l'intérieur de la tombe ou près d'elle, sur la pierre ou la coupe de couverture, dans l'urne cinéraire, parfois dans la fosse mais autour de l'urne. Dans la t.17 de Alba des tessons étaient sur la dalle de couverture de la tombe.

Bien que de nombreux traits en commun ont été envisagés dans tous les sites funéraires du territoire nord-occidental, il y a quand même des éléments qui nous paraient caractéristiques de chaque nécropole. La variété des structures tombales semble typique notamment de la Lombardie occidentale, les éléments de construction de la tombe étant rares dans les nécropoles de la Lomellina et du Piémont, à l'exception de la t.17 de Alba. À Gambolò, le matériel d'accompagnement caractérise notamment les tombes les plus anciennes, les parures des tombes du Bronze Récent étant représentées notamment par les épingles. Par contre, une grande variabilité de structures et du matériel d'accompagnement a été constaté à Canegrate. S'agit-il d'une différence de rituel ou faut-il penser à une différenciation entre communautés? La chronologie et la typologie du matériel étant les mêmes, l'idée que nous avons affaire avec des communautés ayant un rôle différent (richesse majeure? contrôle sur le territoire?) mérite d'être proposée, bien qu'avec prudence, en l'absence de données suffisantes concernant les habitats.

Bibliographie

BASERGA, G.

1936 Tombe dell'età del Bronzo a Locarno. *Rivista Archeologica dell'Antica Provincia e Diocesi di Como* XV, fasc.111/113. Côme: Società Archeologica Comense: 33-40.

BERNABO' BREA, M.

1997 Il limite sud-occidentale dell'area terramaricola e le facies culturale dell'Appennino emiliano occidentale nell'età del bronzo media e recente. *Terramare* 1997: 433-435.

1998 I rapporti tra l'area terramaricola e le facies occidentali nei territori a sud del Po durante l'età del bronzo media e recente. *"Preistoria e*

[14] La pratique du décharnement avant incinération est très difficile, voire impossible, de déterminer pour les ossements brûlés, bien qu'elle peut être supposée (Dedet 1992, p.71). Aucune preuve ne peut cependant être apportée à ce sujet pour les nécropoles ici prises en compte.

[15] Tombes à épée: 6 (groupe 2), 25 (groupe 3), 92 (groupe 9), 153 (isolée). Tombes à poignard: 75 (groupe 2), 110 (groupe 1), 129 (groupe 9), 132 (isolée), 144 (groupe 10), 146S (isolée).

[16] B. Dedet a proposé des raisons matérielles (le bris dû au passage sur le bûcher, ou objets incomplètement récupérés) ou spirituelles (le bris parce que la vie dans l'au-delà est conçue comme l'inverse de la vie sur la terre ou parce que la part équivaut au tout) (Dedet 1992, p.272). L. V. Grinsell a proposé des interprétations diverses: "libérer" l'esprit de l'objet qui doit accompagner le défunt dans l'outre-tombe; prévenir les risques de vol en rendant inutilisables les objets de valeur; empêcher une réutilisation des objets appartenant au défunt; le sacrifice rituel des objets; etc. (Grinsell 1975).

[17] En état "fonctionnel" étaient aussi les vases accessoires en céramique fine, jamais trouvés intentionnellement brisés.

Protostoria del Piemonte", Actes XXXII Riun. Sc. IIPP. Florence: 433-441.

BIETTI SESTIERI, A. M. (éd.)
1992 *La necropoli laziale dell'Osteria dell'Osa*. I, II. Rome: Quasar.

BOCQUET, A.
1997 Le peuplement préhistorique des Alpes françaises du nord. *"La Valle d'Aosta nel quadro della preistoria Protostoria dell'arco alpino centro-occidentale", Actes XXXI Riun. Sc. IIPP*. Florence: 425-453.

CANETTA, E.
1970 Nuovi ritrovamenti della cultura di Canegrate a Premeno. *Sibrium* X: 91-93.

CASTELFRANCO, P.
1875 Necropoli di Rovio nel Canton Ticino. *Bullettino di Paletnologia Italiana* I, II. Parme: 21-24; 57-64.
1891 Tombe di Monza dell'età del bronzo. *Bullettino di Paletnologia Italiana* XVII, ns. II, tomo VII. Parme: 34-47.
1909 Sepolcreto della Scamozzina presso Albairate in provincia di Milano. *Bullettino di Paletnologia Italiana* XXXV, I-IV. Parme: 1-12.

CHAPMAN, R., I. KINNES and K. RANDSBORG
1981 *The Archaeology of Death*. Cambridge: Cambridge University Press.

CRIVELLI, A.
1943 *Atlante storico della Svizzera italiana*, Bellinzone.

DAVID ELBIALI, M.
1997 Il Vallese (Svizzera) nel quadro dell'arco alpino centro-occidentale durante il Bronzo recente: alcuni elementi d'affinità meridionale. *"La Valle d'Aosta nel quadro della preistoria e Protostoria dell'arco alpino centro-occidentale", Actes XXXI Riun. Sc. IIPP*, Florence: 487-497.
1998 Rapporti tra Piemonte e area alpina nell'età del bronzo. Table Ronde *"Preistoria e Protostoria del Piemonte", Actes XXXII Riun. Sc. IIPP*, Florence: 461-467.

DEDET, B.
1992 Rites funéraires protohistoriques dans les Garrigues languedociennes. Approche etho-archéologique. *Revue Archéologique Narbonnaise*, Supplément 24 CNRS, Paris.

DEGANI, M.
1962 La necropoli terramaricola della "Montata" (Reggio Emilia). *Preistoria dell'Emilia Romagna, VI Congresso Internazionale di Preistoria e Protostoria, Documenti e Studi a cura della Deputazione di Storia Patria per le province di Romagna* vol. VII, I, Bologne: 63-110.

DEL LUCCHESE, A., G. ODETTI and R. MAGGI
1989 Le Bronze moyen en Ligurie. *"Dynamique du Bronze Moyen en Europe occidentale", Actes du 113e Congrès National des Sociétés Savantes, Strasbourg-Haguenau*, Paris: Editions du C.T.H.S. 459-472.

DEL LUCCHESE, A., R. DE MARINIS and F. M. GAMBARI
1991-92 L'Italia settentrionale. I. L'Italia nord-occidentale. "L'età del Bronzo in Italia nei secoli dal XVI al XIV a.C.", *Rivista di Archeologia* 10: 31-38.

DE MARINIS, R. C.
1981 Appunti sul Bronzo Medio, Tardo e Finale in Lombardia (1600-900 a.C.). *Actes I Convegno Archeologico Regionale*, Milan, 173-189.
1994 Preistoria e Protostoria del territorio di Lecco. Casini S. (éd.), *Carta Archeologica della Lombardia. IV. La provincia di Lecco*, Modène: 19-80.
1998 L'età del bronzo nell'Italia nord-occidentale. Questioni di cronologia e di terminologia. *"Preistoria e Protostoria del Piemonte", Actes XXXII Riun. Sc. IIPP*, Florence: 443-451.

DOMANICO, L.
1998a Problematiche del Bronzo Recente in Italia nord-occidentale. *Actes XIII Congrès UISPP*, vol. 4, Forlì: 379-388.
1998b Tomb structures in North West Italy during the Recent Bronze Age. *Actes XIII Congrès UISPP*, vol. 4, Forlì: 515-520.
2000a Dinamismo culturale e fenomeni di scambio in Italia nord-occidentale durante la Tarda età del Bronzo. Rapporti con l'Etruria e suddivisione cronologica del Bronzo Recente. Negroni Catacchio N. (éd.), *"L'Etruria tra Italia, Europa e Mondo Mediterraneo", Actes Quarto Incontro di Studi Preistoria e Protostoria in Etruria*, Milan: 59-81.
2000b L'âge du Bronze Récent (XIV-XIII s. av. J.-C.) en Italie nord-occidentale. *Thèse de doctorat, Ecole Pratique des Hautes Etudes, Sciences historiques et philologiques*, Sorbonne, Paris.

FORMENTINI, U.
1941 Tomba a cremazione scoperta nel territorio di Rossano. *Notizie Scavi II*, 7-8-9: 173-175.

GAMBARI, F. M.
1998 L'età del Bronzo in Piemonte. *"Preistoria e Protostoria del Piemonte", Actes XXXII Riun. Sc. IIPP*, Florence: 64-84.

GHISLANZONI, E.
1932 Il sepolcreto di Appiano Gentile. *Rivista Archeologica della Provincia di Como*, Côme: 3-20.

GRINSELL, L. V.
1975 *Barrow, Pyramid and Tomb*. Londre: Thames and Hudson.

HERTZ, R.
1907 Contribution à une étude sur la représentation collective de la mort. *Année Sociologique*: 48-137.

PRIMAS, M.
1975 Zur Interpretation weiträumig verbreiteter Kulturelemente in Norditalien und dem alpinen Gebiet während der Jungbronzezeit. *Kolloquiums "Geschichte des 13 und 12 Jh.v. Chr.", Jahresbericht des Instituts fur Vorgeschichte der Universität Frankfurt a.M.*, München: 46-56.

1998 I rapporti culturali transalpini. *"Preistoria e Protostoria del Piemonte", Actes XXXII Riun. Sc. IIPP*, Florence: 469-472.

RITTATORE VONWILLER, F.

1953-54 La necropoli di Canegrate. *Sibrium* I: 7-43.

1956-57 La necropoli di Canegrate (Scavi 1954-1955-1956). *Sibrium* III: 21-35.

1975 La cultura di Canegrate. *"La cultura protovillanoviana", Popoli e Civiltà dell'Italia Antica* vol. IV: 253-262.

SIMONE, L.

1990-91 La necropoli della tarda età del bronzo di Gambolò (PV). *Sibrium* XXI: 89-147.

1992-93 Nuovi ritrovamenti nella necropoli di Gambolò (PV). *Sibrium* XXII: 19-42.

Terramare

1997 Bernabò Brea M.- Cardarelli A.- Cremaschi M. (éd.) Le Terramare. La più antica civiltà padana. Catalogue exposition, Modène.

VANNACCI LUNAZZI, G.

1988 Le tombe a cremazione di Cilavegna (Pavia): nuovi dati sull'età del bronzo in Lomellina. *Rivista Archeologica dell'Antica Provincia e Diocesi di Como* 169, Côme: 5-35.

VENTURINO GAMBARI, M. (éd.)

1995 Navigatori e contadini. Alba e la valle del Tanaro nella preistoria, Alba.

VENTURINO GAMBARI, M., M. GIARETTI and R. OBERTI

1995 Alessandria, loc. Cascina Chiappona. Rinvenimento di sepoltura a cremazione della media età del Bronzo. *Quaderni Soprintendenza Archeologica del Piemonte, Notiziario* 13: 302-303.

VITAL, J.

1998 À propos des rapports culturels entre l'Italie nord-occidentale et le sud-est de la France du Bronze Moye au Bronze Final. *"Preistoria e Protostoria del Piemonte", Actes XXXII Riun. Sc. IIPP*, Florence: 473-481.

1999 Identification du Bronze moyen-récent en Provence et en Méditerranée nord-occidentale. *Documents d'Archéologie Méridionale* 22: 7-115.

HOLY FIRES AND PROFANE FIRES IN PROTOHISTORIC ETRURIA. NEW DATA FROM THE SETTLEMENT OF *SORGENTI DELLA NOVA* (VITERBO - ITALY)

Nuccia Negroni-Catacchio, Massimo Cardosa, Laura Guidetti and Monica Miari

Introduction

The territory of central Italy on the Tyrrenian Sea, between Rome and Florence was in the 1st millennium B.C. the residence of Etruscans.

Actually this area is rich with far more ancient archaeological evidences, beginning with several finds attributed to the Early Palaeolithic (about 700,000 B.C.). The testimonies of the Neolithic era are also very rich, starting with the first examples of the impressed pottery and of the Eneolithic period. The Rinaldone Culture, developed in this area during the Eneolithic period, is particularly important for its relationship with the Aegean and populations living in Anatolia.

During the Late Bronze Age (Final Bronze Age XII-X century B.C.) a new cultural *facies* called "Protovillanovan" developed. It is characterised by wide settlements placed in naturally defended rises, heralding the rising Etruscan cities. In fact, the inhabitants of these protovillanovan villages colonized the large tuff planes and founded the future cities.

The settlement of *Sorgenti della Nova* has been excavated for 25 years, therefore it is the best known of these proto-historic settlements. Therefore it has been possible to analyse the general settlement pattern and the typology of each living and service structure.

This paper will describe the use of fire and discuss its interpretation. In some cases it has been possible to interpret these evidences as having religious and ritual significance, whilst in others it is probable that their use for practical purposes related with everyday life.

The Settlement of Sorgenti della Nova

The settlement of *Sorgenti della Nova*[1] lies on a tuff-pumice crag on the northwestern borderline of the Viterbo province, to the north of Rome, just on the boundary between Tuscany and Lazio (Fig. 1). The site is naturally fortified thanks to the steep cliffs surrounding it, carved by the two bounding streams.

Fig. 1. Position of settlement of Sorgenti della Nova.

A spring with a considerable flow pours from a grotto bored at the foot of the hill. This picturesque spot has been visited for years by tourists for its particular evocative power.

The first survey took place 1938, but scientific excavations began in 1974, when a quarry of pumice stated to operate and partially destroyed the settlement. The investigations have been carrying on non-stop since that year.

[1] The publications about this site are numerous, the complete bibliography can be found in Negroni Catacchio, N. (ed.) 1981 e Negroni Catacchio, N. (ed.) 1995. The last publication is: Negroni Catacchio, N. and Domanico, L. c.s.

Digs have gone over some wide sectors (I-IX) placed on the northern and southern slopes of the crag and on the "acropolis" and they brought to light a broad proto-Etruscan settlement, nearly a proto-urban, and a quite interesting medieval village.

The whole area, occupying the space between the two ditches, extended to about 15 hectares, while the zone of the naturally fortified *plateaux*, residence of the main part of the village, covers now an area of about 5 hectares. It has been calculated on the basis of the dwellings found in relation with the excavated area that about 1,500 inhabitants must have lived there.

The excavations revealed a wide centralized and intensively "urbanized" settlement, with houses and service facilities very close to, each other; the rational organisation of space and the presence of a general setting up plan seem to imply a centralized programming. (Fig. 2)

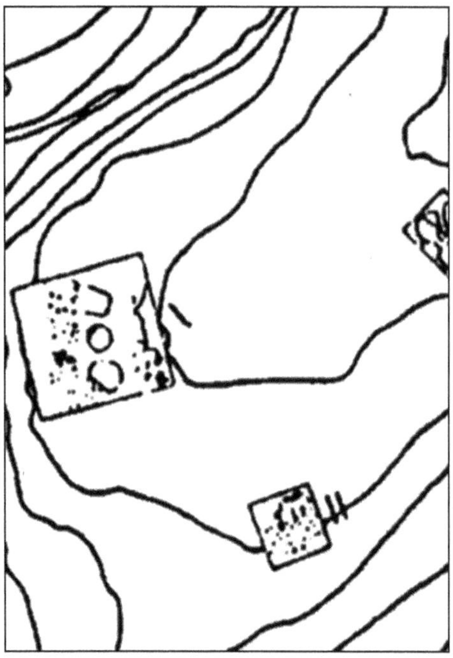

Fig. 2. Plan of Sorgenti della Nova with the dig sectors.

On the top of the hill dwellings were rock-embedded base huts, with sub-circular or quadrangular plan, often with a porch leading to the entrance; the walls and the roof were made of perishable materials. These huts were not very spacious and for this reason they were suitable for nuclear families with a narrow circle of members.

The sides of the crag are occupied by artificial terracing, which involves impressive diggings and a great labour force. Finally *large elliptical plan huts* were built on the terraces. These dwellings were large enough to lodge more than one family or at least enlarged families. Also these houses, with foundations on postholes and gullies had the walls and the roofs made of perishable materials.

Several grottoes were opened in the rock face behind the huts, some of them were used for habitation, others as cult places, and others had service functions. Small cavities in the rock in fact gave place to large hearths or, at least in three cases, clay ovens for cooking.

Along the lower tableland, in the southern sector, there has been brought to light a *wide structure embedded in the rock*. At first analysis it seems to be a large ditch over 30 m in length and about 2.5 m in depth, but it has not been completely dug, yet. Probably it was utilized in the first phase of occupancy of the site and used to settle dwellings partially underground. In fact it has been filled full and covered with a platform of stones still in use during the Final Bronze Age.

Finally, at the eastern corner of the top plateau, a staircase with high steps carved in the rock, called holy stairway, seems to identify a ritual place belonging to the whole community of *Sorgenti della Nova*.

The settlement was born around mid of IX century B.C., during the flourishing of the protovillanovan *facies*, which in Etruria comes immediately before the villanovan Iron Age and the urban phase of the Etruscan civilisation. Other settlements similar in size and importance are known In this period, which probably controlled the territory from which they got their sustenance. Likely they were ruled by aristocratic classes in whose wealth the presence of mineral sources took probably a part and which, at some point, were possibly in struggle for power.

The major part of these sites were abandoned between the end of X c. B.C. and the beginning of IX c. B.C. and *Sorgenti della Nova* had the same fate. Just in those decades the future great Etruscan cities (Vulci, Tarquinia, Cerveteri, etc.) were rising. Their most ancient materials are entirely identical to those of the Final Bronze Age settlements. Consequently, the hypothesis that the protohistoric communities of *Sorgenti della Nova* and other similar centres founded the future Etruscan cities, according to the plan of getting together and occupying the same wide *plateau* near the coast, is becoming more and more persuasive. If it is right, the inhabitants of this large centre could be considered, at least from the second half of the XI c. B.C., "Proto-Etruscans".

The Medieval Settlement

After the protohistoric phase the site has been reoccupied only in the Medieval period: a quite large settlement arose whose core was a castle standing on the upper part of the cliff. Nowadays only the tower and traces of the foundation are left. Also at the top, at the western side of the castle and separated by an artificial ditch, the remains of a church, in use at least until the XIV c. A.D, have been partially brought to light. A great ossuary grave belonging to the medieval era and containing more than 150 individuals was dug right next to the church. The University of Gottingen is now carrying on the physical anthropological analyses on the skeletons.

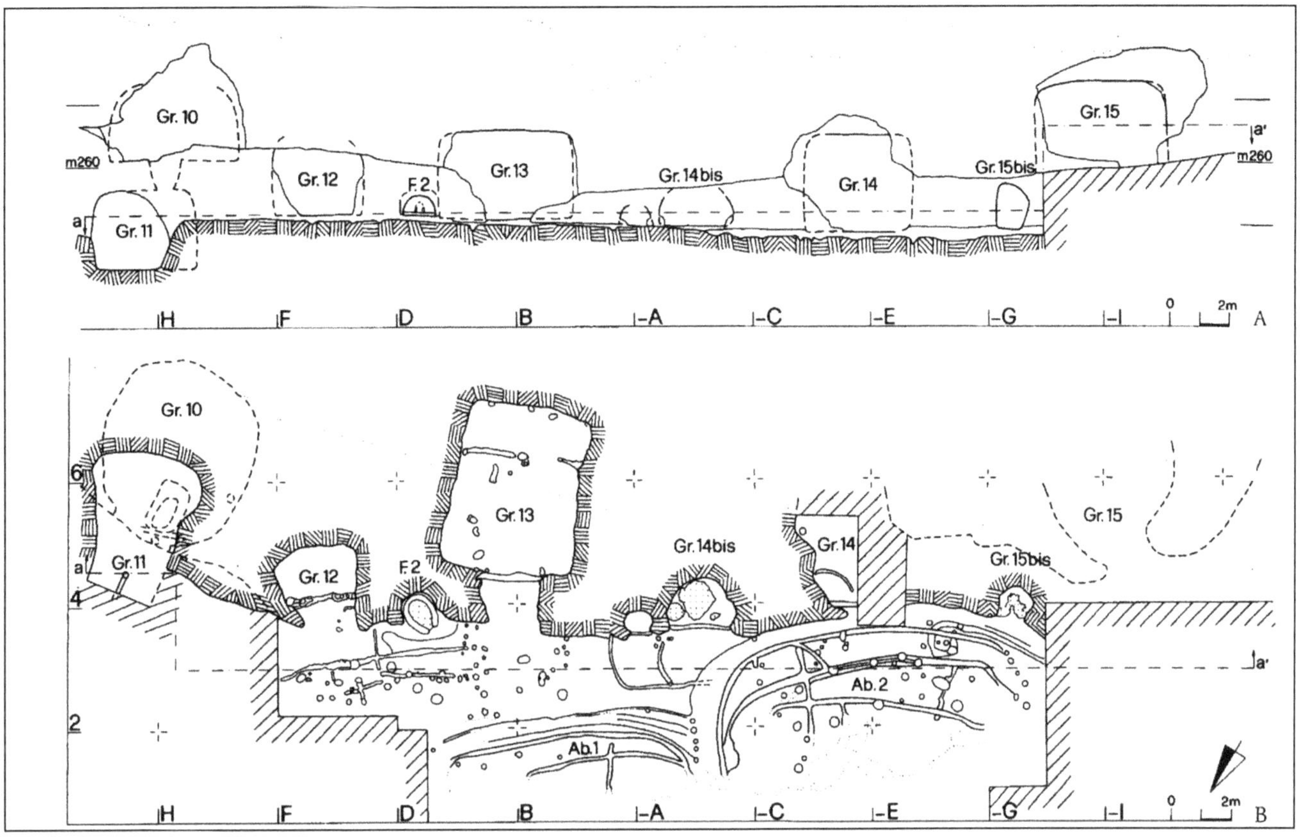

Fig. 3. Sector III. A: prospect; B: plan

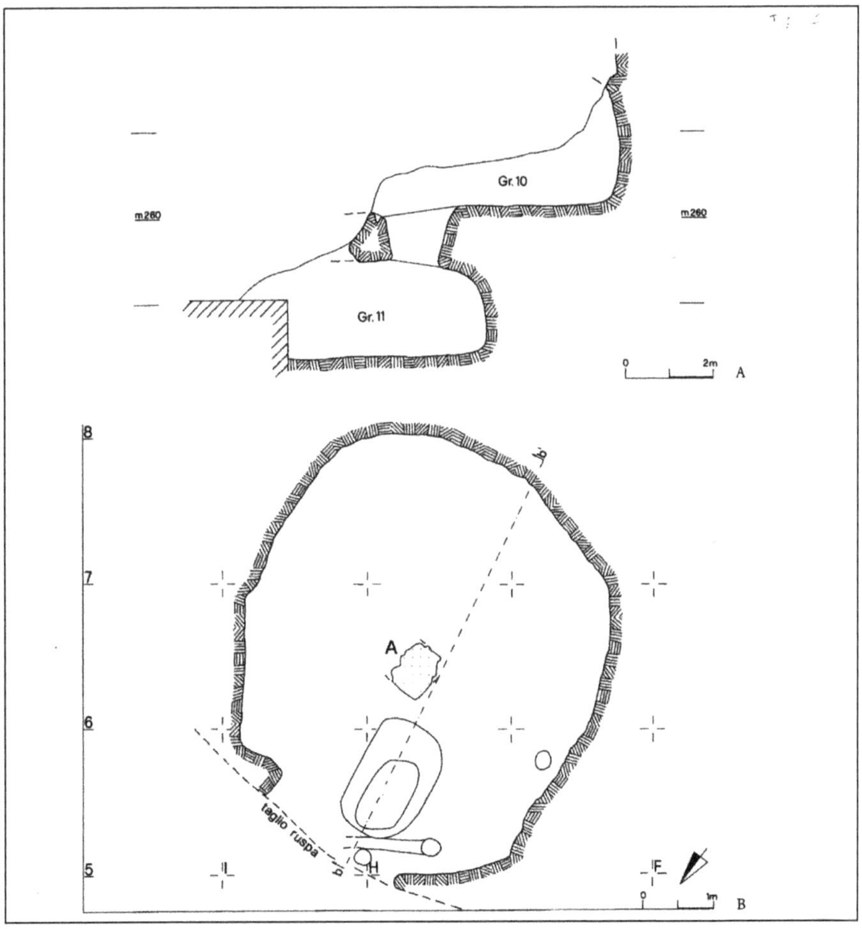

Fig. 4. Sector III. A: section of grotto 10, grotto 11 and the hole between them; B: plan of grotto 10 with the burnt clay layer of the hearth (A).

The real settlement, smaller than the Bronze Age one, is characterised by a series of caves artificially carved into the rock. The grottoes are different in dimension and some are simply an enlargement of the protohistoric ones. It is a real rupestrian village as many others in this region.

Wood dwellings with foundations of postholes and gullies carved in the rock, and others with foundations carved in the rock and small dry-stone walls are also present. The roof and the walls were made out of wood and on one side there is often a small room used as pigsty or sheepfold.

The medieval settlement was abandoned around the XIV c. A.D. and the site was never reoccupied again, showing a very different situation in comparison with many others centres of the region, which have been inhabited up to the present days. For this reason the archaeological layers were kept untouched and allow us an in-depth study of settlement models and style. (NNC)

Ritual Structures and Holy Fires

Sorgenti della Nova is one of the few protohistoric settlements where it has been possible to single out and recognise several actions related with rituals and cult within a chronological and typological common context. This gave the idea for a consideration about the "public" or "private" connotation that such acts could have within the settlement (Negroni Catacchio et al. 1989-90; Domanico 1995). In such sacred contexts the fire has very often an important role, directly as method for the consecration of the offering or indirectly as the role and symbolism inherent to the hearth, centre of the domestic life, but also of the social life with the banquet or the eating of shared meals.

In the III sector, placed halfway up the northern slope, a ritual complex that can be certainly considered one of the more interesting ritual complexes within the Protovillanovan Culture has been brought to light. (Fig. 3) It is composed by a large grotto (*grotto 10*), communicating through a hole in the floor with an underlying room (*grotto 11*). (Fig. 4)

The upper chamber is almost completely occupied by a large hearth carefully made up: three thick layers of carbonised materials and burnt clay are alternatively put on top of as many layers of preparation made out of gravel and crumbled pumice compacted. The layers of frequentation contained several newborn pigs' bones, suggesting important clues for the interpretation of the whole complex (De Grossi 1995 [1998]; Cardosa and Milanesio 1997 [2000]; De Grossi and Minniti [in print]).(Fig. 5)

The sacrifice of the baby pig, in fact, becomes, in the Classic period, characteristic of the cult of Demeter. The most popular feasts in the Goddess's honour were the *Thesmoforie*, celebrations exclusively reserved to women. During these feasts shared meals were eaten, according to special rituals, and new-born pigs were sacrificed and thrown in underground hollows.

Fig. 5. Sector III. Hypothetical reconstruction of grotto 10 with the ritual pyre.

Then the remains, collected and mixed together with seeds, propitiated the women's fruitfulness. The "prehistoric" character of these feasts is usually recorded in the ancient texts: Diodorus, referring to the Sicilian *Thesmoforie*, asserts that they are celebrations made "to imitate the ancient life", and the fact that the Athenian women during the feast had to sleep in leafy branch huts was explained with the same interpretation (Cardosa and Milanesio 1997 [2000]: 381). Therefore shared meals, the sacrifice of new-born pigs, the presence of an underground room connected with the ritual place are common characteristics both to the rituals related with the *grotto 10*, and with the thesmophoric cults in honour of Demeter. In this connection an information from Herodotus is particularly interesting (The History II, 171, 2-3). His opinion was that such rituals were taught to the women of the Pelasgians by the daughters of Danaos; as is well known, Greeks believed that the Pelasgians were a population living in central Italy before the Tyrrenoi/Etruscans, or sometime even confused with them (Cardosa and Milanesio 1997 [2000]: 383). Anyway their memory seems to be often and in more generic terms in association with Mediterranean usage and populations preceding the migration of the Dorians. Therefore it is likely that the similarities are connected with a cult of fertility, belonging to the Mediterranean tradition, common to Greece and Etruria, which after the Dorian invasion would have evolved in the cult of Demeter, Mother Goddess par excellence.

On the contrary, the ritual context brought to light in a preliminary dig in 1974, in an area strongly affected by the quarry, has characteristics related to the private life. Here, on the floor of an embedded structure almost entirely destroyed and so, therefore, difficult to understand, inside a large hearth a small hollow covered by a stone has been identified. Burnt broad beans and barley seeds had been placed inside the hearth (Domanico 1995: 344) evidently as consecration rite.

A similar situation, in a previous period of time, has been recognised in the Fiora valley in the Ponte S. Pietro Valle settlement. Here a small hollow had been dug in the clay floor of a hut and then filled with burnt corn, vetch and *althea*, a medicinal herb (Negroni Catacchio 1981: 133). Likely, in this case, it concerns rituals to consecrate the living space, through a symbolic offering of the earth's fruits "eaten" by the fire.

A very similar situation has probably been noticed also in the Protovillanovan layers on the Tarquinia's Civita, where in an area with post holes there has been identified a shallow pit lined with clay and burnt soil and partially covered with a stone (Bonghi Jovino 1986: 84). (MC)

The Service Structures and the Fire

Sorgenti della Nova stands out for the high degree of technical and functional specialization concerning the service structures connected with the use of fire. It is possible, in fact, to recognise a greater variety in comparison with the contemporary settlements of different structures for cooking (in particular *kilns* and *hearth-niche*) in addition to the constant presence of *hearths* for heating and lighting houses[2].

The frequency of some features allows to infer a developed ability of planning both according to the close relation between function and building techniques of the fire-structures and their position related with the spatial organisation of the whole settlement.

First of all, most of the dwellings had an inner hearth, which could consist of a simple layer of ashes and charcoal, sometimes surrounded by stones, or could be a real surface of burnt clay rather regular in shape (Miari 1995: 283-290).

Hearths made up of a simple layer of ashes and charcoal have generally narrow dimensions and have been found only inside houses. In fact they are placed in the grotto of the sector Va, in the grotto 13 of the sector III, in the dwelling 3 of the sector I and on the floor of the wide embedded structure in the sector Vb. The exiguous thickness and simplicity of manufacture allow us to assume that they could be fireplaces for inside lighting or simple cooking with the aid of clay stoves. The postholes found beside the hearth in the house 3 (sector I) and in the sector Vb also suggest the use of small spits (see also Sheffer 1981: 92-98).

Sometime this simple kind of fireplace without any preparation layer occurs also outside the dwellings, but in that case it presents a series of stones placed all around the layer of ashes and charcoal. Examples of this kind have been found in the sector Vb[3] and in the area in front of the grotto 13 in the III sector. The hearth laid on the re-occupation layers of the house in the sector Ve[4] and the fireplaces of the grotto 7 in the IV sector, and the one found at the entrance of the niche in the grotto 7 (Negroni Catacchio 1995: figs. 54-55) show the same protective pattern. In many cases they are in connection with a second utilization of some structures. (Fig. 8)

A clear intention of planning the inside living space can be seen in the more complex structure of the hearths with base made up of burnt clay laying directly on the rock. They were carefully and expressly made since the beginning phases of the dwellings. At *Sorgenti della Nova* this kind of hearths have a rectangular shape, no more than 1 m in length and they are usually placed in the middle of the inside room (e.g.: grotto 10, sector. III, grotto 7, sector IV and embedded dwelling, sector Ve). The fireplace (a plate of burnt clay) in the dwelling 6 - the hut with elliptical plane recently found in the IX sector, is the only one placed near the longer side wall (Dolfini 1996-98: 18).

In the grotto 10 the base of the hearth laid directly on a preparation layer made of cobbles and potsherds. (Fig. 6) When discovered, this grotto was thought to have a grotto/kitchen function (Miari 1995: 292-294), but after the analyses carried out on the bones, it has been found that it was related to ritual practices, connected to the roasting of pig meat. Therefore it could no longer be considered as a service structure, but had to be counted among the cult place of the settlement (see Cardosa, *supra*).

The same sacral significance has been conjectured for the grotto 7 in the IV sector, where the third burnt clay fireplace of the settlement has been found, right in the middle. The importance and the singularity of the destined use of these two structures, in two out of the three of which this particular kind of fireplace has been found, could not be of any significance related with the presence of choice criteria in planning their position and construction.

---

[2] The study of the service structures has been approached in detail in the general publication about the settlement of Sorgenti della Nova (Miari, M. 1995), to which it is possible to refer for more details. With particular attention to the reconstructive hypotheses: Negroni Catacchio, N., Miari, M. 1992 (1995).

[3] In the sector Vb two different phases of frequentation have been clearly recognised: the more ancient corresponding to the use of the embedded structure, and the following filling of it with a phase of frequentation as outdoor area equipped with various facilities - Miari, M. in Negroni Catacchio, N. (ed.) 1995, 181-196.

[4] See M. Cardosa in Negroni Catacchio, N. (ed.) 1995, 221-234 for a detailed analysis of the phases recognised inside the dwelling of the sector Ve.

Fig. 6. Sector III, grotto 10. A: the potsherds stratum under the burnt clay layer, B: the hearth.

The burnt clay base, with or without underlying preparation *stratum* of cobbles and potsherds, is a common functional element of every structure for cooking.

In accordance with a tendency of planning functional areas for different activities, that was certainly applied at Sorgenti della Nova, that has a pattern of well organised and advanced settlement, shared spaces with characteristc structures expressly devoted to specific activities were built for cooking, baking bread and flat bread and toasting seeds.

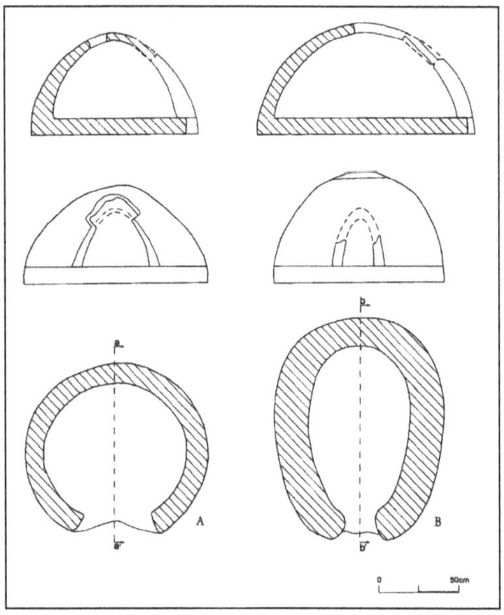

Fig. 7. The burnt clay ovens. A: sector III destroyed, section, prospect, plan; B: sector III, section, prospect, plan.

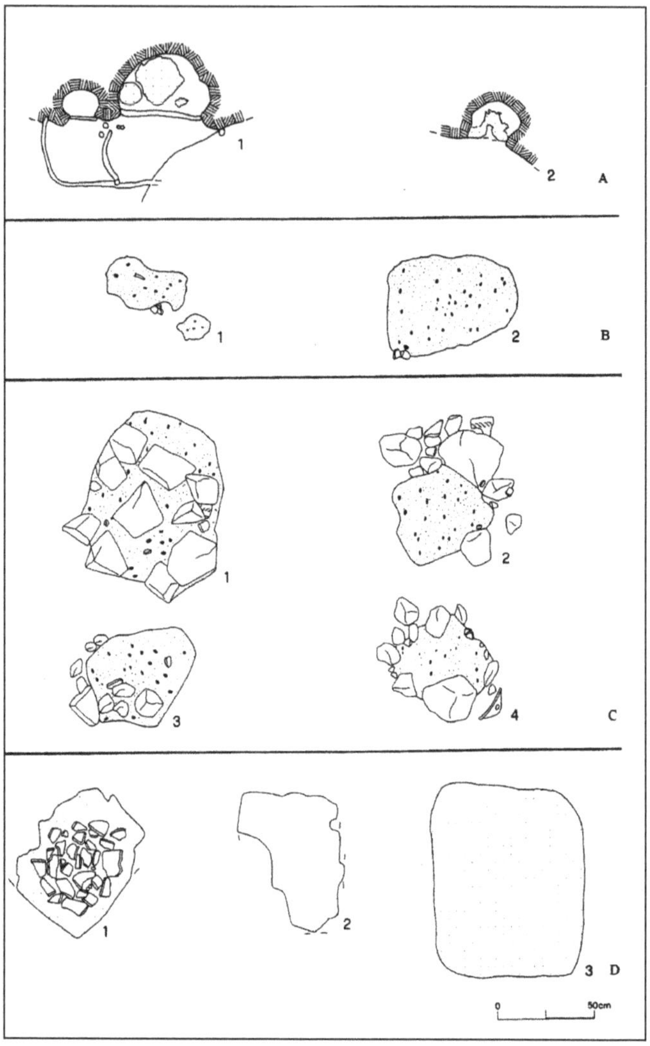

Fig. 8. The hearths. A: hearths in structure dug in the rock. Sector III, grotto 14bis (n.1); grotto 15bis (n.2). B: hearths without preparation base. Sector Va, grotto (n.1); Sector Vb, wide structure embedded in the rock (n.2). C: hearths with base surrounded by stones. Sector Vb, US 92 (n.1): Sector III, area in front of grotto 13 (n.2); Sector Ve, hut (n.3); Sector IV, grotto 7 (n.4). D: hearths with burnt clay base. Sector III, grotto 10 (n.1); Sector Ve, hut (n.2); Sector IV, grotto 7 (n.3).

Wide hearths made of burnt clay with preparation layer in cobbles and potsherds took up the whole area of the niches 14bis and 15bis, respectively 3 m^2 and 1 m^2 in width, dug in the tuff wall of the III sector (Miari 1995: fig. 114 A). Charcoal, ashes and animal bones were spread on the whole floor and the front areas. The lack of a roof must have had the purpose of letting the smoke go out of the room, while the wall at the back offered a natural shield from winds and helped to reach higher temperatures. The good functioning of this solution is confirmed by the frequency of this pattern even in the Etruscan settlement of Acquarossa (Sheffer 1981), where fragments of clay stoves testify the "domestic use" of this kind of structures.

Fig. 9. Sector III: hypothetical reconstruction of the area in front of the grottos. From left grotto 12, the burnt clay oven, grotto 13 and the area of the grotto 14bis.

Fireplaces for craftsmen's activity (furnaces for ceramic or smelting and casting metals) have not been found until now at Sorgenti della Nova[5].

Finally it is very likely that also the terracotta kilns were used for cooking and baking; in fact the analysis of the technical characteristics and the comparison with similar situations in prehistoric and protohistoric contexts allowed to exclude the possibility that they were furnaces (Miari 1995: 276-283). In particular, the relatively rather small inside width, the dome shape with a hole in the back and on the top for the escape of the smoke and the steam, are particularly related to cooking and baking purposes. The bases are usually elliptic in shape. (Fig 7) Also this type of structure has been in used in the historic period, as we can see at Tarquinia (Bonghi Jovino 1986), Acquarossa (Sheffer 1981; 1987) and Cures Sabini (Guidi-Alfonsetti 1988).

However, the settlement of *Sorgenti della Nova* differs from the other known examples for the number of kilns and their position. In fact they were always placed within complex and well-organised residential areas, with dwellings with elliptical plane, service structures, grotto and room carved in the rock. Three new kilns have been discovered in the last few years in the IX sector (see Guidetti *infra*), which can be added to the already known two belonging to the sectors III and III *destroyed* (Miari 1995). It is interesting to notice that the kilns 4 and 5, unfortunately in very bad condition, were placed inside a recess in the back wall of the terrace, therefore reflecting the same pattern found in other areas of the settlement. On the contrary, the kiln 3 was built inside the house 5. This unusual position, along the northern wall of the hut recalls the situation identified in the house 6, in the same sector (Dolfini, A. 1996-98: 18). Therefore it can not be excluded that these different patterns are related to the history of this sector. (Fig. 9)

In conclusion, the numerous service structures of the protohistoric settlement of *Sorgenti della Nova* present constant characteristics, therefore it is possible to draw some common patterns.

First, the close relation among the technical and the primary purpose characteristics can be considered proved with the presence and the strengthening, just in this period, of highly specialised typologies for cooking and baking.

Second, the location of fireplaces with base of burnt clay – kilns, hearth-niche and hearths - implies a precise planning both while building the dwellings and when more important and incisive works for the realisation of the terraces on the tuff hill holding the protohistoric village were needed.

The new excavations, inside the IX sector, are another confirmation of the results, even with some new suggestions.(Fig. 11) In particular, considering the kiln which has come to light inside a house with elliptical plane, the problem of the planning of the common spaces for the everyday activities and all the structures related to them (kiln and heart-niches) should be studied in more depth in relation with the structure utilised by a single living unit. (Fig. 10)

[5] On the contrary, in the near settlement of Scarceta, have been found important traces and structures connected with the processing of metals - Poggiani Keller, R.1991 (1993).

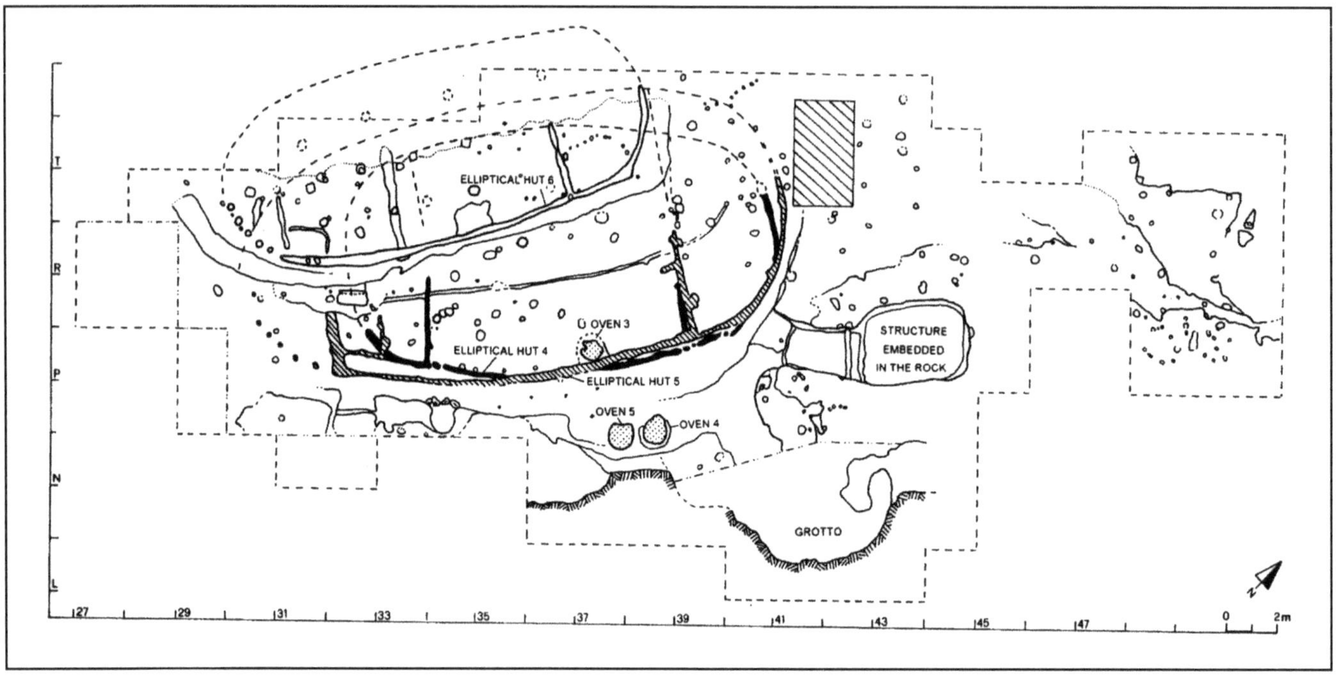

Fig.10. Sector IX: plan.

Fig. 11. Sector IX: area of oven 4 and 5.

Finally the areas, which have been excavated in the last 25 years, are essentially devoted to houses and domestic activities, and this fact does not conflict with the lack, inside the settlement, of traces of metallurgical activities or ceramic furnaces. Unfortunately, the incompleteness of data does not allow to make any hypothesis about the function of the large burning plane of burnt clay, surrounded by a layer of soil, cobbles and potsherds, found in the northern slope, during the first excavation, in 1974[6]. The fireplace, remarkable in width and thickness – at least 5.30 m in length and more than 40 cm thick – is very different from all the other structures connected with the use of fire and could be related to other productive or craft activities. (MM)

The Kilns of the IX Sector

In the last years the excavations carried out in the IX sector, placed on the northern slope of the crag, brought to light three new terracotta kilns, two of which (kiln 4 and 5) are located in a deep niche dug in back side of the upper terrace, whereas the third is placed in one of the three dwellings which succeeded in the life of this area. The two outdoor structures are in a very bad state of repair. In fact of the *kiln 4*, situated in the northern part of the niche, only the bottom and the beginning of the walls are preserved; of the other one (*kiln 5*), right in the middle of the recess, only the bottom plate is left, but wide fragments of baked clay were still in place when it was brought to light. Probably they were part of the dome, pulled apart when it was abandoned. They have not been excavated but they have been restored in common accord with the Superintendence for Southern Etruria, depending on a general plan of revaluation of the whole area.

Anyway the investigation carried out in the neighbouring layers allowed to assume that they were in use simultaneously, at least for some time, and they belong to the last living phase of the sector. This part of the settlement (sector IX) stretches in two artificial terraces of more than 20m in length, carved in two successive chronological phases. On the upper part the traces of two dwellings, following each other in the same place, have been recognised

[6] In the same zone there was a second hearth, similar for building techniques, but smaller in dimension – Negroni Catacchio, N. (ed.) 1995, fig.58.

(*house 4* and *5*) belonging to the first occupation of the area[7]. Then the first terrace was cleared out and probably transformed in a service area for food processing and conservation. At this stage of the settlement life, the niche was probably carved and the two kilns were built. As far as it is possible to judge by the first analyses, they were realised with the same technique and materials as the former kiln placed inside the house 5.

Actually the *kiln 3* is the only one in the whole settlement realised inside a dwelling. For this reason and to make sure that it was really connected with the foundations of the *house 5* and not built after its demolition, it was decided to dig it out entirely.

It was covered by a wide area of burnt soil and clay, only the bottom part of the dome-shaped walls were still in placed, but the plate was perfectly preserved. Therefore the removal of the layers was carried out with extreme care and caution, just to comprehend exactly the building methods and the stratigraphic relation with the foundation of the two dwellings, to understand not only if it was placed inside one of them, but also inside which one.

At a first glance the cooking plate seemed to be formed by a unique layer directly covering the preparation *stratum*, but then it was found that it was divided in two different beds: one on the top red in colour and an other one underlying and black. The preparation bed was composed by a thick unit packed with cobbles and potsherds laying directly on the rock, in a shallow hollow. Its maximum diameter (102 cm) can be compared to the dimension of the kilns recovered by the other sectors of the settlement (III and III *destroyed*) and allows us to assume that the general shape must also have been quite similar (See Miari *supra*).

After the realisation of the shallow cavity a circular platform made of river cobbles and potsherds (even some wide stove fragments) was placed and stuck together with sandy soil, then a narrow kerb of clay was put all around the plate in order to curb and bound the fragments. Above the preparation bed a first layer of clay was spread and after this operation a fire was lighted in order to seal the underlying bed. Eventually the cooking plate was built: a second layer of red clay placed on top of the burnt black clay.

Having built the base, the dome-shape walls, preserved to a maximum height of 20 cm and a width of 19 cm, were raised. The walls presented a hole in the backside still closed with a small stone, probably an inlet placed just in front of the door that must have been built looking out on the valley.

The excavation permitted to make sure that the kiln was really built inside the *house 5*, near the backside wall.

This kiln of *Sorgenti della Nova* is comparable to one similar in shape, construction material and methods and position inside the dwelling belonging to the house of Area III - Trench 31 (Final Bronze Age-Early Iron Age) at Morgantina (Leighton 1992). Here, however, the kiln was protected by rock to a depth of about 20 cm, the dwelling having an embedded base. On the contrary, at *Sorgenti della Nova*, it was placed close to the wall made out of wood poles and leafy branches, but covered with clay. Inside the foundation gullies in fact fragments of baked clay with reed impressions have been found. This finding - rare at least until now at *Sorgenti della Nova* - proves that the internal wall was lined with clay. It cannot be a coincidence that the only dwelling with an inside kiln is also the only one that certainly had a lining to protect the leafy branch walls (Cardosa 1996-98: 17). (LG)

[7] For more detailed information see *Notiziario 1996-1998*.

References

BONGHI JOVINO, M.
1986 Gli scavi dell'Università degli Studi di Milano (campagne 1982-1985 sul Pian di Civita). In: Bonghi Jovino, M. (ed.): *Gli Etruschi di Tarquinia*. Modena, 81-140.

CARDOSA, M.
1996-98 L'abitazione 5. In Negroni Catacchio, N. (ed.): *Preistoria e Protostoria in Etruria. Notiziario 1996-98*: 12-14.

CARDOSA, M. and M. MILANESIO
1997 (2000) La grotta 10 di Sorgenti della Nova (Farnese – VT): affinità di rituali tra Etruria protostorica e Grecia postmicenea, In Negroni Catacchio, N., (ed.): *Preistoria e Protostoria in Etruria. Atti del Quarto Incontro di Studi*. Milano: Manciano-Montalto di Castro-Valentano. 2000, 379-390.

DE GROSSI, J.
1995 (1998) Analisi dei resti faunistici da alcune strutture di Sorgenti della Nova. In Negroni Catacchio, N. (ed.): *Preistoria e Protostoria in Etruria. Atti del Terzo Incontro di Studi*. Manciano-Farnese 1995. Firenze 1998, 169-180.

DE GROSSI, J. and MINNITI, C. (in print)
Testimonianze di pratiche cultuali nella Grotta 10 di Sorgenti della Nova: recenti analisi sul materiale osteologico. In Negroni Catacchio, N. (ed.): *Preistoria e Protostoria in Etruria. Atti del Quinto Incontro di Studi*.

DOLFINI, A.
1996-98 L'abitazione 6. In Negroni Catacchio, N. (ed.): *Preistoria e Protostoria in Etruria. Notiziario 1996-98*: 17-19.

DOMANICO, L.
1995 Aree sacre e indizi di culto. In Negroni Catacchio, N. (ed.): *Sorgenti della Nova*. Firenze, 339-346.

GUIDETTI, L.
1996-98 Il forno 3 dell'abitazione 5. In Negroni Catacchio, N. (ed.): *Preistoria e Protostoria in Etruria. Notiziario 1996-98*: 14-17.

GUIDI, A. and ALFONSETTI, D.
1988 Cures Sabini. In: AA.VV., *Problematiche di scavo delle strutture abitative dell'età del Ferro*.

Quaderni della Soprintendenza Archeologica del Lazio 1: 41-52.

LEIGHTON, R.
1992 The Protohistoric Settlement on the Cittadella. *Morgantina Studies* IV, Princeton.

MIARI, M.
1995 Tipologia delle strutture di servizio. In Negroni Catacchio, N. (ed.): *Sorgenti della Nova*. Firenze, 275-300.

NEGRONI CATACCHIO, N. (ed.)
1981 *Sorgenti della Nova. Una comunità preistorica e il suo territorio nell'Etruria meridionale*. Catalogo della mostra, Roma: CNR.
1995 *Sorgenti della Nova. L'abitato del Bronzo Finale*, "Origines", edited by Istituto Italiano di Preistoria e Protostoria, Firenze.
2000 *Notiziario 1996-1998 Preistoria e Protostoria in Etruria. Notiziario*. Milano.

NEGRONI CATACCHIO, N. and L. DOMANICO (in print) L'abitato protourbano di *Sorgenti della Nova* (VT): dagli spazi dell'abitare all'organizzazione sociale. In *From huts to house. Transformation of Ancient Societies*, Proceedings of the International Seminar edited by the Norwegian Institute at Rome and the Swedish Institute of Classical Studies at Rome. Rome 1997.

NEGRONI CATACCHIO N., L. DOMANICO and M. MIARI
1989-90 Offerte votive in grotta e in abitato nelle valli del Fiora e dell'Albegna nel corso dell'età del bronzo: indizi e proposte interpretative. Proceedings of the Conference: *ANATHEMA: Regime delle offerte e vita dei santuari nel Mediterraneo Antico*, Roma 15-18 giugno 1989, in *Scienze dell'Antichità* 3-4: 579-598.

NEGRONI CATACCHIO, N. and M. MIARI
1992 (1995) La ricostruzione grafica delle strutture protostoriche in Etruria meridionale: problemi e metodi. In: *The Fifth Conference Of Italian Archaeology. Settlement and Economy, 1500 B.C. - A.D. 1500.* Oxford 1992. Oxbow Monograph 41 1995, 521-530.

POGGIANI KELLER, R.
1991 (1993) L'abitato di Scarceta (Manciano, GR) tra Bronzo Medio e Bronzo Finale. In N. Negroni (ed.): *Preistoria e Protostoria in Etruria. Atti del primo incontro di studi*, Manciano-Farnese 1991, Milano 1993, 301-310.

SHEFFER, C.
1981 Acquarossa, Vol.II,1: Cooking and Cooking stands in Italy 14000-4000 B.C.. *Acta Inst.Rom.R.Sueciae*, 4,38,2,1.
1987 Forni e fornelli etruschi di età arcaica. In: AA.VV., *L'alimentazione nel mondo antico. Gli Etruschi*. Roma, 97-105.

MAN - FIRE - STONE – FIRE-CRACKED STONES AS A CULTURAL EXPRESSION

Phyllis Anderson Ambrosiani

Introduction

Knowledge of sites containing fire-cracked stones in Sweden has changed dramatically during the course of the 1900s. The few examples known in the early years of the century are today only a fraction of the thousands of fire-cracked stone monuments and sites documented in many parts of the country. (Fig. 1) Twentieth century Bronze Age research has maintained generally the view of this material as the waste of habitation sites, and fire-cracked stones have been tied to a context where settlement, hearths and food preparation have given this material its subsistential meaning and chronological framework.

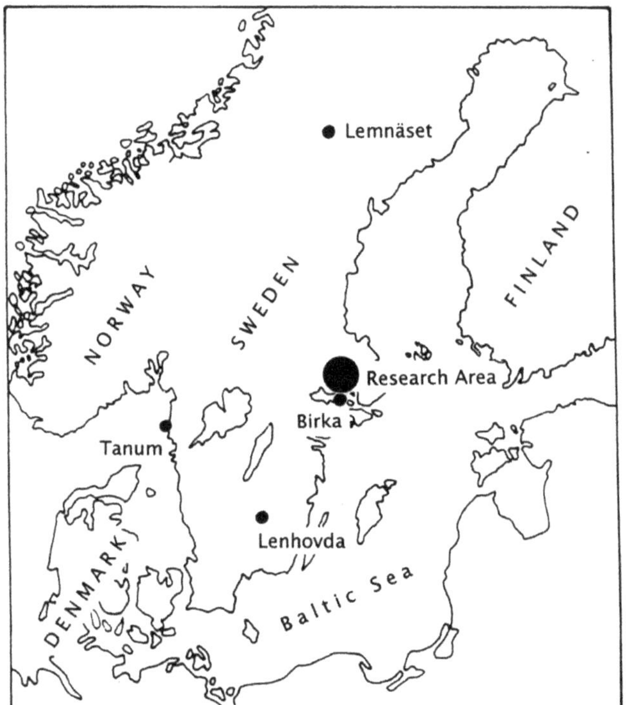

Fig. 1

Since the 1960s settlement archaeological studies have tended to use the ancient monument category "heap of fire-cracked stone" as the characteristic physical marker for Bronze Age habitation sites, pointing out specific locations of settlement in the modern landscape. The high degree of generalization in these studies has left much of the detailed information about this material unused and unexplored. The general assumption that fire-cracked stones is waste cleaned from the hearths of settlement sites has locked this complex and varied stone material to a singular context, causing it to appear unstructured and apparently trivial. It has been difficult to reach greater knowledge about Bronze Age society using fire-cracked stones and fire-cracked stone remains. Fundamental questions and phenomenon characteristic of this comprehensive material have remained unanswered and difficult to approach in traditional research. This situation poses a particular problem, among other things, because the fire-cracked stone material as it is recognized today constitutes a sizeable part of the total archaeological material and the most characteristic remains of some areas. This encourages further research and suggests also that renewed discussion and new perspectives are needed in approaching this problem. But if fire-cracked stones are not settlement waste, how can this material be understood and what potential information does it hold about past cultures?

Background

Today all too little is known about many of the "classical" problems often related to these remains. These concern mainly aspects of their 1) distribution – the reasons behind the differences in the settlement pattern anticipated in the fire-cracked stone remains and the settlement pattern understood to be reflected by, for instance, graves (for example Kjellén and Hyenstrand 1977; Larsson 1986: 150ff; Jensen 1989: 139; Johansen 1993: 101); 2) status – the relation between heaps of fire-cracked stones and traditional grave monuments, for example, the reason why fire-cracked stones are found in traditional grave monuments and why burials are found in heaps of fire-cracked stone (for example Nylén 1958; Rentzhog 1967: 70f; Hyenstrand 1968: 75f; and more recently, for example, Kaliff 1997), 3) relation to other traditional monuments – for example, why do finds of fire-cracked stone remains appear near, and sometimes overlie, rock carvings (for example, Bellander 1938: 93f; Kjéllen 1939: 82 and 94f; Rentzhog 1967: 70f), and 4) variation in deposition – why, for instance, are fire-cracked stones found collected together in piles or heaps as well as spread out in flat layers (for example Gustawsson 1949; Rentzhog 1967: 74ff).

But more direct questions should also be asked of this material as well. Why do the number and size of heaps of fire-cracked stones appear to vary between obviously contemporary sites, and what is the reason for the notable differences in amount of fire-cracked stones between sites? What do quantity and volume represent (Anderson [Ambrosiani] 1984 and 1986)? Why do some areas with traditional Bronze Age remains, as graves and rock carvings, lack heaps of fire-cracked stone? And, indeed, if fire-cracked stones are to be plausibly explained as waste from food preparation, why don't all Bronze Age settlements have such heaps (Anderson Ambrosiani 2001a)?

The consequence of a habitation-waste and food-related line of reasoning would suggest that a central problem should concern food preparation and differences in culinary traditions and how these may be reflected in fire-cracked stone remains. Do the notable variations in the fire-cracked stone material correspond to contemporary variations in this tradition? And, of course, then why do the settlement and habitation sites of earlier and later, Stone Age and Iron Age, cultures lack heaps of this stone material? What are the causes or conditions behind these differences? These questions, in my experience, have not been discussed by the research that has so unanimously supported the interpretation of this material as this kind of waste. As heaps of settlement waste, this material at least should be expected to throw light on the strategies of waste production, accumulation, disposal and deposition practiced by Bronze Age societies. This, however, has not been established.

Studies of the 1990s have shown renewed interest in explaining the existence of this stone material from the perspective of ritual, cult and religion, retaining, to a greater or lesser degree, the interpretation of settlement waste but supplementing it with explanations founded in religion and, in some cases, also ideas founded in popular folk belief (for example, Lundqvist 1991; Ullén 1994; Olausson 1995; Kaliff 1997; Widholm 1998; Karlenby 1999; Runcis 1999). This has resulted in the not so small task of having to determine in which cases fire-cracked stones and fire-cracked stone remains belong to the different spheres of "profane" and "sacral" respectively, or to which degree they are to be divided between these two contexts. Is the fire-cracked stone material in the context of graves, rock-carving locations, bronze working sites, or even in itself, "profane", "sacral", neither, or both perhaps?

Earlier Research, Terminology and Definition

Our present picture of this stone material and current ideas about what it represents have developed from a combination of experience gained through archaeological fieldwork, of information from its identification and documentation, and from our own constructed definitions and terminology. In other words, it is the accumulated and collective result of more than a century of work by professional archaeologists, students and initiated amateurs. Some of the trends that characterize this development (Anderson Ambrosiani 2001a: 115-119) are discussed below.

Some of the earliest interpretations of archaeological fire-cracked stone in Swedish contexts build on contemporary ethnographic traditions from other parts of the world (Almgren 1912) and comparative West European examples from archaeological sites outside the main Scandinavian Peninsula, from Denmark and Germany (Sarauw and Alin 1923). In 20th century Bronze Age research fire-cracked stones have partly been seen as the result of burning (for example Almgren 1912; Sarauw and Alin 1923: 264; Stenberger 1964: 289; Janzon 1991: 205), but also as the result of cooling, as a rule, through contact with a liquid, often water (for example Gustawsson 1949; Stenberger 1964: 289; Meschke 1968: Chapter 3.4.4.). The dominating idea that fire-cracked stones are settlement waste seems to be founded in two very similar chains of association. One builds on the connection between hearths, fire-cracked stones, and food preparation using cooking pits (for example Almgren 1912) and the other on the combination of hearths, fire-cracked stones, and food preparation by boiling in containers using heated stones (for example Gustawsson 1949). Nevertheless, already in the first half of the 1900s fire-cracked stones also had been observed in the contexts of graves, rock carvings and bronze working.

Through the years, a number of terms, mainly focusing on the physical state and appearance of this material, have developed. Their meaning has varied during this development, but contemporary variations in their use appear also. Most notable are two terms; *skörbränd sten*, translated brittle-burnt stone, and *skärvsten*, literally, split stone, although it is often translated as fire-cracked stone. *Skörbränd sten* appears earlier in the archaeological literature than *skärvsten*, which became more common around the mid-1900s at a time when these two terms are used interchangeably. But in the mid-1900s, yet another term, of a different nature, is introduced for this same material, namely, *koksten*, translated cooking or boiling stones, a term which also explicitly expresses function (Gustawsson 1949). This term is used often, but not exclusively, for the fire-cracked stone material in north Swedish contexts.

This suggests that perhaps some of the difficult questions that seem inherent to this material may have developed as a result of the way in which it has been approached in earlier research. In light of this, some questions, regarding the kind of information it may yield about prehistoric cultures and what earlier research has expected of it, may be interesting to reconsider.

For example, the strong tie between fire-cracked stones and cooking hearths, which has been at the center of the interpretation of this material in prehistoric contexts, is founded in the idea that all fire-cracked stone has had one single common origin, in hearths, regardless of whether it has been explained as the result of heating or cooling. A key question, then, is what happens in the case that fire-cracked stones prove to have other origins as well?

Another example is the development of ideas and attitudes about fire-cracked stone in the 20th century, which may be characterized as cumulative, but simultaneously also oversimplified, complicated and complex. This development appears to have produced a homogenized and idealized picture, a stereotyped cliché, which is widely spread and widely accepted today – for what fire-cracked stones *are*, is, of course, everyday knowledge for archaeologists. The core of this idealized picture is composed of a few common characteristics which appear and reappear in discussions and interpretations. But to this core are an additional number of various, less well-established and less well-defined, characteristics. Some of the problems in interpreting this material may have been caused by difficulties in using these characteristics in identification, and successfully distinguishing between similarities and differences in this mass material. A key question is, then, whether the fire-cracked stones found in the heaps in one location, for example, are actually comparable to the fire-cracked stones in heaps from another location? What happens if they are not? How and why do they differ?

While the definitions of fire-cracked stone and fire-cracked stone remains have concentrated on describing the role of this material and where and in what context it should appear, the non-verbal information in these idealized pictures also have regulated the conditions for its existence and explanation in archaeological contexts. Conclusions based on this are interesting, because they may be used to create a different point of departure for research on this material, approaching it from a perspective not exclusively tied to an ideology of waste.

Changing Research Perspectives

An important pre-understanding of the fire-cracked stone material lies in the conditions for its appearance. Fire-cracked stones are the result of human activity, whether they are found in connection with graves, bronze working, settlement or ritual. For example, naturally split stone, through natural catastrophes or weathering, falls outside the archaeological definition and understanding of this material. The general Swedish term for this material, *skärvsten*, does not explicitly voice the central cause for its creation, which is exposure to fire. The English equivalent, fire-cracked stone, does. By definition, sharp-edged stones produced by mechanical crushing, splitting or formation fall also outside the archaeological definition and understanding of this material.

Thus it may be concluded that, although fire-cracked stones are found in connection with many kinds of remains and in many various contexts in prehistoric (Fig. 2), historic (Fig. 3) and modern times, and appear in varying density in different parts of the country, they seem to share a mutual condition of origin. And fire-cracked stones may be understood as the physical manifestation of man's use of fire – where similarities and variations in appearance, find context, quantity, size and distribution are all important evidence of ideas about what fire and stone could, and should, be used for – and sources of information about the technical knowledge of the societies which used it (Anderson [Ambrosiani] 1992: 18 and 1995).

Fig. 2

Fig. 3

Therefore, man – fire – stone lies at the heart of the origin of this material, but the meaning of this combination in this paper's title may be understood in several ways. It expresses the combination of *human activity* and *energy* in interaction with *raw materials* and *resources* (EAA catalogue, 2001). But in a broader sense it defines the foundation of a new approach or *research perspective*. It appears to embody also an *idea* so basic to this archaeological source material, independent of its location, function or chronological framework, that it may be understood as expressing one central *concept* of the fire-cracked stone material as such. Together, man, fire and stone constitute also the fundamental *precondition* for its appearance in archaeological contexts; recognizing from an archaeological standpoint an inherent assumption of its meaning, namely, that fire-cracked stones are the result of human activity, not natural causes, making them a kind of "artifact" and defining at the same time the fundamental *technology* behind its formation – the human use of fire (Anderson Ambrosiani 1992 and 2001).

Identifying Fire-cracked Stones and Preliminary Results

The majority of fire-cracked stone remains in Sweden have been registered through the systematic ancient monument survey by the National Heritage Board (formerly, The Central Board of National Antiquities) in Stockholm, where the identification of fire-cracked stones was the main criteria for its registration as, for example, a heap of fire-cracked stone (*skärvstenshög*), or as an embankment with fire-cracked stones (*skärvstensvall*). Although the method of identification itself has not been recorded in the surveyor's handbook (Selinge 1969), two generally recognized characteristics of these stones were: 1) fire-cracked stones emit a soft and low, "short" thud or thump when they are hit against another stone, and 2) fire-cracked stones crack apart easily when struck against another stone. To my knowledge, these criteria are still today generally accepted as identifying features of fire-cracked stone.

However, the important point here is that these two criteria are not characteristic for all fire-cracked stone. In the above discussion and in earlier archaeological literature, among others, it is clear that there are a variety of different characteristics of this stone material, and these vary. In field survey, the criteria for identifying fire-cracked stone is characteristic for brittle-burnt stone or *skörbränd sten*, but not for *skärvsten* generally, which is the less hard burnt, split stone, more angular in form, sharp-edged and firmer in structure. The written documentation of ancient monuments and sites shows that the distinction between fire-cracked stone and natural stone was not always easy, and not easily understood. The problem of how hard-burnt the fire-cracked stones of different monuments were, was not, and has not been, a central question for their registration by the survey. This is important as it implies a biased registration for remains in favour of the brittle-burnt stone material. Because of this the distribution picture based on survey results may not be completely representative, showing instead an "idealized" distribution of heaps. Along with these are other stone variations only partly, or not at all, included in the register.

Control visits in 1998 and 1999 to a number of selected sites throughout my research area in SW Uppland, an area whose ancient monument survey I also participated in 1980, verified these uncertainties and inconsistencies in the registration of fire-cracked stone remains of the area. Remains with hard-burnt stone have been registered, and to a greater extent than other remains with the sharp-edged "firmer" fire-cracked stone. In some cases, remains with naturally geometrically - shaped stone also have been registered as heaps of fire-cracked stone.

On many sites erratic boulders, still *in situ*, and in some cases also bare rock surfaces with clear traces of burning are evident. These were not registered by the 1980 survey and, to my knowledge, traces of fire-affected rock of this kind have not been recognized in this area earlier. Fieldwork in connection with my current doctoral research on these questions has shown a correlation between this evidence and ancient monuments and sites from the Bronze Age and Early Iron Age but not with, for example, Late Iron Age cemeteries in the area. (Fig. 4) Among other things, the distribution and character of damage evident on these boulders, and the lack of fire-damage to other nearby remains and landscape features speaks against the idea that this has been caused by naturally set fires.

Fig. 4

Further discussion of this question lies beyond the limits of the present paper, but there also seems to be a connection between the appearance of fired boulders and areas that have been cleared of stone within these sites. Some stone-cleared areas, understood as characteristic evidence of Bronze Age activity, were registered by the 1980 survey. But a connection between these and fire-cracked stone remains, to my knowledge, has not been recognized earlier in the archaeological literature. This suggests that, at least for some of these sites, fire-cracked stones appearing in connection with clearance cairns and areas with traces of early agriculture may find their explanation in the process of clearing areas from boulders and rock using fire. Fire-cracked stones in these contexts earlier have been generally assumed as subsequent cultivation of abandoned settlement sites. The low number of registered areas with clearance cairns particularly in the area of SW Uppland (Hermodsson 1996) may also reflect an aspect of the problem of identification and how this material has been perceived (Anderson Ambrosiani, forthcoming).

Fig. 5

Concluding Discussion

Remains with fire-cracked stone are a broad source of new knowledge about prehistoric life and technology, ideas and traditions. Although Bronze Age research has traditionally viewed fire-cracked stone remains as a material characteristic of that period particularly, and subsequently to a certain degree also for the earliest Iron Age, fire-cracked stones as such exist in contexts far beyond these narrow chronological limitations, appearing instead over a period of many thousands of years (Anderson Ambrosiani 1995). The nearly total fixation in the past c. 50 years on the explanation of fire-cracked stones as settlement waste and the stereotyped picture that has been built up of this material in the course of the 20th century, have left much of the potential of this material as a source of knowledge about past cultures unrecognized. Because fire-cracked stone remains often have been analyzed from similar prerequisites as other traditional archaeological remains as prehistoric graves, many aspects of the problems related to this material have been left unexplored (Anderson Ambrosiani 2001). (Fig. 5)

In my research in southwest Uppland, the remains I have studied, among other things, have shown that hearths are not the single origin for all fire-cracked stone material. The picture of these remains, from the ancient monument survey of 1980, is not fully representative for the distribution of the varied fire-cracked stone material of the area. Evidence that fire has had a broader use than imagined earlier, is apparent in purposely burned boulders and bare rock surfaces from many sites. Evidence points to the activity of purposely exposing boulders and bedrock outcrops to fire in order to produce suitable stone material for the construction of graves, both cairns and stone settings, and areas of the terrain have been cleared from rock using fire. And on this basis, there appears to be two main kinds of fire-cracked stone, one which is the heavily brittle-burnt fire-cracked stone and the other which is the sharp-edged, but less hard-burnt, fire-cracked stone.

The research perspective expressed in man-fire-stone makes a strong statement about this material. Through this, another status of the fire-cracked stones than waste may be identified, allowing it to be understood as a cultural expression within a broader context of material culture and society. The fire-cracked stone material and associated remains, using different approaches from a variety of areas, appears to hold rich research potential as a source of information about past cultures, traditions, social development and change that today is essentially untouched. One key area of this research is, of course, the complex of questions related to the mentality and technology of fire - the ideas, concepts and uses of fire - and stone by prehistoric society. From this perspective, fire-cracked stones and fire-cracked stone remains may perhaps not only be understood in the sense of a cultural expression, but it may offer a new approach to looking at expressions of culture and to identifying social and cultural differences (Anderson Ambrosiani, forthcoming).

References

ALMGREN, O.

1912 En uppländsk bronsåldershydda. *Fornvännen*, 7: 32-51.

ANDERSON (AMBROSIANI), P.

1984 Frequency diagrams and size-volume calculations. In Å. Hyenstrand (ed.) *Bronsåldersforskning –*

kring aktuella projekt. Arkeologiska rapporter och meddelanden från institutionen för arkeologi vid Stockholms Universitet, 16: 61-75.

1986 Frequency diagrams and size-volume calculations. A wider perspective. Det 4. Nordiske bronsealder-symposium på Isegran 1984. Oslo: *Varia* 12: 124-132.

1992 Pyroliths. In Å. Hyenstrand and A. Thomasdotter Jakobsson (eds.), *Forntid i förändring. Aktuell Arkeologi III*: 15-21. Stockholm Archaeological Reports 25.

ANDERSON AMBROSIANI, P.

1995 Skärvsten – ett mångtusenårigt fenomen. *Populär arkeologi* 13 (4): 30-31. Luleå.

2001a Människa – eld – sten. In: Bolin, H., A. Kaliff and T. Zachrisson (eds.) *Mellan sten och brons. Uppdragsarkeologi och forskning kring senneolitikum och bronsålder*, OPIA 27 & Stockholm Archaeological Reports No. 39: 114-131.

2001b Man-Fire-Stone. Aspects on fire and society in Bronze Age Sweden, *7th Annual Meeting, EAA, Final Programme and Abstracts Esslingen am Neckar, Germany*, Stuttgart: Verlagsburo Wais and Partner, p. 92.

BELLANDER, E.

1938 Bålrösen – Offerrösen. *Kulturhistoriska Studier tillägnade Nils Åberg 24.7.1938*, Stockholm: 91-100.

GUSTAWSSON, K.-A.

1949 Kokstenshögar. *Fornvännen*, 44: 152-165.

HERMODSSON, Ö.

1996 *Fornlämningar i Uppsala län. En nyckel till fornminnesregistret.* Stockholm: Riksantikvarieämbetet.

HYENSTRAND, Å.

1968 Skärvstenshögar och bronsåldersmiljöer. *Tor* XII: 61-80.

JANZON, G.

1991 Skärvsten. In: *Sten- och bronsålderns ABC*, 205. Historia i fickformat. Statens Historiska Museum. Stockholm.

JENSEN, R.

1989 Bronze Age settlement patterns in the Mälaren basin – ecological and social relationship. In Nordström H-Å. and A. Knape (eds.) *Bronze Age Studies, Transactions of the British-Scandinavian Colloquium in Stockholm, May 10-11, 1985*, Stockholm: Statens Historiska Museum Studies 6: 133-150.

JOHANSEN, B.

1993 Skärvstenshögar och sörmländsk bronsålder. In Andræ T. et.al. (eds.) *Arkeologi i Sverige*. Ny följd 2: 99-118. Stockholm: Riksantikvarieämbetet.

KALIFF, A.

1997 *Grav och kultplats. Eskatologiska föreställningar under yngre bronsålder och äldre järnålder i Östergötland.* Diss. Aun 24. Uppsala.

KARLENBY, L.

1999 Deposition i skärvstenshögar. En studie kring avfallshantering och religion under äldre och yngre bronsåldern i sydvästra Uppland. In Olausson, M. (ed.) *Spiralens öga. Tjugo artiklar kring aktuell bronsåldersforskning*, 117-128. Avdelningen för arkeologiska undersökningar, Skrifter nr. 25. Stockholm: Riksantikvarieämbetet.

KJELLÉN, E.

1939 Nyupptäckta hällristningar i sydvästra *Uppland. Upplands Fornminnesförenings Tidskrift*, 46: 81-100. Uppsala.

KJELLÉN, E. and Å.. HYENSTRAND

1997 *Hällristningar och bronsålderssamhälle i sydvästra Uppland.* Uppsala: Upplands Fornminnesförenings Tidskrift, 49.

LARSSON, T. B.

1986 *The Bronze Age Metalwork in Southern Sweden. Aspects of social and spatial organization 1800-500 B.C.* Diss. Archaeology and environment 6. Umeå.

LUNDQVIST, L.

1991 Undersökta skärvstenshögar I Västsverige. *Arkeologi I Sverige*, Ny följd 1: 43-60. Stockholm: Riksantikvarieämbetet.

NYLÉN, E.

1958 Pryda, skydda, binda? *Gotländskt arkiv.* Meddelanden från föreningen Gotlands fornvänner 1958, 23-38. Visby.

OLAUSSON, M.

1995 *Det inneslutna rummet. Om kultiska hägnader, fornborgar och befästa gårdar i Uppland från 1300 f.Kr. till Kristi födelse.* Diss. Riksantikvarieämbetet. Arkeologisk undersökningar Skrifter 9. Stockholm: Riksantikvarieämbetet.

RENTZHOG, S.

1967 Om skärvsten och skärvstensrösen. *Tor* XI (1965-1966): 61-82. Uppsala.

RUNCIS, J.

1999 Den mytiska geografin. Refliktioner kring skärvstenshögar, mytologi och landskapsrum i Södermanland under bronsålder. Spiralens öga. Tjugo artiklar kring aktuell bronsåldersforskning (M. Olausson, ed.), 129-158. Avdelningen för arkeologiska undersökningar, Skrifter nr. 25. Riksantikvarieämbetet. Stockholm.

SARAUW, G. and J. ALIN

1923 *Götaälvsområdets fornminnen.* Skrifter utgivna till Göteborgs stads trehundraårsjubileum genom jubileumsutställningens publikationskomitté, III. Göteborg.

STENBERGER, M.

1964 *Det forntida Sverige.* Stockholm.

ULLÉN, I.

1994 The Power of Case Studies. Interpretation of a Late Bronze Age Settlement in Central Sweden. *Journal of European Archaeology* 2: 249-262.

WIDHOLM, D.

1998 Rösen, ristningar och riter. Diss. Acta Archaeologica Lundensia. Series Prima in 4, 23.

Unpublished Sources

Ancient Monument Register.
Riksantikvarieämbetet (National Heritage Board). Stockholm.

MESCHKE, C.
1968 *Hur skall boplatserna publiceras? Ett förslag till publikation av Umeälvens inventerade boplatser av stenålderskaraktär, med definition av detta begrepp*. Lic. avh. Stockholms universitet.

SELINGE, K.-G. (ed.)
1969 *Inventering av fasta fornlämningar*. Fornminnesinventeringen arbetsföreskrifter 1969. Stencil. Riksantikvarieämbetet. Stockholm.

TOMBS, TEXTS AND IMAGES: FIRE IN GREEK FUNERAL RITUALS

Vasilica Lungu

All things are an equal exchange for fire and fire for all things...
Herakleitos, ed. H. Diels, frg.90

On Fire and Ritual

For the Greeks, fire was one of the essential elements of the cosmos; it transformed the material world, sometimes as instrument, sometimes as material structure (Aristotle, Metaphysica: A.984.a.5). Heraclite believed that fire represented both the material substance, and the active factor of qualitative transformation of life. Empedocle (ed. H. Diels, frg. 96), who may be considered the 'grand-father' of structuralism, postulated that 'the bones contain two parts of earth, others two of water and finally four parts of fire'. One could infer that similar 'materialist' ideas should be determined their preference for a complete transformation of the corpse by fire. Their epoch is very close to that of the rise of cremation in the Greeks necropoleis, and then they cannot be ignored. It is now commonly believed that the late 8th to 6th centuries B.C. and later witnessed a normative use of fire in funeral rites and rituals discernible in art, literature and archaeological finds.

Numerous scholars have studied the cremation rite in the Greek world, each following different or common approaches. Several among them consider that the tombs with cremation covered under a tumulus no doubt have their predecessors in an ancient tradition, which was started in the Bronze Age (Nilsson 1955; Coldstream 1977; Kurtz and Boardman 1971; Musgrave 1990; Morris 1987; 1992). In different regions of the Greek world cremation became more usual about 700 B.C., as well as at Athens (Cook J. 1953; Cook R. 1960; Morris 1987, 1992, 2000), the treatment of the corpse being usually correlated to the nature of the society and the state (Oestigaard 1999).

The Greek literature suggests that the consummation of the corpse by fire was an usual behaviour at the aristocratic level. Its heroic character was largely spread by the elitist poets (Homer, *Iliad*: 23 and 24, Andronikos 1968) or by tragedy, all promoted in public representations before a numerous audience. It is in the light of this tradition that cremation seems to have been predominant in Roman Hellenized upper classes to the time of Augustus, in the first century B.C. (Toynbee 1971: 39-42). Ralph Merrifield (1987: 60-61) is convinced that "the belief in the virtues of the physical remains of the dead cannot be reconciled with the practice of cremation". In his opinion, fire "is necessary to separate completely the dead from the living". If burning is received as a general, dogmatic rule, why did many Greek people ignore it? Vidal-Naquet (1986: 139-140) concluded that fire, especially when seen as a main manifestation of Greek religion, involved the contrast between nature and culture.

According to archaeological data, it is evident that rites, cremation and inhumation alternated or often were practiced side-by-side in Greek necropoleis. But beginning on the 8th – 7th centuries B.C. the Greeks used more cremation for adults. Not only the Homeric influence on burial customs could be considered there, but also the social and historical evolution of Greek societies would have produced diffusion of the use of fire in funerary practices. At Athens, Morris (2000: 294) noted that "cremation was a striking element in a visual system which looked back to the past, recalling the simpler religious practices and social relationship of the Dark Age". In the Geometric Period cremation was practised in Rhodes, Cyclades, Euboea, Attica, Crete and Cyprus (Snodgrass 1971, 189). I suggest that the cremation was spread more in Greek World at the beginning of the Archaic period not as an act of "reversion" but as a result of turning to East and admission of an orientalizing style of life. We don't forget that the Greek spiritual elite raised before and after 600 B.C. from East Aegean Regions. Moreover, the beliefs of certain philosophical sects or the practices of person initiated of different mysteries or cults had certainly their influence on the funeral customs (Bruit Zaidman and Schmitt Pantel 2000: 72-79). They were able to support general explanations for this phenomenon, but call attention to regional variation in the pattern. An example could be the distribution of cremation over the map of west Greece dominion, where Carter (1998: 103) pointed to the distribution of the tombs with cremation related to the origin of colonisers, and noted that this rite "was particularly true of the colonies sent out from Euboea where cremation predominated; it was much less common in the Dorian colonies". The long tradition of the Greek cremation in the West Greece would have provided a strong background for the development of burning rituals in ones of the colonies (at Siris, Berlingo 1986: 120-121). The involvement of the cremation in a funeral was of course an ancient practice at most of the Aeolian region (at Pitane, Akurgal 1987). On the Black Sea shores, the necropoleis of the Greek cities demonstrate a common practice of rites, cremation and inhumation, typical of archaic city-states (de Polignac 1995: 81-88). Paradoxically, however, it is evident more strength of inhumation on the north and east shores, and

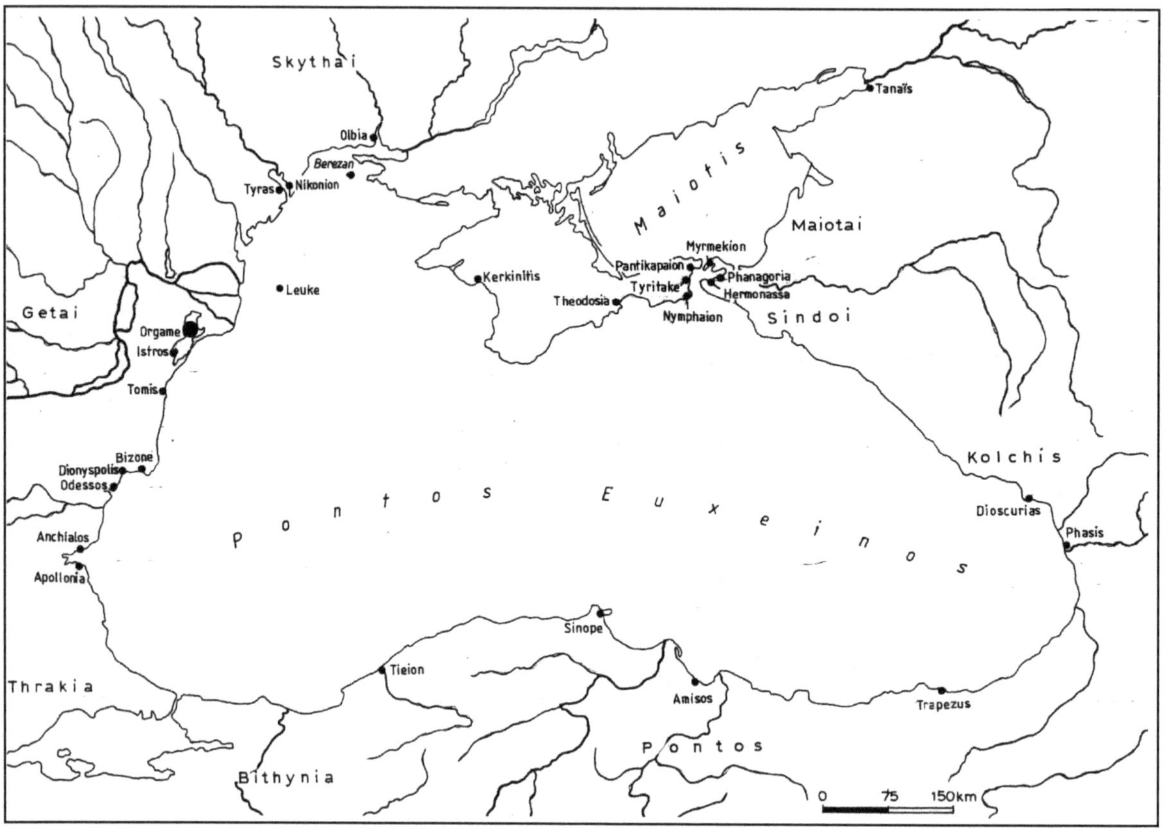

Fig. 1. Map of Miletus Colonies in the Black Sea area.

a sure domination of cremation on the west side, both dating from the Archaic to Hellenistic period. It may not be insignificant to take here in account that it may be influenced by the different religious traditions during the cohabitation between Greeks and new communities in foreign lands. Cremation seems to have been predominant in west indigenous peoples (Thracians, Getes), and inhumation in north and east indigenous necropoleis (Scythes). In these regions, the cremation rite spread more widely there after the first phase of Greek colonisation, with the intensification of the communication between Greeks and Non-Greeks, particularly they placed around the shores west of the Black Sea.

The colonisation process generated a particular Greek cultural fragmentation into "barbarian" lands, which requires that the new institutions and partitions of space be established in the Greek new *poleis*. In this disruptive process another kind of social cohesion is exemplified by the perspective of the relationships with new peoples. Often they were hostile. All these generalisations are more or less applicable to the Greek *poleis* established in the Thracian lands, where there are good reasons to believe that the local peoples did not appear friendly to the Greek colonists, the peoples lived around the Black Sea being notorious for their bellicosity (Malkin 1987: 55). There is such evidence to consider that the conversion to cremation rite revealed in Greek colonies on the West shores of the Black Sea could be explained in order to this familiar feature to Thracians. They are also mentioned in front of the Trojan gates (*Iliad* IX, X), where they known directly a possible common experience with many peoples from East Aegean. After such centuries, they receive a considerable migration of Ionian peoples toward their territories, with a wide spread of the Homeric epic, and this process seems to make acceptable an intense practice of cremation rite in a new Greek diffusion. It is important to note that cremation rite was not restricted only to Greek colonizers of this region; Herodotus (2.5.8) reports that the Thracian peoples, and the Getes between those, widely used cremation in their necropoleis. The continuity of rite over time may be achieved even by this common experience. The use of fire seems to be particular for this region that the Homeric model is imposed as a general rule in funeral customs, particularly related to the warrior burials (at Athens, see Antonaccio 1995: 222). Appealing to certain common meanings of the Greeks and Thracians religion, I suggest that this important 'coincidence' reflecting a similar behaviour in the ideological field. It is well known the mythical association between the religious leader of Thracians peoples, Salmoxis with Cronos (FHG III, frg. 23), Hestia and Pythagoras (Ziegler 1967: 2302–2303,with a summary of sources).

In the light of this similar/different features, I try here to establish a new approach where different ways of interpretation are put together to analyse the Greek funeral experience. Our attention is focused on the excavations of a Greek necropolis at Orgame (Argamum; Jurilovca village, Tulcea County), a Milesian colony (see Hecataios from Miletos, *apud* Stephanus from Byzanthium: Οργάμη πόλις ἐπί τῷ Ἴστρῳ), situated in the west Pontic area (Dobroudgea, Romania) (fig. 1). There, the remains of different kinds of fire customs in, and around the tombs, have recently been investigated. It is also reasonable to assume that their description would be substantiated by the data of literary and iconographic evidence. By its content, the present study is an on-going critical approach to some information regarding the use of the fire in funeral rituals underlying this Greek custom within the colonial area of the Black Sea. It is based on a flexible methodology that estimates the complexity of relationship between three important sources: archaeology, ancient texts and Greek vase-paintings, though it is not often that they 'speak with the same voice' (Murray 1993: 239). The aim is to point out the importance of the use of fire within Greek tombs and religious life.

Study Case: the Orgame Necropolis

The necropolis of Greek Orgame, identified in front of the ancient city, to its west side, provides exceptionally clear evidence of the use of fire in funeral customs. They are associated with a tumular necropolis, which revealed a considerable anthropological complexity (Lungu 2000a; 2000b). Systematic excavations on some sectors of this necropolis defined the significance of an urban structure, where the cremation of the dead was the main rite, similar to Istros necropolis (Alexandrescu 1966).

The earliest burials took place about the third quarter of the 7th - early of the 6th centuries B.C., and they widely spread in a common area, at a short distance west of the city. By the 6th century B.C. the cremation appears as a regular norm well define in tombs of tumular shape. Cremation persists in the following period as well, being the most frequent treatment of the dead, both in the Classical period and later. From the Archaic times, burials started to be organized in a large space to west and to south of the city. The bulk of our archaeological evidence comes out from two large periods, the 7th – 6th centuries B.C. and the 4th – 3rd centuries B.C., and it provides that the demography at Orgame reached its peak in its last stage. The lack of adequate evidence for a large period of time from the 5th century to the early 4th century B.C. in the excavated areas has not handed to understand the development of the same rites. I believe that present excavation is responsible for this hiatus.

Particular for the necropolis at Orgame is the very large number of cremation tombs and other auxiliary pyro-ritual activities related to the tomb. There were common features with others Greek colonial necropoleis, as Istros. Both Greek necropoleis emphasis the presence of traces of a complete burning, sometimes the only pieces left being some bits of human bones. This was possible because of the important natural fuel reserves of the forests existing in Antiquity. Apollonios from Rhodes (*Argo*. 4.310), and Pausanias (3.19.11, who describes the Island Leuke, nearly the mouth of the Danube), Pseudo-Scymnos (frags. 786-787: Πεύκη δὲ λέγεται δὶα τὸ πλῆθος ὥν ἔχει πευκῶν) gave references about the abundance of forests in the vicinity of the Orgame area (cf. Paskovski 1967; Dihoru and Donita 1970; Ivan 1992).

a. Primary Cremations

The earliest example of a cremation burial in the necropolis of Orgame occurs in the 7th century B.C. in the North side of the tumular necropolis, positioned on a top of a hill. Here the author recently investigated a large sector of an archaic necropolis related to Greek colony from Orgame. The most part of the present text is focused on the largest burial example, the tumulus-TA95 (fig. 2).

Fig. 2. The tumulus-Heroön TA95, and a part of Archaic necropolis.

The tumulus is topped on the ancient level with limestone slabs without any ornamentation preserved, oriented to the west in line with the front of the Greek settlement (Lungu 2000a; 2000b). It has the most impressive proportions by comparison with all the other excavated tombs in the necropolis, its overall structure covering more than 42m in diameter by 1.75-1.80m height. Moreover, this is one of the biggest Archaic monumental tombs in Greek colonial area from the Black Sea. Considering the chronology and the Ionian origin of the most of the finds which appear for the first time both in this tomb and in the civil settlement, one may postulate that the earliest use of the cemetery with cremation such as general rule coincides with the earliest stage of the colony.

The tumulus evidence revealed traces of a large pyre correlated to a remarkable Greek Iron Age heroön, i.e. a hero's tomb (Lungu 2000b). It is a central pyre formed around a cremation pit cut in the rock and protected by a very impressive stone ring. Finally it was protected by a tumulus and surrounded by a ritual ditch cut in the rock and used widely for centuries for the ritual offerings to the dead. The stone mass excavated at the beginning from this ditch has been used at the ring erection to surround the central pyre. A lot of limestone shows altered features indicating that the pyre in the central pit was extinguished in the last stage with the stone slabs. At Halos, in Thessaly, Ferguson (1991: 148) points out the same practice from the 8th-7th centuries B.C.

In tumulus TA 95, the archaeological evidence provided that the deceased body was placed directly on the top of the pyre built over a pit. On the east side of the pit a short ditch cut in the rock to produce a good air-drought gives the T-shape for the complete pit. The amount of fuel used was enough to complete consume the human body. Short fragments of the bones were left on the pyre place, covered with a substantial ash layer indicating a strong fire produced by the combustion of oak wood. A similar case was described by Homer (*Iliad*: 23, 24) at the funerals of Patroclos and Hector, where the wood was consumed in very considerable quantities. There is more evidence that the ritual at the Orgame hero-tomb has been influenced by Ionian funerary tradition (Lungu 2000b). On the pyre level, to the east side of the pit, some cremated animal and bird bones were found on a circular surface burned to red; these must be connected with the animal offerings brought κατ'ευχήν (following the custom) at the time of the human cremation. It should be noted that it seems to be a common custom in certain areas (i.e., Ofeltes' tomb at Nemea, Pausanias, 2.15. 3; Antonaccio 1995: 176) where the hero cult was practised. Very close to the pyre, and surrounding it, lay the fragments of many ceramic vases associated with offerings to the dead, the cremation having been formed under the remains of many logs from the pyre (fig. 3).

Fig. 3. Detail of the oak log fallen from the pyre.

The ritual of vase deposition has been attested by Homer (*Iliad*: 23) when he described the Patroclos pyre. Unfortunately, archaeological finds at Orgame show the absence of the main part of the human bones, the central pit of the tomb containing only a small quantity of cremated remains of one dead male. One cannot clearly explain the lack of the most part of the cremated bones, but judging by the ancient traces of the recurrent plunders, the removal of bones acts as a possible explanation. In this case, one can propose as a model the Homeric ritual regarding the deposition of cremated bones into an urn and buried at the place of cremation (*Iliad* 23: 105-257). This assumption may be supported by the presence in a robber trench of a short thin fragment of gold (fig. 4), which may be considered as being an ornament of a cyst made of wood or other material. It is in this context that we should understand the translation from the pyre of the bones of hero by reason of a safe preservation, frequently mentioned in archaic and classical Greece (Herodotus, *Histories*: 1.68; Plutarch, *Cimon*: 8). Another explanation for the absence of the bones is suggested by an example at Taras (Tarentum, Italy), where the ashes of Phalanthos were dispersed over the land of the new city after his death, in this way the founder protecting the colonists (Malkin 1987: 216ff.). Certainly, the field of the hypothesis could be expanded, but archaeological finds in Orgame, as the presence of robber pits from ancient times, sustain the first interpretation regarding the plundering of the urn.

Fig. 4. Gold piece from the tumulus- *heroön* TA95.

To reconstruct the image of the pyre from the hero-tomb at Orgame, consideration of some *signa priscae artis* in pictorial contexts is useful. On an amphora painting attributed to Myson (Paris: Louvre, inv G 197; ARV^2 729, 4; Simon 1976: 107-108, fig. 133) (fig. 5) and dated about 500/490 B.C., Croesus, the famous king of Lydia, is burned on a pyre built in front of a bronze-walled courtyard (Bacchylides, *Ode* 3). The artist depicted the King sitting upon his throne pouring alone his last libation on his own pyre. The pyre looks impressive, exhibiting a stratified wooden construction. Many logs, probably by different essences, the painter using different colours to differentiate them, are arranged in a grid-like-pattern with large spaces left free to draught the flames. Some details connect this image to Patroclos's pyre in Homeric description (*Iliad*: 23) such as the impressive quantity of the wooden material, the dimensions of the pyre etc. Both evidence, literary and iconographic, help us to study the original shape of the pyre from the tumulus-*heroön* TA95 starting from the remains of the archaeological excavation.

The image represented on the Myson amphora is a reflection of the historical event of the Croesus death after the Cyrus triumph over the Lydian king in 547 B.C. Through this representation of Croesus, the painter has created an image of a social order in which the Lydian king, as a metaphor of the world of men, is subject to divine authority. The presence of the sceptre in the hand of Croesus gives him the religious authority to make the ritual of libation at his own tomb. The meaning of his gesture could be envisaged as punishment,

Fig. 5. The image of Croesus' cremation on pyre, on an amphora by Myson Painter, 5th century B.C. (Paris, Louvre G 197).

closer to the mystic-Herakleitian combination of cosmology with ethics. Croesus must pay for his unlimited desire for wealth shown in the attack over the Persians. But he seems to choose his death according to the principle known at the famous meeting with Solon, preserved by Herodotus (1.29-33). In Solon's opinion, happiness depends on the ending (τέλος) of life. Croesus prefers the death to slavery. His burning on the pyre emphasizes the process of transformation by fire of mortal body, according to Heracleitian significance of fire as being the principal agent of the universal exchange. There has been a tendency to recognize the religious significance of the change by fire of the human body into divine spirit. Croesus' divine authority, reflected by his last human act of libation, provides that his cremation could evoke a sacrificial access to the gods (Morris 2000: 293).

It is important to note, in the same representation, the name of the second character, Euthymos, who ignites the fire on the pyre with two torches. His name could be related to the family of the Greek verb θύω = *to burn offerings for the Gods*, (Chantrain 1968: 448, see θύω 2) and it was put intentionally to add sacredness to the burning process of the pyre, and to transfer the burning of the Croesus' human body in an abstract, ideal or heroic level. This iconography of Myson amphora suggests that Croesus was placed on the pyre according to Greek mystery and myth. The large, fleshy figures with their significant gestures have a very rich, almost eloquent quality to have the power of the *logos* for the reception of the universal values of Greek religion. There is a symbolic function of vase iconography indicating that the ceramic had a great cultural value, and that looking was in fact much more influential in social life, the visual images have an important role to spread the ethic values in ancient communities of Greek *poleis*.

The Myson amphora obviously suggests that the representation of a human cremation provides a communal sensorial experience in the eye of the observers in order to exalt the memory of the deceased. Moreover, the dead's cremation customarily happened at night, as has been described for the burning of Hector (*Iliad* 24: 781-782). As described in the Homeric text, the moment of the burning action was more spectacular, and the pyre become the most important centre for the assistant's attention, fire receiving an aesthetic value. In the morning, the people from the funeral cortege extinguished the pyre with wine, and covered the remains with an earth mound. Archaeological evidence supports often this custom: it was mentioned in princely tombs from Salamis (Coldstream 1977: 350) or from Pithecusai (Ferguson 1991: 148). In women's tombs, the pyre may be quenched with water from hydriai, as we can see on a bell-krater depicting Alkmena pyre, and signed by Python Painter (British Museum, Cat Vases F 149, 350-340 B.C.).

Although, I recognise that the interpretation of iconography as well as of literary sources can involve a great degree of subjectivity, nevertheless I think that the relationship between image, text and archaeological evidence could be reciprocal, and the discoveries in Orgame support the above iconography of the Greek vases. The hero-tomb construction is achieved by focusing on a particular cult. After the death, the burned corpse, being perhaps of a community leader or even its founder (*oikistes*), aquieres a symbolic meaning very important for express the group solidarity. It was employed in the consecration of the sacred fire, and his tomb became a sanctuary (*heroön*) where the pyre of the altars were regularly feed by means of the offerings brought there. Therefore, we may regard the hero's commemorative as part of a ritual process that guaranteed the identity of a new foundation. So the presence of his tomb (*heroön*) is performed to initiate a new phase of the same importance like the transfer of sacred fire in the new *polis* or the setting up of altars in a new place. According to literary testimony from Herodotus (1.146.2), Graham (1983: 25) noted that 'it was a rule for the colonists to take with them fire from the sacred hearth (Hestia) of the mother city, in order to kindle with it the sacred hearth of the colony'. For Greek colonists at Orgame there was an important social duty connected, not only with the construction of an historical memory to preserve their own physiognomy among the 'barbarian' peoples, but also with the cosmogonic implications to preserve the link with their motherland.

The description of the tomb at Orgame infers a rather complex ritual. The sherds filling the central pyre points to a date from the early third quarter of the 7th century B.C., although, the pottery from the exterior ditch dates from the 7th century B.C. and continues till the middle of the 3rd century B.C. The latter consists of drinking wares, as amphorae and jugs fragments, few crater fragments, and drinking cups suitable for a commemoration of a hero. The majority were imported vases, with a few local hand made wares. The presence of the Greek pottery, tiles, ashes, cremated bones, etc, is a clear evidence for sacrifices and feasts for the dead, which could be identified in a common Greek practice of burned offerings (θύα, Pindar, frag. 129; see, also an inscription from Priene, 2nd century B.C., *SIG* 1003). These offerings were no doubt simply set on the ground or on special stone assemblages. Their remains identified as ash or cremated bones show that the dead received a regular supply of roasted flesh in ritual circumstances and, once ritually consecrated, the hero was inextricably integrated within the ideological framework of a Greek *polis* through the rituals related to fire. Apollonios from Rhodes attested us that 'for those who have arrived in a foreign land the custom was to sacrifice to the local gods and heroes' (= ἐνναέται τε θεοϊς : τοῖς γάρ ἐπιόημοῦντας ἔθος ἦν ἐν τῇ ἀλλοδαπῇ θύειν θεοϊς ἐπιχωρίοις καὶ ἥροσι, Scholia in Apollonium Rhodium Vetera, ed. C. Wendel 1974; see also Nilsson 1950 : 2, note 2; Guthrie 1968: 29; Malkin 1987:193). Greek ancient literature tells us much more about the practice of the burning of animal offerings in the hero-cult (Herodotus 6.38; Pausanias 2. 20).

The symbolic role of the ditch surrounding the tomb could be assimilated to one *temenos* (sacred precinct), as it shows a distinct preference for separating sacred and profane areas, this practice being frequently reported to tombs (Herodotus: 9.116). This division of spaces corresponds to the idea of separation of sacred and profane areas similar to those existing on the *acropolis* in Greek *poleis* (see Plato, *Laws*: 745 b). The offerings present inside the ditch conferred sacredness to this area, and the heroön was reserved as a focalising centre for the titular of the cult. The fire to burn the offerings on the altars, the wine or the water to make the libations at the tomb could be seen as the elements by which one could recognize the religious unit.

It is commonplace at Orgame that other grandiose tumuli with primary cremation revealed certain similar features in funeral rituals, like the tumulus T IV at Orgame (Lungu 1995: fig. 5), where the tomb's components are set in the same systemic circular configuration. Here, one horizontal layer of stones extends the exterior side of the central ring, and surround the mound added over the pyre remains. Similar to the heroön mentioned above, the tomb's components are organized around a central pyre. In the middle of the pyre was installed a wooden bed (2.20 m by 1.60 m), and the body of the dead was placed on it and cremated together with the funeral offerings. There are many burned parallel wooden beams attributed to the funeral pyre from tumulus T IV (fig. 6). Their presence is request evidence in the Orgame necropolis. The cremated bones discovered there were very small, and the ash layer covered the pyre remains to a depth of about 15-20 cm. The fragmentation of the bones established that the amount of the fuel was enough to completely consume the body. After burning, the pyre deposit was covered with a thin layer of brown earth and overlapping it there were many unburned amphorae, which had evidently been used for putting out the flames of the fire (at Orgame, see Lungu 1995; see also, Canarache 1957). Finally, it was covered by a very large mound of soil and stones. It seems obvious that the prominent mound was erected to protect the remains of the most important member of a family (*oikos*, Sirsa 1986: 163-194). This person especially honoured could be the leader of one family or its founder; his family kept his memory alive by regular ceremonies identified by the offering left, and his tomb was associated with particular installations for offerings after burial, like *bothroi* or altars (Lungu 2001). Other members of the same family were cremated outside their tombs, and they were buried on reserved places, which surrounded the leader's tomb. Such funerary disposal reveals a social model of the Greek colony established at Orgame: the social group was organised into families, each *oikos* acquiring its own funeral space, and it is there that the family's members were added one by one.

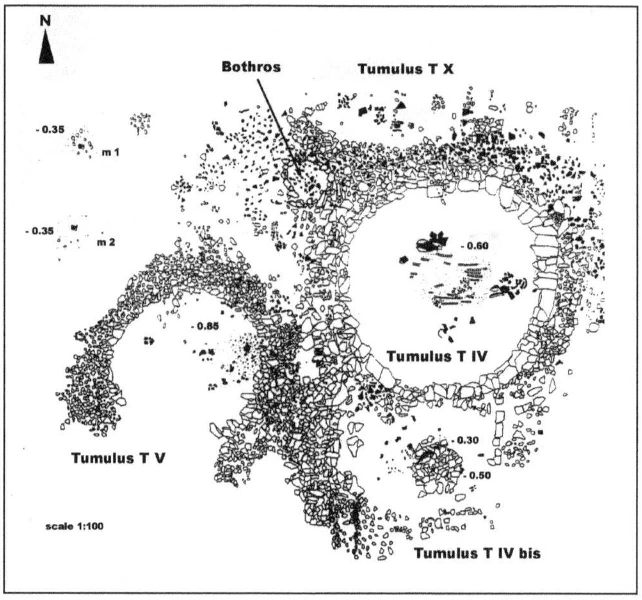

Fig. 6. Familial funerary complex at Orgame (tumuli T IV, T IVbis, T V, T X).

b. Secondary Cremations

Secondary cremations were more widely identified in familial plots at Orgame. They include human and offering remains burnt on a pyre and buried elsewhere in a container or on the soil, in a different place to that tomb. Structurally the toms with secondary cremation are normally distinct from the principal tomb, but their circular plan is also typical. Their status seems to have been of secondary order compared to the first type, and inferred a different social position for the members of the same family (fig. 5: tumuli T IVbis, TV TX).

In Orgame necropolis, the secondary cremation tombs are built on the ancient soil surface with stone rings surrounding the mound of earth, which covered the burial remains. The whole tomb's area was surrounded by a stone ring, and covered by earth and stones, analogous with the tombs with primary cremations. All the tombs of this kind are smaller than those with primary cremations. In plan they occupied usually much less than a half to a third of the total area of the principal tomb. Even if the upper part of many tombs is missing, there are sufficient archaeological data to understand their plan. At least 8 cases of the familial plots display these features, and show the persistence of this general pattern of the start of the necropolis from the Archaic to Hellenistic times (fig. 6).

The earliest burials with secondary cremations are dated from the 7th century B.C., and they contain small quantities of cremated human bones and offerings placed in the middle of the tomb. An important number of this type of tombs was excavated in the familial plots dating from 4th – 3rd centuries B.C., when the deposition of cremated remains in urns became popular. The use of a narrow-necked vessel (like amphorae, *hydriai*, *oinochoai*) as a burial container for ashes was then typical for the Greek sites where cremation was practicised outside the tomb. A common practice become to deposit only the ashes and cremated bones in these containers. An example from the tomb T Bm2 (Lungu 2000 b: Figs. 5.4; 6.3) infers that the corpse may have been transported on a wooden bed or in a sarcophagus decorated with ivory ornaments and cremated with it, but there are not other data that this was a regular norm at Orgame. More numerous seems to have been the tombs with simple deposition of the corpse on the wooden pyre. There is much evidence that the combustion of the human body was quite complete, and this seems to have been dependent on the fuel used to burn. Similar to the tombs with primary cremations, the bones found here are very small, but these remains could be associated with ordinary people, who did not have an important place in social life of the city.

One can infer that beside adults, children, man and women, were also cremated and buried in the familial plots. Child cremation could be identified particularly from 4th and 3rd centuries B.C. tombs, when this rite could be considered as an ordinary custom by the Greeks (Rudhardt 1963: 10-20; Brisson 1978: 114; Juvenal 15.140; Plinius N.H. 7.72, for Roman Period). The presence at Orgame of an *enchytrismos* (child' deposition in a amphora) dating from the Archaic period (middle of the 6th century B.C.) suggests that from this epoch, infants were buried inside the necropolis, at the limit of a reserved familial area. Ideas concerning the status of the children as social non-persons (Kurtz and Boardman 1971; Morris 1987: 105; 1992: 61) have survived in this example. Ialissos, from Rhodes, (Clara Rhodos VIII 1936: 12), and Eretrie, from Euboea, (Bérard 1970: 50; Vidal-Naquet 1974: 147), are two other regions where inhumation was the current child burial custom. Moreover, Berard and Vidal-Naquet noted that inhumation was the normal rite for individuals up to the age of sixteen. Lewis Binford (1971: 22) insisted that the corporate involvement in the mortuary rites of "lower status sub-adults" was limited to the familial level.

Within the context of the Orgame necropolis, the familial plots comprise a burgeoning structure on a same area, the distances between tombs usually being very close. Each structure is articulated from many different elements connected to the primary and secondary cremations, both being practiced by Greeks from this colony displaying deep interrelations within their own social structure, as well as with the prevailing economic realities. The presence of regular offerings suggests that the necropolis was organised like a sacral place for the members of the Orgame community who at death were buried in accordance with specific customs. The particular status of human remains is reflected often in structures of Greek necropoleis. With one exception (the plot of tumuli VI,VII,VII bis, and IX), the most part of the discoveries from Orgame Greek necropolis display a very common pattern issued from the association of primary and secondary cremations, i.e. a realistic image of a homogenous elite into a 'citizen cemetery' (Morris 1999). A mental projection of the funeral space reveals an organisation based on the tomb intra-relations as well as inter-relations of sacral spaces that forms the necropolis: a principal tomb, usually the biggest, including a central pyre, all other tombs with secondary cremations of the members of the same family surrounding it. A similar model comes out from the Istros necropolis, another Milesian colony further south along the coast (Alexandrescu 1966: 133 - 294). To judge from the organisation of the sacral area of the necropolis and the attendant rituals that were practiced there, the conclusion is an elitist pattern that probably explains the social origin of the colonists, known at Istros as ἀινεῦται (commercial aristocracy, Plutarch, *Quaest.Gr.* 32, 28 c; Alexandrescu 1986: 63-64; Avram 1990: 243). Despite the absence of texts referring to the presence of ἀινεῦται at Orgame, the spatial configuration of tombs and the complexity of rituals suggest their existence.

A particularity of the Orgame Necropolis is the Ionian model of the tumulus-heroön (TA 95), whose structure suggests a certain instance for one of the most significant concepts of the Milesian philosophy known by many sources from the 6th century B.C. Parmenides describes, according to the cosmology promoted by different "schools" of Hesiod, of the Ionians, and of Pythagoras, the organization of the space in concentric circles (Burnet 1970: 215). This mental projection is familiar also to the Martin's (1975) theory related to the organization of civic space. Inside the Greek cities the sacred fire (Hestia) was placed in its centre, just as the funeral fire inside the hero-tomb. In my perception the model of space configuration is based on the main principles of the Greek cosmology regarding the systemic vision of the Universe (see West 1971).

Starting with the Greek colonisation from the 8th – 6th centuries B.C. a new world order was instituted, and this process probably engendered a revolution in burial rites in Greece, and in the new dominions. Two main factors, outside certain philosophic ideas that also contributed, seem to emerge quite clearly during this age, i.e. the belief in the survival after death of personal individuality, which prevailed in Homeric epic, and a strategy of foundation of Greek colonies in foreign lands. Both archaeological evidence and ceramic iconography do, in fact, reveal that here was in this epoch an increasing conviction that the depart from their home involved a danger, and their survival, and more fortunate life than that experienced into foreign land could be home involved a danger, and their survival, and more fortunate life than that experienced into foreign land could be protected hereafter by the souls of the departed. It is in the light of these beliefs about the nature and the state of the human soul after death that it was regarded as powerful and received the successors' veneration. Hence the recurrence in the picture language of Greek vases of such themes, as departures of warriors, combats, tombs, pyres, representations of *prothesis* (the laying-out ceremonial), or *ekphora* (the funeral cortege), sacrificial offerings on the altars of Gods or heroes, real or mythological, spread more in colonial areas. One can suppose that the imagery of Greek vases preserves this memory and it becomes a basic vehicle of public information in order to continue social inter-relationships and religious customs. According to this quality, Greek representations on vase-paintings add complementary information to the archaeological evidence.

It is true that these visual pattern have relatively much more explicit correspondences in surviving literature (Homer, Hesiod, Herodotus etc.). Most of the tombs where they occur belonged to the Greek communities who forcefully expressed their identity. It is also reasonable to assume with Thucydides (2.43) that it is more important to preserve a sacred place with burial goods and customs for its safekeeping than to have a real tomb. According to it, one can think that the use of cremation by Greeks in funerary customs was supported by two main processes: the rise of warfare and the phenomenon of colonisation in the Iron Age.

In the light of recent excavations, it seems that the cremation was widely practiced along the western coast of Black Sea, from the Archaic to Hellenistic period. This was recorded from those necropoleis which have been extensively excavated, such as Istros (Alexandrescu 1966), Tomis (Bucovala 1967) and Orgame (Lungu 1995, 2000a, 2000b), all these belonging to the Miletus motherland. For the large presence of cremation in these necropoleis we can propose certain possible sources of external influence. Similar features in the practice of contemporaneous cremations or rituals can be noted in the Aeolian area, in the necropolis of Pitane (Akurgal 1987). Moreover, the presence of the Aeolian pottery was attested on the west coast of Black Sea (Dupont 2000). The possibility of influence from Phrigians, who used widely large mounds, wooden construction inside the tombs, and cremation, should be taken into account. In this case, the idea of a mixed origin of the colonisers under the flag of Miletus has a good support. I have also cited above the cremation customs in the Thracians and Getes necropoleis. Similarities with Greeks were obviously attested in an emphatic presence of the elite tombs. In the light of the recent interpretations concerning the hero cult (Antonaccio 1995), one could infer that the cremation and some of the pyro-rituals particular there could be assimilated in a common tradition. In this perspective, there must have been some connection between the role of fire in Hesiodic and Homeric epic and the adoption of the cremation ritual as the Greek funeral norm. In reality, as well as in poetry or iconography, the image of the corpse consumed by flames transformed the memory of the dead in a tangible and credible legend, i.e. history.

The burning of the hero's body and the burial in Orgame Greek necropolis, conceived as a delimited space outside the city-walls, provides precious information on social status and religious symbolism. During the process of archaeological analysis, it is possible to follow in some aspects the construction of tombs, as in the case of the principal buildings in a Greek city. The erection of the hero-tomb was preceded by sacrifices and prayers, and burnt offerings were also performed after the burning. The same practices were usually for the buildings placed around the *agora* and the temple of the patron deity. The similarities with the living space in the city entertained the Greeks believe in a future life of the soul after burial. Connected to the ancient Greek funerary rituals, fire could be associated with an ethic standard by which the behaviour of the dead are judged, the tomb becoming the mirror of their moral kind. The fire had the role to transform the human's bloody flesh into exemplary memory of the soul. It suggests that the cremation could be felt a respectful way of honouring the body. The community's visual experience of the pyre kindled becomes complementary of contemporaneous people's life, and it assured the communality of the Greek *polis*. It was a main purpose of the colonisation process. The archaeological evidence of the west Black Sea colonies provides that the preference of cremation was, moreover, too particular, and it has to be perceived as regards the creation of a historical memory.

Abbreviations

ARV^2 = BEAZLEY, J.D. 1963, *Attic Red-Figure Vase-Painters*, 2nd ed., Oxford.
SIG = *Syllogae Inscriptionum Graecorum*

Bibliography

AKURGAL, E.
1987 *Grechische und römische Kunst der Türkei*, München.

ALEXANDRESCU, P.
1966 Necropola tumulara. Sapaturi 1955-1961. In Condurachi, E. (ed.), *Histria* II: 133-294.
1986 Aristotel despre Constitutia Histriei, *Studii Clasice* 24, 63-70 = 1999, L'aigle et le dauphin. Études d'archéologie pontique, Bucarest – Paris, 108-116.

ANDRONIKOS, M.
1968 Totenkult, *Arheologia Homerica*, III, Göttingen.

ANTONACCIO, C. M.
1995 *An Archaeology of Ancestors. Tomb Cult and Hero Cult in Early Greece*, Lanham, Maryland.

AVRAM, A.
1996 Modes de contacts entre Grecs st Gètes à Histria à l'époque archaïque, in O Lordkipanidze, P.Lévêque (eds.) Sur les traces des Argonautes. Actes du 6e Symposium de Vani (1990): 241-251.

BELL, C.
1992 *Ritual Theory, Ritual Practice*, Oxford.
BÉRARD, Cl.
1970 L'Héroon à la porte de l'Ouest, *Erétrie* III, Bern.
1980 Éretrie, l'organisation de l'espace et la formation d'une cité grecque. In Schnapp, A. (ed.) *L'archéologie aujourd'hui*, Paris.
1983 L'Héroïsation et la formation de la cité: un conflit idéologique. In *Actes du Colloque International organisé par le Centre National de la Recherche Scientifique et l'École Française de Rome*, Rome 2-4 decémbre 1980, Rome: 43-59.
BERLINGÒ, I.
1986 La necropoli arcaica di Policoro in Contrada Madonelle. *Siris-Polieion*, Policoro: 117-127.
BINFORD, L. R.
1971 Mortuary practices: their study to potential. In Brown, J. A. (ed.), *Approches to the Social Dimensions of Mortuary Practices. Memoirs of the Society for American Archaeology* 25: 6-29.
BOARDMAN, J.
1998 *Early Greek Vase Painting*, Oxford.
BRISSON, L., (ed.)
1978 Aspects politiques de la bisexualité. L'histoire de Polycrite, *Hommages à M. Vermaseren*, Leiden.
BRUIT ZAIDMAN, L. and P. SCHMITT PANTEL
2000 Religion in the Ancient Greek City, (English edition) Cambridge.
BUCOVALA, M.
1967 *Necropole elenistice la Tomis*, Constanta.
BURNET, J.
1970 *L'aurore de la philosophie grecque*, Paris,.
CANARACHE, V.
1957 Importul amforelor stampilate la Histria, Bucharest.
CARTER, J.C.
1998 The Chora of Metaponto: The Necropoleis, University of Texas Press.
CHANTRAINE, P.
1968 *Dictionnaire éthymologique de la langue grecque*, Paris.
COLDSTREAM, J.N.
1977 *Geometric Greece*, Oxford.
COOK, J.M.
1953 The cult of Agamemnon at Mycenae, in Geras Antoniou Kerapoullou, Athens: 112-118.
COOK, R.M.
1960 Archaeological argument: some principles, *Antiquity* 34: 177-179.
DIHORU, G. and N. DONITA
1970 *Flora si vegetatia in Podisul Babadag*, Ed. Academiei RSR, Bucharest.
DÖRPFELD, W.
1905 Über Verbrennung und Bestattung der Toten I, alten Griechenland, *Ethnologie* 37: 538-541.
DUPONT, P.
2000 Trafics méditerrannéens archaïques: quelques aspects, in Fr. Krintinger (ed.), Die Ägäis und das westliche Mittelmeer, *Archäelogische Forschungen* 4: 445-460.
FERGUSON J.
1991 *Fra gli dei dell'Olimpo. Un'indagine archeologica sulla religione della Grecia Antica*, (1989), édition traduite par Martino Menghi, Bari.
GRAHAM, W. K. C.
1968 *The Greeks and their Goods*, 2nd edition, London.
GUTHRIE, R.
2000 Reciprocity and ritual. Homer and Tragedy in the Developing City-State, 2nd edition, Oxford.
IVAN, D.
1992 *Vegetatia Romaniei*, Bucharest: Ed. Tehnica.
KURZ, D. and J. BOARDMAN
1971 *Greek Burial Customs*, London.
LUNGU, V.
1995 Une tombe du IVe siecle av.J.-C. dans la necropole tumulaire de la cite d'Orgame-Argamum, Peuce IX, 231-286.
2000a Pratiques funéraires et formes d'organisation sociale dans la nécropole de la cité grecque d'Orgamé', in G. Simion and V. Lungu (eds.), *Tombes tumulaires de l'Âge du Fer dans le Sud-Est de l'Europe, Actes du IIe Colloque International d'Archéologie Funéraire, Tulcea 1995:* 101-118.
2000b La tombe d'un héros et l'organisation de la nécropole d'une cité milésienne du Pont Euxin', in: V. Lungu (ed.), *Pratiques funéraires dans l'Europe des XIIIe-IVe s.av.J.-C., Actes du IIIe Colloque International d'Archéologie Funéraire, Tulcea 1997*: 67-86.
2001 Rites funéraires et éspaces sacrés dans la nécropole d'Orgamé, dans M. Kolia (ed.), *Actes du colloque international de Rhodes*, in print.
MALKIN, I.
1987 *Religion and Colonisation in Ancient Greece.* Leiden-New York-København-Köln.
MARINATOS, S.
1970 Further discoveries at Marathon, *Archaiol. Analekta ex Athenon* 3: 349-363.
MARTIN R. P.
1975 Problèmes de topographie et d'évolution urbaine, in *Contribution à l'étude de la société et de la colonisation eubéenne, Cahiers du Centre Jean Bérard* 2, Naples: 48-52.
MERRIFFIELD, R.
1987 *The Archaeology of Ritual and Magic*, London.
MORRIS, I.
1987 *Burial and Society. The Rise of the Greek City-State*, Cambridge.
1992 *Dead-Ritual and Social Structure in Classical Antiquity*, Cambridge.
1999 Iron Age Greece and the meanings of "princely tombs", in P. Ruby (ed.), *Les princes de la protohistoire et l'émergence de l'État*, Napoli.
2000 *Archaeology as Cultural History*, Bodmin, Cornwall.
MURRAY, O.
1993 Death and the symposion, *Archaeologia e storia antica* 15, Napoli: 239-254.

MUSGRAVE, J.
1990 Dust and Damn'd Oblivion: A study of cremation in Ancient Greece, *The Annual of British School of Archaeology 85*, 271-286.
NILSSON, M.
1950 *The Minoan Mycenean Religion and its survival in Greek Religion*, 2nd edition, Lund.
1955 *Geschichte der Griechischen Religion*, München.
OESTIGAARD, T.
1999 Cremations as transformations: when the dual cultural hypothesis was cremated a carried away in urns, *European Journal of Archaeology* 2, no 3: 345-364.
PASCOVSKI, S.
1967 *Succesiunea speciilor*, Bucharest.
POLIGNAC, Fr. de
1995 *Cults, Territory, and the Origin of the Greek City-State*, Chicago.
RUDHARDT, J.
1963 Sur quelques bûchers d'enfants découverts dans la ville d'Athènes, *Museum Helveticum* XX, 1, 10-20.
SEAFORD, R.
2000 *Reciprocity and ritual. Homer and Tragedy in the Developing City-State*, 2nd edition, Oxford.
SIMON, E.
1976 *Die griechische Vasen,* Berlin.
SIRSA, G.
1986 La famille dans la cité grecque (Ve-IVe s.av.J.-C.), in Burguière, A., Ch. Klapish-Zuber, M. Segalen, Fr. Zonabend (eds.) *Histoire de la famille*, Paris, 163-194.
SNODGRASS, A. M.
1971 *The Dark Age of Greece. An Archaeological Survey of the Eleventh to the Eight Centuries,* Edinburgh.
TOYNBEE, J. M. C.
1971 *Death and Burial in the Roman World,* Baltimore and London.
VIDAL-NAQUET, P.
1974 Le cru, l'enfant grec et le cuit', in *Faire de l'histoire* 3.
VIDAL-NAQUET, P.
1986 *The Black Hunter*, Baltimore.
ZIEGLER, K.
1967 *Paulys Realencyclopädie der classischer Altertumswissenschaft*, Stutgart IX A2.
WEST, M. L.
1971 *Early Greek Philosophy and the Orient*, Oxford.

CREMATION OUTSIDE THE GRAVE: SEARCHING FOR THE TRACES OF A FUNERAL PILE

Nicholas Petrov

The overwhelming majority of the cremations in the Russian North-Western area during the pre-Christian Medieval period (from the 5th till the 10th centuries AD) were made outside a grave. As usual such cremations are present in the archaeological record as congestions of cremated bones which were placed in ceramic urns, or in pits on a natural surface, or simply in a mound of a barrow or on its top. No traces of a funeral pile (annealing of sand, a lot of charcoal pieces and so on) were found during the excavations of such graves.

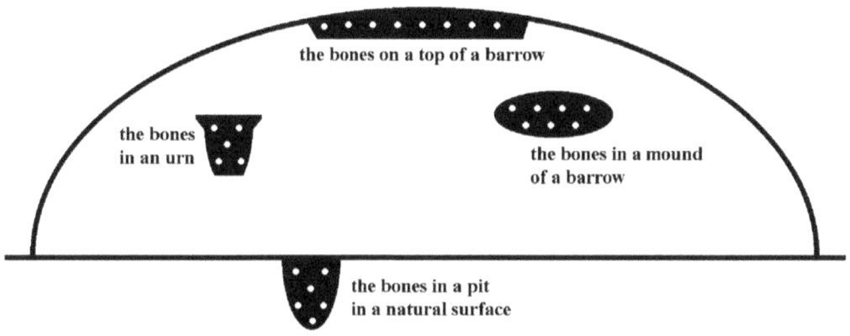

Fig. 1. The positions of the graves with the cremations in a barrow.

This circumstance raises great difficulties for the analysis of the excavated data on one hand and — on the other — for the correct interpretation of the funeral ritual. As for the data themselves, we can not reconstitute for example the complete set of the grave goods in such cases as we don't know whether part of the goods were left on the location of a funeral pile or not and if they were — what these goods were? Thus when we study such pre-Christian Medieval cremations we must deal with incomplete data wittingly. For example, the grave goods from the barrow 47 (located at the barrow cemetery "St'opanovo-3" and excavated in 1997) include 5 bronze badges-"shells", an iron rectangular girdle clasp, 2 small fragments of melt glass, 2 bronze rectangular plates with a point stamped ornament and a fragment of some bronze object. The grave itself discovered in the top part of the barrow contained the compact cremated bones of at least two persons, which were buried here. The clasp testifies that one of them was a man and the plates that the other was a woman. These grave goods do not seem to be a complete set of metal and glass elements of the funeral dress of the cremated persons.

Sometimes it is obvious that all (or almost all) bones were transferred to a grave from a funeral pile as they weighted only a few kilograms. On the other hand, sometimes only several small fragments of the bones were present in a grave and it is obvious that only some symbolic part of the cremated bones was placed in a grave, and the other part was left on a pile. But in the majority of the particular cases is it impossible to approve unequivocally, according to the weight of the bones, whether all the remains of a cremation were transferred to a grave or not? Even the anthropological analysis cannot help in such situation very often, as the integrity of these bone fragments is very poor.

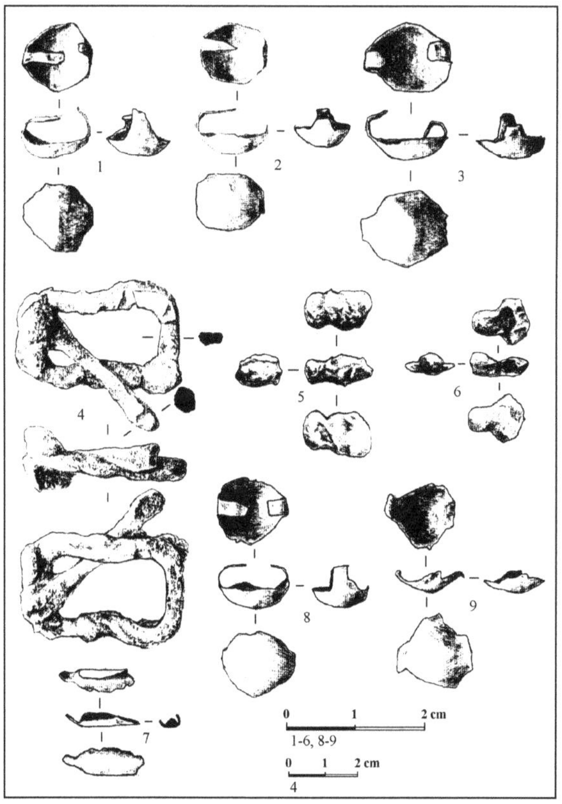

Fig. 2a. The grave goods from the grave No 1, barrow 47, "St'opanovo-3" (1–3, 8–9 — bronze badges-"shells"; 4 — iron rectangular girdle clasp; 5–6 — the fragments of melt glass; 7 — the fragment of some bronze thing.

The other main point of the difficulties in these researches is the correct understanding of the funeral ritual and it is possible to continue the previous question. What ritual actions connected a funeral pile to a grave, how were the cremated bones taken out from a pile and how were they transferred to the grave?

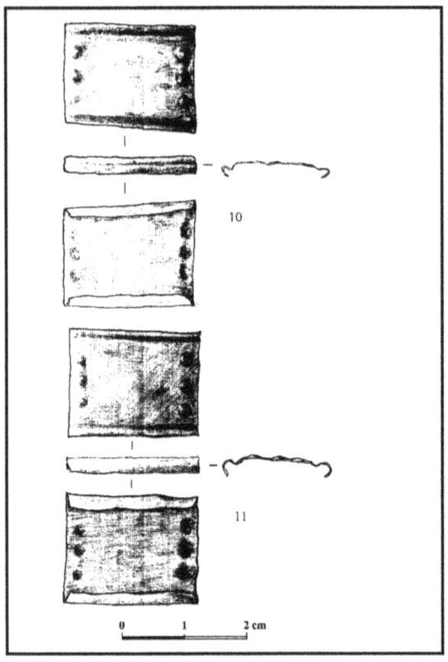

Fig. 2b. 10–11 — bronze rectangular plates with the point stamped ornament).

Thus the searches and the excavations of the locations of the funeral piles is an important direction of the field works. But such searches are complicated by the absence of some clear attributes of these objects on modern ground surface. The main barrier here is not so much these difficulties, but our poor knowledge about the steps of the funeral ritual and its "technology" before the placing of cremated bones into the grave.

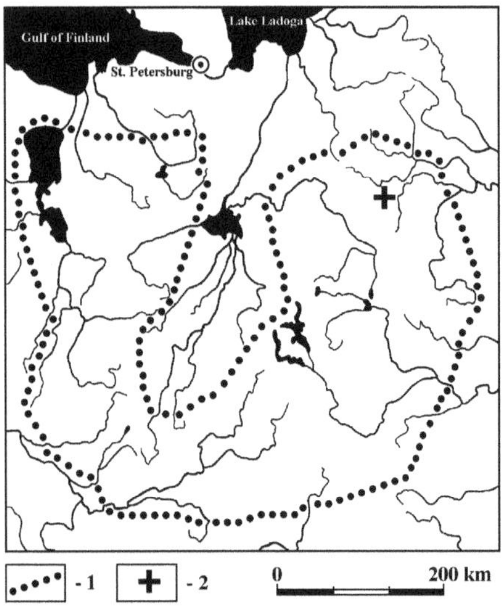

Fig. 3. The group of the barrows "St'opanovo-3" at the area of archaeological culture of Pskov's long barrows (1 — the border of archaeological culture of Pskov's long barrows; 2 — location of the group of the barrows "St'opanovo-3").

An indicative situation was revealed during our excavations in 1999 of the mentioned barrow cemetery "St'opanovo-3". This cemetery belongs to the archaeological culture of Pskov long barrows that occupied the lager part of North-West modern Russia and is dated generally during the 5th–10th centuries AD. It is compared now most likely with the native Finnish tribes the most convincingly (Ligi 1993; Lebedev 1994; Panchenko, Petrov and Selin 1994). The researches of the last two decades show that this culture is based on the local antiquities of the previous period (the so-called "Early Iron Age") which also reflect the culture of Finnish population. A main cultural innovation, which divides the Finnish cultures there before and after the 5th century, is the barrow with cremations (Lopatin and Furasyev 1994; Bashenkin 1995). The barrow cemeteries of the Pskov long barrows culture include not only the long barrows themselves (which are typical for this culture) but also round, rectangular and square barrows.

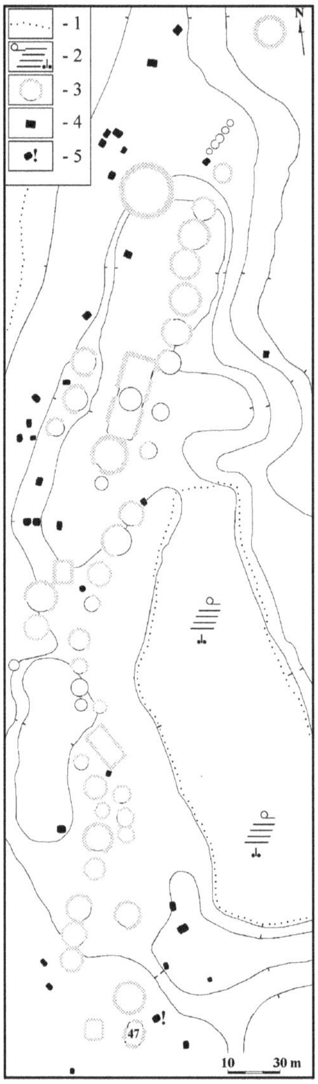

Fig. 4. The plan of the group of the barrows "St'opanovo-3" with the locations of the pits, equidistance — 1 m (1 — the contours of the forest, 2 — thin forest on a mossy swamp, 3 — a ditch around the base of a barrow, 4 — a pit, 5 — the excavated pit).

Since 1995 to present I have managed the excavations of the group of 50 barrows "St'opanovo-3" located on a sandy mound on the left marshy coast of the river Kushavera, near the village St'opanovo in the Novgorod region of Russia (Petrov 1996; 1998–2001). In order to find a location of a funeral pile we excavated one of the 33 pits, which were traced on the modern ground surface near the barrows.

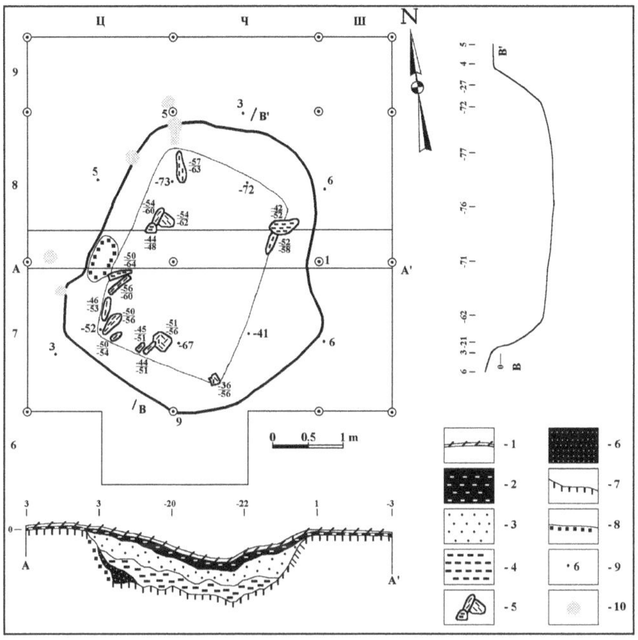

Fig. 5. The pit near the barrow 47, "St'opanovo-3": plan, section and profile (1 — turf, 2 — light gray sand, 3 — brown sand, 4 — black humous, 5 — the fragments of the carbonized wooden logs, 6 — dirty-yellow sand, 7 — yellow natural sand, 8 — red annealed natural sand, 9 — the marks of the level of natural sand from the zero mark, 10 — the large stumps and roots of the modern trees).

Fig. 6. The fragments of the carbonized wooden logs in the north-west bottom part of the pit near the barrow 47, "St'opanovo-3" (view from the East).

This pit was located right near the barrows in the southern peripheral part of the cemetery. The grave goods discussed earlier belong to the grave in this barrow. The sizes of the excavated pit were 3.8 x 3.3 m, the depth — up to 0.8 m. A number of 15 fragments of carbonized wooden logs (figs. 6-7) were found in the bottom part of its infill — in the layer of a black humous soil. (The upper part of the infill containing brown sand was connected probably with the slump of the edges of the pit when its original function was finished.)

All logs were located in horizontal position; their length was up to 42 cm and their thickness from 2 cm up to 20 cm. The thickness of the black humous soil between these wooden remains and the bottom of the pit was from 5 up to 24 cm. It is very important to note that the central part of the north-west slope of the pit was annealed — the natural sand had not a yellow (as usual) but a red shade here and was separated from the black humous soil by a layer of dirty-yellow sand. Thus one could interpret this as the result of some strong fire lit in this pit, the remains of which are these fragments of logs.

Fig. 7. The fragments of the carbonized wooden logs in the the south-west bottom part of the pit near the barrow 47, "St'opanovo-3" (view from the North-East).

But there were no cremated bones or any other finds discovered in the pit. On the other hand, the ^{14}C-date of the charcoal sample from the pit[1] corresponds to the 8th–10th centuries AD and thus falls under the period of the archaeological culture of Pskov long barrows — the 5th–10th centuries AD. Moreover this date correlates with the date of the barrow 47, located near this pit. The bronze rectangular plates with point stamped ornament among the grave goods in this barrow are typical for the archaeological culture of Smolensk long barrows, dated 8th-10th centuries AD (Yenukov 1990: 54–55, 58, 61, 66, 73–79, 201–205, 215, fig. 14–16, 25).

[1] The ^{14}C-date of the charcoal sample from the pit was analyzed in the laboratory of the Institute for the History of Material Culture, Russian Academy of Sciences (St. Petersburg). The code of the sample is Le-5837.

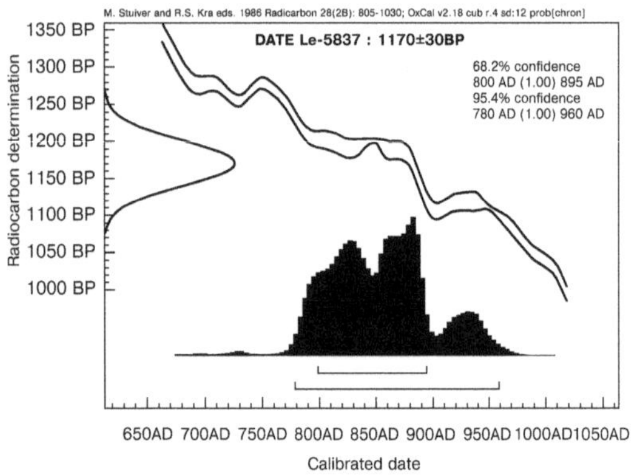

Fig. 8. The calibrated ^{14}C-date of the charcoal sample from the pit near the barrow 47, "St'opanovo-3".

Therefore, taking into consideration the date, the location of this pit and the presence of the traces of a strong fire, the function of this pit are very probably ritualistic. Unfortunately, its interpretation is difficult because of the very poor information about the existence and location of a funeral pile in the Early Medieval texts. In Medieval Russian annalistic texts (which are relevant in general to the region and to the period of the excavated object) one can find only a few mentions on cremation as a pagan funeral ritual. And it is difficult to understand whether the described cremations were made as a pile on the ground surface or as a pile in a pit dug in the ground surface. In the preamble for *Povest' vremennikh let* ("Tale of the Bygone Years"), which is dated by the beginning of the 12th century A.D., the process of a pre-Christian ritual of cremation is described in the following way: "they [the pagan tribes from Eastern Europe] made a great heap of logs, put a dead person on this heap and cremated it" (quoted according to the Laurentian Chronicle (1997: 14) — the earliest text which reflects this tale and is dated by 1377).

From a mere "technological" point of view, a cremation in a pit is a better method than a cremation on the ground surface since it is obvious that a fire in a pit and not on an open surface offers the best conditions for a funeral pile consuming.

Nevertheless, because of the absence of cremated bones in this pit, the particular character of its function is not clear yet. On the other hand, the absence of the bones is not the unequivocal certificate that the pit was not the location of a funeral pile. The size of the pit corresponds to such a function. But if the temperature of the fire inside the pit had not been very high the bones could not have cracked in small fragments and then they could have been recovered from the pit completely. And of course any variant of a final conclusion on this point can be reached only after the excavations of several such pits and — may be — after some experimental archaeological research.

Bibliography

BASHENKIN, A. N.

1995 Cultural and historical processes between the rivers Mologa and Sheksna during the end of the 1st millenium BC – 1st millenium AD. *Slavyano-russkiye drevnosti,* 3: 3–29.

LAURENTIAN CHRONICLE

1997 *Polnoye sobraniye russkikh letopisey,* I. Moskva.

LIGI, P.

1993 National romanticism in archaeology: The paradigm of Slavonic colonization in North-West Russia. *Fennoscandia Archaeologica* X: 31–39.

LEBEDEV, G.

1994 Slavs and Finns in Northwest Russia revisited. *Fennoscandia Archaeologica* XI: 89–95.

LOPATIN, N. V. and A. G. FURASYEV

1994 About the role of the sites of the 3th–5th centuries in the formation of the archaeological cultures of Pskov's long barrows and Tushemlya-Bantzerovshchina. *Peterburgsky arkheologitchesky vestnik (Petersburg Archaeological Herald)* 9: 136–142.

PANCHENKO, A. A., N. PETROV and A. SELIN

1994 "Language replacement" by Priit Ligi. *Fennoscandia Archaeologica* XI: 96–99.

PETROV, N.

1996 The early medieval archaeological sites in the middle reaches of the river Kushavera. pp. 99–104. In *Drevnie kul'turi i tekhnologii. Novie issledovaniya molodikh arkheologov Sankt-Peterburga.* Sankt-Petersburg:

1998–2001 "Dr. Nicholas Petrov: personal web-site" — http://members.fortunecity.com/nickpetrov/

YENUKOV, V. V.

1990 *Ranniye etapi formirovaniya smolensko-polotzkikh krivitchey (po arkheologitcheskim materialam)* [The early stages of formation of the tribe *krivitchi* at the Smolensk-Polotzk region (according to the archaeological data)]. Moskva.

THE FIRE OF TJELVAR: GUTNIC SYMBOLS ON PICTURE-STONES

Christer Westerdahl and Michal Artzy

In prehistoric iconography it is hard to distinguish between the portrayal of the sun and that of fire. In this study we wish to show that a circular motif found on picture-stones on the island of Gotland in the Baltic sea, which are thought of as being of the sun, might actually be depicting fire. In order to achieve our goal, we chose two aspects of the cultural heritage distinct to the island of Gotland. One is the picture-stones of the Iron Age, dated AD 200 to 700 and the second is the ancestral myth of the of the Gotlanders or Gutes (Swed. Gutar). The first written attestation of this myth is from ca. AD 1200, some five hundred years after the dates of the stones. Although the time difference between the two sources should be kept in mind, the proportions are minor in comparison with some other efforts at ethnoarchaeology or related fields.

Questions relating to possible depictions of fire were recently treated, in passing, by Raimond Thörn at the meeting of the EAA in Lisbon, Portugal, in 2000. The liminal aspect of the long rows of hearths, the "great rows of fire" of the Bronze Age, was compared by him to the border-shaping rows of cup marks of certain rock-carvings, such as those of the great Vitlycke rock of Tanum, Bohuslän, Western Sweden.

Fire must have appeared more volatile to prehistoric man than the sun both in reality and in the cognitive world. In our case the properties appear transitory between elements and transform, both the dead body of a man and the stone itself.

The Gotlandic picture-stones are unique as was pointed out by Anders Andrén in his brilliant study. (Andrén 1991) As a matter of fact Gotland possessed other unique phenomena, a well-known fact among Nordic archaeologists. Thus, this is a suitable starting point for a discussion on a specifically Gotlandic or Gutnic historical and cultural identity and the specifically Gotlandic symbols. As an island, Gotland can be viewed as a society or community, which can be examined as a multi-edged social, cultural and religious maritime "laboratory".

According to Nylén/ Lamm (1987) there are 440 registered picture-stones of which 111 are to be assigned to the older types. Those often display rowing ships and whirlwheels on the carved front side (Fig. 1). There are some curb stones, and of these are only 7 are presented with ships although wheels are more prevalent. This is true for Bro No 27, Hangvar Austers No 126, Lärbro No 175, Martebo No 195, Sanda No 215, Stenkyrka No 274, Väskinde Björkome No 335. At least on about 50 older stones, whirlwheels or other circular figures appear. There may be many others of the older type bearing this motif, but the decoration is fragmentary or cannot be ascertained.

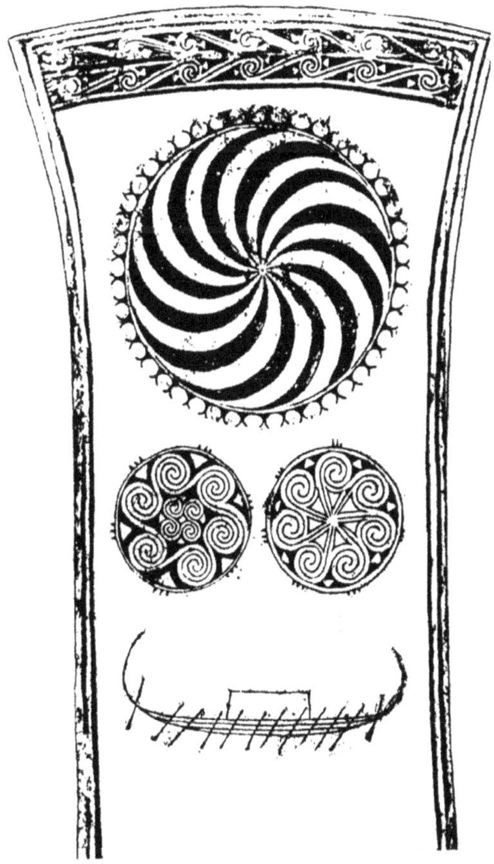

Fig. 1.

The ships are often provided with two steering oars in each stem or stern. Sometimes only the oars are marked, and no rowers. This is found at Väskinde Björkome and Bro I. Sometimes the ship is propelled by rowers, whose tension in their backs is plainly visible, e.g. at Sanda church and churchyard where the tallest stone of was found. There are also other circular forms with an ornamental `fill', as in Väskinde churchyard, Bro I, Hablingbo and Havor II. Some support one or more snake figures, placed often at or on the edge. We find this at Bro Ka II, No 24, Hablingbo Havor II, Hangvar Austers I and Martebo K. (Fig. 2)

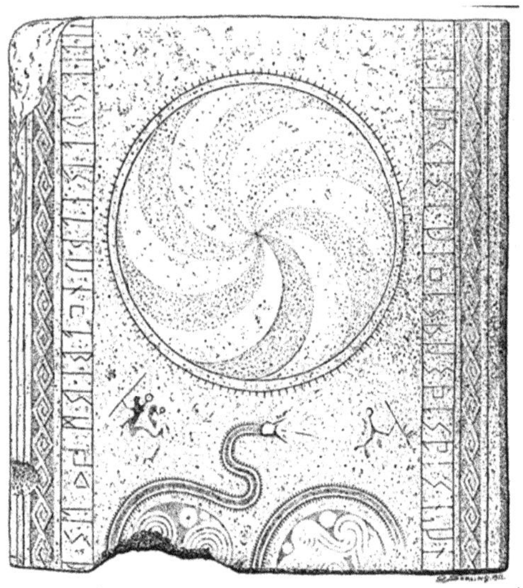

Fig. 2

There are examples where no ships are represented, but only whirlwheels and/or deer or other terrestrial quadrupeds, in particular horses (see Fig. 5). There are also depictions of armed men or warriors (see Fig. 9). Some examples are Vallstena Vallstenarum, and Hablingbo Havor. Mostly on the later types of stones ducks or geese appear in connection with the ships (Fig. 3). They are dated to the Late Iron Age, with Broa VII in Halla parish as an example. In another example at Smiss I, Garda, a figure resembling a picture stone with an axe-form appears to be put at the place of the mast and sail. (Fig. 3). The upturned stem and stern of this ship indicates that it is to be viewed as a sailing ship. We place them within the 8th century. It is highly probable that in Gotland, an island, we see an early appearance of the sail among in the northern area. (e.g. Westerdahl 1995).

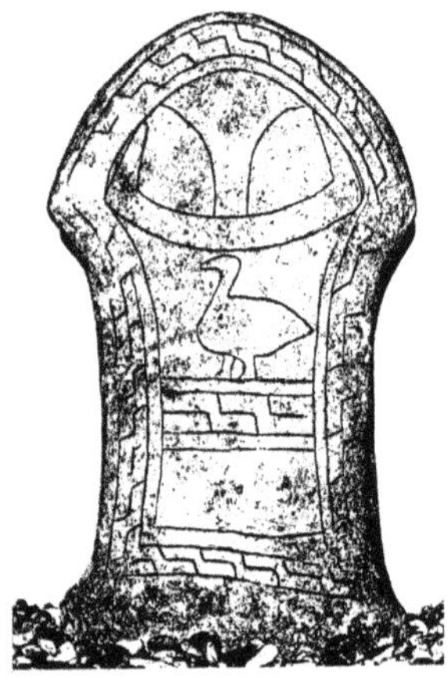

Fig. 3.

These iconographic representations, individually, do appear in Northern Europe, although rarely in the close vicinity of Gotland. It is the combination of the themes which is only found on the Gotlandic picture-stones. The form of the early stones is quite distinctive, reminiscent of an axe-blade with the edge turned upward, like the sail -or possibly cargo? As seen on the Smiss I Garda stone (above). They do not, necessarily, conform to the "doors to other worlds" which Andrén compares with a reasonable prototype to the doors of the Late Iron Age longhouses.

These, unfortunately, are the only extinct representations where ship, wheel and snake pictures appear together. It is quite possibility, that other, less permanent, materials were decorated with the same motives, e.g. textile for wall decoration, cloth, shields and wooden details in buildings. Their absence limits our perspective, albeit unconsciously. With this lack of other direct evidence in mind, one might argue that the picture-stones have a direct relationship to burial and graves. The combinations of motives could therefore have a sepulchral, ritual and/ or mythical significance. Thus limiting the appearance of the combined motifs to the particular stones.

Nydam

A most remarkable find is in the new excavations at the Nydam bog in Sundeved, close to Sønderborg in southern Jutland, Denmark. Great amounts of sacrificed weapons and ship detail have been discovered. Some of the weapons and mountings of leather straps for swords are veritable treasures. A few early runic inscriptions found there accentuate our relationship with the owners whom became less anonymous, especially the chiefs, Harkilar and Wagagastir. That is if not the latter is the name of the battle-axe, on whose shaft or handle the inscription appears (cf. the runologist Marie Stoklund in Rieck *et al.* 1999 in print).

So far three sacrificed ships were located in Nydam, of which the oldest is an iron-clinkered oak ship destroyed by hewing, and is dated c. AD 250. A pine ship, ca. 19 m long, which has been dated by means of one of the shields deposited, to ca. AD 300. This later ship was partly used as firewood after the excavations of the 1860's and is only fragmentarily preserved. The third and later vessel was built some years after AD 310 and is a large oak ship, 23 m long, which has been preserved and partly reconstructed in the Schleswig-Holsteinisches Landesmuseum in Schleswig, N. Germany.

The steering-oar or rudder of the pine ship is intact. An anchor and anchor-stock, the oldest in Northern Europe, is among the finds, but it is unclear to which of the ships they belong. Three large wooden sculptures of male heads belonging to the decoration in the stem section of the large oak ship were located, but they have obviously been placed on both sides as mooring posts, turned outward, not crowning the stem itself, like later ships with dragon heads and the like. These grim faces further connect us, in a personal manner, to the Late Roman Iron Age.

The most fascinating find, relating to this study, was a well-worked, large, semicircular wooden piece of unknown function and asymmetric form, ca. 1 m long and 5 cm thick

(Fig. 4). Presumably, it is connected to the pine ship, where it must have served as a part of some constructive feature. It is made of lime-wood on which were found a clearly carved whirlwheel, a flower cross in rosette-form put in a circle and a somewhat damaged snake figure which runs along the edge. The snake figure appears somewhat atypical in relationship to the Gotlandic snakes, e.g the picture-stones of Hellvi, L. Ire I or Havor Hablingbo IV (Fig. 5). Flemming Rieck, who directed the excavations, has recently been able to ascertain the existence of two bird figures, of which the larger part belongs to a separate fragment, now united with the principal wooden piece. (Fig. 4).

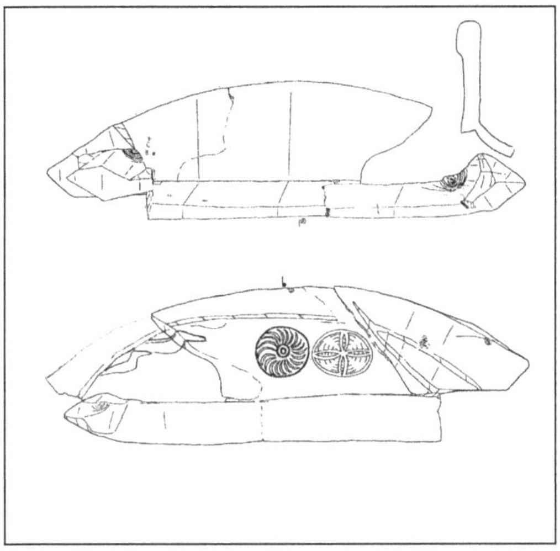

Fig. 4.

Fig. 5.

Why is this remarkable wooden object in the Nydam bog? Has it really been attached directly to a ship? It is probable that it is derived from the weapon sacrifice of the pine ship, although it can not be proven. In the first discussions Rieck pondered the possibility that it may have been part of these remarkable "baldaquins" or roofs of some of the picture-stones (Figs. 1, 2), sometimes provided with rings or shields at the edges. It is, as has been already pointed out, asymmetrical and does not conform well in size. Concurrently, it has to be remembered that the ship depicted on the picture stone is totally unrealistic as a man-of-war, even if the Gutnic picture stone ship as such is close to the ship types of Nydam.

The decorated part is fastened to a massive beam with slots laid crosswise at both ends for some kind of connection. It might have belonged to a stem rib and/or the strakes in the stem. This at least is a possible explanation of its asymmetric form. In a more specific manner, the asymmetry might have arisen because a rib strengthened was to receive the fastenings of a side-rudder on the starboard side. The preserved side-rudder probably belonging to the 'pine ship' and is 1,8 m long, pre-supposing a strengthened rib, of which, however, no trace remains.

Already at an early stage, Flemming Rieck has pointed out the similarity between the combination of motives on this wooden piece and on the Gotlandic picture-stones (Rieck 1994, 1997, 1997). Let us point out again that, indeed, the combination of these motives found on this piece and as a group on the picture-stones of Gotland, although, some of the motives, individually, might appear in other contexts. The excavator also points out to the existence of a flower cross of the same type as on the wooden piece, which has been carved into one of the oar-holes of the pine ship.

The similarity mentioned above does not have to surprise us. During the Late Iron Age period C 2 archaeologically dating to AD 250/260-310, the provenience of the weapon sacrifices in Denmark appears to be the southern Baltic area, including central Sweden and the larger islands in the Baltic. This has been shown conclusively by Jørgen Ilkjær in his exposition of the large find from Illerup Ådal of central Jutland. The Nydam find consists of at least five depositions of which the last must belong to the end of of the 5th century.

One possible explanation is that the invading crew of the pine ship came from Gotland. At any rate since the pine was very rare in the Danish area, ships constructed of pine indicate contact to the coasts of the southern Baltic (e.g. Wagner Smitt 1984). The rich pine forests in Gotland have not only supplied timber for ships' construction in historical times, but also great amounts of high-quality sojde, Gotl. `pit-extracted tar'. Bearing this in mind, we might have an absolute dating of some of the motives on the Gotlandic picture-stones. (Fig. 6)

Fig. 6.

It seems apt to look at Nydam in a wider religious perspective, with reference to death, ritual and graves as well as myths and not necessarily only as a site of sacrifice . The religious and magical sphere, in martial contexts, has been dealt with by Otto Höfler and his cultic Männerbünde (Höfler 1934). This may have some relevance to the war ship units of the time, as displayed by the magnificent weapon sacrifices.

The Guta Saga

We wish now to present the myth of the origins of Gotland and of the Gotlanders, as presented in the Guta Saga (e.g. Säve 1859, Peel 1999). A man, named Tjelvar discovered Gotland. At that time the island was so bewitched that it sank by day and rose up at night. But following Tjelvar's introduction of fire to the island, it never sank again. Tjelvar bore a son, named Hafdi, who married Huitastierna (White Star). These two were the first to settle in Gotland. On their first night she dreamt that three snakes were coiled in her womb, and felt as they crawled out of her lap. She related the dream to Hafdi, her husband, and he interpreted it as follows:

> 'Everything in rings is bound.
> Inhabited this land will be;
> We shall beget sons three.'

The translation is by Peel (1999), and is presented here with a slight alteration of the names Tjelvar and Hafdi (Peel has rendered these names Thieluar and Hafdhi with the original Anglo-Saxon-type letters th and dh which are difficult to reproduce here). In this mythical legend the island of Gotland was up only in the night and submerged during the day and therefore was not inhabited. This vision appears rather extraordinary, if it can be assumed that this refers not only to the primeval times of Gotland but also of the origins of the known world. In the creation tale of Genesis God separates land and water. In the Guta Saga it is a primeval Man, Tjelvar who lights the first fire and thereby separates the primeval time of incessant land movement and the present time of stable island existence, making both time and space accessible to other humans.

The first mythical settlers of Gotland were Tjelvar's son Hafdi and his wife called White Star. White Star's dream is interpreted as an omen of the impending settlement of the island by the descendants of the three sons. One son, presumably the first, was called Guti who inherited the island, the second was Graipr and the third Gunnfiaun. Unfortunately we know very little of the perpetuity of this myth. A possible comparable example dating to the High Middle Ages will be mentioned below.

The Guta Saga seems to have been written down in its extant form during the first part of the 13th century AD. One of the possible initiators was the famous Danish archbishop of Lund and Primas of Denmark and Sweden, Andreas Suneson, who according to the legend received the Danish cross banner from the sky during the decisive battle at Tallinn in 1219. Some ecclesiastical connection with the clerks of the Swedish diocese of Linköping must also have existed (e. g. Kyhlberg 1991, Peel 1999). In this case, however, we do not note any more significant Christian influence on the original myth.

The association of the myth seems to be specific for the island of Gotland. Its sedimentary structure is rich in marine fossils showing the inhabitants that it must have been formed under water. By this observation follow several dichotomies, day and night, land and water. The isolation of Gotland in the middle of the Baltic may have strengthened the particular identity of the myth, in comparison with other lime-stone islands, like Saaremaa or Oesel and Hiumaa or Dagoe of Estonia. We have no preserved myths of origin in this Fenno-Ougrian area, if we are not to quote the Kalevalan egg swimming in the water.

How long the actual memory of the Tjelvar myth persisted in Gotland is a matter of conjecture. Some ship settings do carry the name 'Tjelvar's grav', `the Grave of Tjelvar'. However, this name may be of a learned source, an attribution made by parsons who had a textbook knowledge of the Guta Saga.

The 'Fire of Tjelvar' on the Stone

A male lights a fire on the limestone. Perhaps this was a male prerogative in mythical as well as in the current times and not only on Gotland. Also the later, Late Merovingian and Vendel period/ Viking Age, picture-stones are erected over dead males. They have been interpreted at times a phallic symbols, although a more likely explanation is that they represented doors to another world, thus the dead male may have the entrance to the beginning of Something Else, the Realm of Death or the After-Life.

The living viewer of these stones/doors is obviously on the outside, the inside of the imagined house behind the stones is excluded from the living. We have reason to believe that the interior is the realm of the women. The home, hearth and the maintenance is, according to Andrén, one of the potent symbols of women, the Gutnic farmer wives.

Hearth-formed graves with smaller picture-stones seemingly marking female graves were noted. Depicted on them are sledges or carts but no ships. However, this gender perspective in attribution to means of transport may appear a little ambiguous. In real boat graves- although not found so far on Gotland itself- several women have been buried, like the mainland Swedis burials at Tuna in Badelunda. The famous grave of Oseberg, Norway, containing two buried women, has got a ship as well as a cart and a sledge.

Also the Gotlandic hearth, where the fire is presumably imagined burning permanently, could be thought of as having a cosmic quality with many facets. It could e.g. signify home and hearth, the inner, protected, against the outer, unprotected, like the journey on land as well as on water. There are fine parallells in a world of belief around the house and the hearth as appearing in the classic work of Fustel de Coulanges (Fustel de Coulanges 1864) as well as in Ian Hodder's ideas on structures of meaning (Hodder 1986). But it also would seem to illustrate the archetype of the *pars-pro-toto*, the part representing the whole, in this case the house and the family.

When the ship is being repeatedly depicted with these specific symbols it could be argued that the situation is evoked in metaphorical form. Thus, the burial involves a journey, both in the Middle and Late Iron Age. Among the Gutnic's heathens it appears to be associated with a dangerous and uncomfortable sea passage. Upon his arrival, the man, just as in life, could be expected to light a fire repeating the feat of Tjelvar. Only then the journey ends and the dead man is assembled to his forebears. In order to help him to symbolically make the transition the fire is depicted as well. This is what we propose may be seen in the whirlwheels of the early picture-stones, and possibly also in other circular forms. That specifically a fire, Tjelvar's fire, is intended, may be indicated by the circular form, the whirl, and possibly other inscribed figures. The squarish grave hearths, which appear later, is being ascribed to the female sphere, according to Andrén as above.

The famous Sparlösa rune-stone of Västergötland is the only mainland monument resembling the later picture-stones of Gotland and with partly the same iconographical set (Westerdahl 1996). On all of its sides is a lengthy runic inscription, with royal connotations. On one of its sides are illustrated, a house, a ship and a mounted warrior with a lion-†or leopard-like follower. The house is a stave-construction and in the opening in the middle a circle is depicted. Since this reasonably could not be a door we presume that the circle represents the hearth. This appears as contemporary evidence that circular figures can signify either the hearth as a construction or as the focus of the house.

It is less probable that the sun would have been depicted in the Gutnic Late Roman Iron Age or Migration Age, even if sun symbols may have had a considerable life span in Nordic prehistory. The supreme fire could no doubt have been identified with the sun, considering the fact that according to the Guta Saga the transformation occurs by way of the night being changed into the day. Only by making the nightly appearance of the island above water permanent, can it be settled.

It should not be denied that it can be difficult to explain picture-stones where several wheel figures appear together and also with more varieties than just the whirlwheel. (cf. the Nydam piece). Is it a question of strengthening repetition or an artistic decoration? Is this a remnant magical quality, which makes the wished-for process more secure?

Fig. 7.

As has already been indicated there are indeed other symbols on the stones which may have been intended as markings of the first fire of Gotland. As for instance, circular figures in general including the Migration Age cross of St. John as of Havor /II/, Hablingbo (Fig. 7; Nylén/ Lamm 1987: 39), cf also the triskele of Smiss in När, where it is directly connected to a female figure, very similar to an Irish Sheila-na-gig. (Fig. 8).

Besides symbolizing the specific mythical fire, the original fire of Tjelvar, it is possible that the fire symbol may have insinuated cremation of the dead man. Both cremation and the more common inhumations are known from the Roman Iron Age and Migration Age. The distinction might have been marked on the stones.

The depiction of the ship should also be considered. The choice of a ship as a symbol could have indicated the rapidity of the transformation. In Ibn Fadlan's Viking Age, c. AD 923, account from the Middle Volga his Rus informant says:

"You Arabs are stupid, because you take those whom you hold dear most and whom you honour most, put them into the ground and soil and worms consume them. We burn them in an instant. So they enter paradise."

(after the Danish translation 1981 by Jørgen Bæk Simonsen). A male lights the fire on the limestone. Is this Gotland in a micro-scale? Many other questions would indeed arise as a consequence. Lime-stone is potentially transformed by fire. Would that transformation lead to other consequences of the myth, e.g. towards funerary ideas?

Fig. 8.

Is it possible, that the Tjelvar myth was known c. AD 400? What is known of other myths which have been put down in writing as late as during the Middle Ages? We could surmise, that the trappings of the Odin myth, or at any rate what we know of as the later attributes of Odin, were depicted on gold `bracteates. Bracteates are brooche-like ornaments partly modeled on Roman coins (after Karl Hauck et al.), dated to the Migration Age (AD 300-600).

It is indeed curious why no reference is found to this specific Gotlandic myth on the later picture-stones, which are much closer to the time when the Guta Saga was first written. On these picture-stones the mythic world seems to be taken from the common Nordic and common Germanic heritage with Sigurd Fafnesbane and the Volsungs starring. Is this an effect of the Gutnic integration, politically and cognitively, into a greater community, the Svea realm of Central Sweden?

The ship depicted on the picture-stones can have concurrent meanings: a transporter or "transformer" -possibly with direct connection to a divine power and as a symbol of a specific god. (Ellmers 1995, Crumlin-Pedersen e.g. 1995). The well-known symbolic function of the snake in the North as the protector of the hearth and thereby the luck of the household was recently described†in the work of Flemming Kaul. With this metaphor -of safety?- we find ourselves once more inside the ambiguous character of gender and cosmology:

man/ woman
at home/ under an open sky
inside/ outside
inmark/ outmark (Sw. utmark)

The inmark is†the cultivated and homely area immediately adjacent to the houses, the outmark being on the other hand what belongs to the woods beyond the farm. The snake belongs to both, either in myth or in reality. Perhaps the ship and also the snake symbolized a kind of transition or journey between these aspects. However, this vague reasoning does not answer the question why we find the same combinations of symbols on Gutnic picture-stones as in the sacrificial bog at Nydam. Do they express the same Gotlandic unicity, perhaps also of a secular kind as well. A maritime symbolic ethnicity?

Gutnic Unicity Once More?

The question as to why are there no boat graves or ship settings from the Iron Age as on the mainland should be placed alongside the query as to why are the picture-stones of graves are found especially on Gotland? One answer lies in this particular identity which is created by the Gutes. It is still evident in many other fields.

The Gutes travelled by ship and always 'abroad', outside the island. Birger Nerman and others have shown colonies of probable mainland Svea and Gutnic origin side by side. The Gutnic identity is expressed by a particular burial inventory, especially those of the females whose ornaments have a very characteristic Gutnic trait, according to authors like Hanna Rydh, Lena Thunmark-Nylén and Anders Carlsson. One example of a special sepulchral identity is the large grave-field at Menzlin in Pomerania, which also incorporated ship settings. Some have been put stem to stern with each other. It may in fact be a reference to the past; the ship settings of Bronze Age Gotland, especially the well-known site of Rannarve in Klinte.

Perhaps re-examination of the pagan past gives new strength to Gutnic identity during the Middle Ages. 'The grave of Tjelvar', a name of several ship settings or the mythical sons, a part of an etymological explanation tale for Havdhem, Gothem and other parish names although not always consistent, but weighty nonetheless. Even the sculpture from the church of Väte, which has been interpreted as Luxuria, one of the seven mortal sins in the Christian world, can be associationed to the Guta saga and the dream of Hvita Stierna. Depicted is a woman from whose breast snakes are emerging, but only two. One is tempted to compare with the snake witch from the picture stone of Smiss No. 3 in När. (cf. Nylén/ Lamm: 40). It should be observed that it also carries the triskele, perhaps a symbol of the same significance as the whirlwheel. (Fig. 8).

Tjelvar's Ship

At this juncture we wish to broach the consequence of the notion that the circular forms might be the first fire of the Gutnic myth of origin, Tjelvar's fire.

Tjelvar seems to be identical with Tjalfi, one of the followers of the god Thor in the myth on the visit of Thor to the home of the giant Utgårda-Loke. In one version he is supplanted by Loke- a light demonical trickster figure of the Nordic mythology, always irritating the gods. Loke appears, in many guises, e.g. presumably as Utgårda-Loke himself, but also as a cultural hero, especially as the inventor of the net in the myth of the hunt for Loke by the principal gods. He then is guises as a salmon and is caught by his own invention. The name of Loke indicates a spider personification (Rooth 1961).

What does Tjalfi or Tjelvar mean?

Tjalfi is very young, almost a child in the myth of Utgårda-Loke. One of his tasks in the competition between the teams of Utgårda-Loke and Thor is to run a heat with Time and of course loses it. Against Time even the gods are powerless. It is notable that another competitor in the races is the Fire. Tjalfi also, unintentionally, splits a bone of one of the immortal bucks of Thor, thereby acquires some of the of Loke's quality which is to irritate gods. Tjalfi in the written Norse tradition is possibly a faded memory of a truly divine or culture hero, and at the same time a semi-divine bricoleur in the vein of Lévi-Strauss. One may even ask whether he has always been important in Gotland and possibly the object of actual cult. One could, thus, consider it a kind of eponymous "ancestor cult. His connection with Thor is interesting in yet another way. Thor was, above all, the thunder-god and his thunderbolt ignited fire.

The ship on the early picture-stones and particularly its association with Tjelvar, might be taken as another symbol. The culture hero, the ancestor of humans in Gotland necessarily reached Gotland aboard a ship, although the Guta saga is not say it explicit on this subject. Yet the ship has an extraordinary importance as a symbol in the North (Crumlin Pedersen/ Munch/Thye /Eds/1995). It would accordingly be surprising if the ship has not been connected to several divinities and mythical personalities other than the the divine family of the Vanir, Njord and Frey (e.g. Crumlin Pedersen 1995). The ship of these gods is called Skidbladnir, meaning 'thin as a hide'. It was special, inspiring good wind and convertible so it could be placed in a pocket. This means that is also dated as a sailing ship from the time of the introduction of the sail, c. AD 750, and forward.

At times Odin lent his horse Sleipnir in order to expedite the transporting of the dead (Ellmers 1995) and in the same way and with the same intention maybe the gods of fertility lent their ship. In an earlier context it has been assumed that Skidbladnir is that special sailing ship on the later Viking Age picture-stones (Westerdahl 1995). The intricate hanfot system on the lower edge of the sail indicates its special character, depicted while catching a good wind (Westerdahl 1997).

The most striking trait the in the picture stone rowed ships from older times is the fact that there are at least two steering oars at both ends, sometimes with the coxswain marked in both places. It is not typical of a real ship, but as a reference to a particular situation. This recurring feature might indicate the approach to a beach like the Gotlandic pall, the limestone shelf edge, where the foremost rower must try the depth and be ready to push the ship sideways. In that case why is a home-coming depicted? Is the Gutnic Realm of Death like Gotland as indicated above?
(Fig. 9).

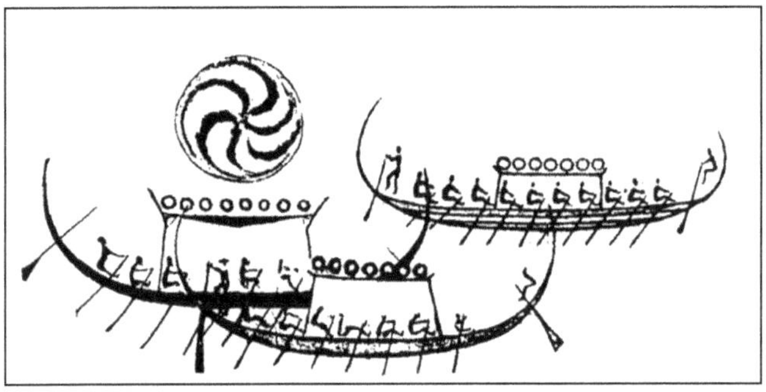

Fig. 9.

The two people wielding the steering-oar could also indicate that it is not only the dead man who runs the ship from the stern, but that he is accompanied. Could it be Tjelvar.

An abstract meaning of 'ship' or the ship 'type'; in the Platonic sense, is less obvious in a basically illiterate society such as the Nordic ones of prehistoric times (for a general reference Ong 1982). A more concrete meaning of an individualized ship is thus extremely probable.

It is attractive to think that this ship could be identified and not maintain a vague notion of a 'ship of death'. From the earliest stone pictures the boat is represented as a rowing ship of the Nydam type. The meaning of the picture would then be that the Original Gute, Tjelvar, directs or transports the dead man back to him. Such a thought could also be extended to other local maritime communities. But in this context Gotland appears to be unique. There are too few parallels on the mainland. As a lonely exception one could refer to the Häggeby stone of Uppland, undated, but very likely Migration Age. On the other hand these early pictorial presentations do not appear as particularly numerous even on Gotland. It should be remembered that the monuments could be very impressive, and possibly of a special importance. Thus, the tallest one of all, from Sanda, is amongst them.

It is attractive to see the circular symbols on the picture-stones as the Fire of Tjelvar, a factor that may have represented microcosmos in the Gutnic picture of the world, the Fire on the Limestone, perhaps even the Devouring Fire of the Grave. But this could of course not be proven, just suggested, in the light of a strong Gutnic identity during the Iron Age and the Middle Ages.

Bibliography

ANDRÉN, A.
1989 Dörrar till förgångna myter- en tolkning av de gotländska bildstenarna. pp. 287-319. In: Andrén, A. (Ed.): Medeltidens födelse. *Symp. på Krapperups borg 1*. Lund.
1991 Förhållandet mellan texter, bilder och ting. pp. 19-39. In: Steinsland, Gro (et al., Eds): *Nordisk Hedendom. Et symposium*. Odense.
1993 Doors to other worlds. Scandinavian death rituals in Gotlandic perspective. *Journal of European Archaeology* 1, Spring: 33-55.

ARVIDSON, S.
1973 *Tjelvars eld*. Uddevalla.

CRUMLIN-PEDERSEN, O. and B. MUNCH THYE, (eds.):
1995 *The Ship as Symbol in Prehistoric and Medieval Scandinavia*. Publications from the National Museum. Studies in Archaeology and History I. Copenhagen.

CRUMLIN-PEDERSEN, O.
1995 Boat-burials at Slusegaard and the Interpretation of the Boat-grave Custom. pp. 87-99. In: Crumlin-Pedersen and Munch-Thye (eds.) *The Ship as Symbol in Prehistoric and Medieval Scandinavia*. Publications from the National Museum. Studies in Archaeology and History I. Copenhagen.

ELLMERS, D.
1995 Valhalla and the Gotland Stones. pp. 165-171. In: Crumlin-Pedersen/ Munch-Thye (eds.) *The Ship as Symbol in Prehistoric and Medieval Scandinavia*. Publications from the National Museum. Studies in Archaeology and History I. Copenhagen.

DE COULANGES, F. and D. NUMA
1979(1864) *The Ancient City. A Study on the Religion, Laws, and Institutions of Greece and Rome*. (orig. La Cité Antique) Gloucester, Mass.

HODDER, I.
1986 *Reading the Past. Current approaches to interpretation in archaeology*. Cambridge: Cambridge University Press.

HÖFLER, O.
1934 *Kültische Geheimbünde der Germanen* I Band. Frankfurt/Main.

ILKJÆR, J.
1994 Illerup Ådal. Mosefundenes kronologi og proveniens, Et resumé. Højbjerg.

INGSTAD, A.-S.
1995 The Interpretation of the Oseberg Find, pp. 139-145. In: Crumlin-Pedersen and Munch-Thye (eds.) *The Ship as Symbol in Prehistoric and Medieval Scandinavia*. Publications from the National Museum. Studies in Archaeology and History I. Copenhagen.

KAUL, F.
1999 Ships on Bronzes. A Study in Bronze Age Religion and Iconography I-II. *Publications of the National Musseum. Studies in Archaeology and History* Vol 3. Copenhagen.

KYHLBERG, O.
1991 Gotland mellan arkeologi och historia. *Theses and Papers in Archaeology* 4. Stockholm.

LINDQVIST, S.
1941-42 Gotlands Bildsteine I-II. KVHAA. Stockholm.

NYLÉN, E and LAMM, J.-P.
1987 *Bildstenar*. Värnamo.

ONG, W.
1990 *Muntlig och skriftlig kultur: Teknologiseringen av ordet*. Göteborg. Orig. Orality and Literacy. The Technologizing of the Word. London etc, 1982.

PEEL, C.
1999 *Guta Saga. The History of the Gotlanders*. Viking. Society for Northern Research. London: University College.

RAUSING, G.
1985 Beowulf, Ynglingatal, Ynglinga Saga. Fiction or History? *Fornvännen* 80: 163-177.

RIECK, F.
1994 Håbet om det overordentlige. *Skalk* 1/94: 7-12.
1997 *Jernalderkrigernes både*. Roskilde.

RIECK, F., E. JØRGENSEN, P. VANG PETERSEN and C. CHRISTENSEN
1999 in print. "...som samlede Ofre fra en talrig Krigerflok" Status over Nationalmuseets Nydamprojekt 1989-97. In: *Fra Nationalmuseets Arbejdsmark* 1999: 11-32.

ROOTH, A. B.
1961 Loki_in Scandinavian Mythology. Stockholm etc.

STEINSLAND, G.
1991 Det hellige bryllup og norrøn kongeidelogi. Larvik. Diss.

STRÖM, F.
1967 Nordisk hedendom. Tro och sed i förkristen tid. Lund etc.

SÄVE, C.
1859 Gutniska urkunder: Guta lag, Guta saga och Gotlands runinskrifter språkligt behandlade. Academ. avh. Uppsala.

WAGNER-SMITT, L.
1984 Wood Species in Viking Age Shipbuilding, pp. 130-137. In: Crumlin-Pedersen, O. and Vinner, M. (eds.) *Sailing into the Past*. Roskilde.

VARENIUS, B.
1992 Det nordiska skeppet. Teknik och samhällsstrategi I vikingatid och medeltid. *Stockholm Studies in Archaeology* 10. Stockholm. Diss.

WESTERDAHL, C.
1995 Society and Sail. On symbols as specific social values and ships as catalysts of social units. In: Crumlin-Pedersen/ Munch-Thye (Eds): 41-50.
1996 Sparlösa-stenen. Symboler och "politiska" attityder i tidig vikingatid. *Västgötabygden* 51(5): 16-21.
1997 Hanfotssystemen på gotländska skeppsbilder –en idé. *Marinarkeologisk Tidskrift* 4/96: 18-19.

www.ingramcontent.com/pod-product-compliance
Lightning Source LLC
LaVergne TN
LVHW071215210926
848482LV00015B/815
* 9 7 8 1 8 4 1 7 1 4 6 7 7 *